IEE ELECTRICAL MEASUREMENT SERIES 5

Series Editors: A. E. Bailey
Dr. O. C. Jones
Dr. A. C. Lynch

Principles of Microwave Measurements

Principles of Microwave Measurements

by G H Bryant

Peter Peregrinus Ltd. on behalf of the Institution of Electrical Engineers

Published by: Peter Peregrinus Ltd., on behalf of the
Institution of Electrical Engineers, London, United Kingdom

Peter Peregrinus Ltd.,
The Institution of Electrical Engineers,
Michael Faraday House,
Six Hills Way, Stevenage,
Herts. SG1 2AY, United Kingdom

British Library Cataloguing in Publication Data

A CIP catalogue record for this book
is available from the British Library

ISBN 0 86341 296 3

Printed & bound by Antony Rowe Ltd, Eastbourne

Contents

x   *Contents*

List of principal symbols

A = amplitude, effective aperture of an antenna, directivity error

α = attenuation constant, phase angle, elevation angle, sensitivity

$a_x, a_0, \ldots$ = unit vectors

a, b = complex wave functions

B_N = noise bandwidth

B = bandwidth, susceptance, effective source match

β = propagation constant

c = coupling coefficient

C = coupling ratio, effective source match, capacitance

$D(\theta, \phi)$ = antenna directivity at angle θ, ϕ

D = directivity

D_M = maximum dynamic range

δ = variance, polarisation phase angle

D, R_c, R_L = directivity and reflection amplitudes

E = electric field vector

E = error ratio, electric field

E_D, E_R, E_S = error terms

e = electronic charge

$\overline{e^2}$ = mean-square-noise EMF

η = power-head efficiency, radiation efficiency, azimuth angle

F = electric-field vector

f = frequency

F = noise figure

ϕ = phase angle, co-ordinate in spherical system

G = conductance, gain

γ = complex propagation constant, gyromagnetic resonance constant

h_s, h_t = source and test antenna height

I = current, intermodulation intercept

i = current

$\overline{i^2}$ = mean-square noise current

J_M = magnetic current density

k = vector wave number

L = loss factor, inductance, length

l = length

λ = wavelength

λ_0 = free-space wavelength

$\mathscr{L}(f)$ = two-sided power-spectral-density ratio

$\boldsymbol{n}$ = unit vector normal to a surface

ω = angular frequency

ρ_L, ρ_D, ρ_c = co- to cross-polar ratios

P = power, port designation

P_{av} = power available into a conjugate match

P_Z = power into a Z_0 match

$P(\theta, \phi)$ = power density radiated at θ, ϕ

p = polarisation loss factor

Q = Q-factor

r = co-ordinate in spherical system, radius, polarisation axial ratio

R = resolution bandwidth, resistance, radial distance

R_0 = minimum resolution bandwidth, characteristic resistance

ρ = reflection coefficient

S/N = signal/noise ratio

$S_{\Delta\phi}(f)$ = one-sided spectral density of phase fluctuation

$S_y(f)$ = spectral density of fractional frequency fluctuation

S = frequency scan

$S(v), C(v)$ = Fresnel integrals

$S_{11}\ldots$ = scattering parameters

$\sigma_y^2(\tau)$ = Allan variance

t = time, transmission coefficient

τ = time constant, time period, polarisation tilt angle

T = periodic time, temperature, transmission coefficient

T_0 = standard noise temperature, 290°K

T_K = Kth forward path in flowgraph

θ = phase angle, co-ordinate in spherical or cylindrical system

θ_B = 3 dB beam width

V = voltage

v = voltage, variable in Fresnel integral

v_p = phase velocity

X_0 = characteristic reactance

x, y, z = rectangular co-ordinates

Y = admittance, Y-factor in noise measurement

Y_0 = characteristic admittance

Z = impedance

Z_0 = characteristic impedance

ξ = intrinsic impedance of free space

Preface

This book is aimed at the postgraduate or final-year honours level, and assumes some knowledge of electromagnetic theory and a general familiarity with microwave components. It should be useful to young engineers, who inevitably bear the brunt of detailed laboratory work, but also to more experienced engineers wishing to update or refresh their knowledge. The guiding strategy is informed by the notion of a young engineer who, on first entering an industrial laboratory, would be struck and perhaps somewhat overawed by the large number of shiny boxes performing apparently quite difficult experiments and obviously based on a lot of complicated theory.

A traditional course on microwave measurements might have begun with transmission lines and a review of the theory and application of microwave components, such as attenuators, standing-wave indicators, ferrite isolators and short circuits, as a necessary prerequisite to a detailed description of some standard techniques and systems. This book takes a different approach and looks at measurements through the instruments rather than through the components, since the reality for most new entrants to the industry is of laboratories stocked with sophisticated instruments, often computer controlled with remote automatic operation, and distinguished on sight only by the number of program select buttons, the type of display and the number of connecting sockets. Again traditionally, sources would have been at fixed frequencies and only spot frequencies would have been selected in the operating band, whereas in modern measurements the ubiquitous swept source is used to obtain a continuous in- and out-band response of reflection, transmission, power level, noise figure, gain and many other results from passive and active devices. This is recognised by beginning with the theory of swept sources before introducing the many instruments that would not exist without them.

A good experimenter, whilst fully understanding the wisdom of rigorous mathematical models, knows that, in a practical subject, theoretical approximations are useful so far as they give a common-sense guide to the processes and the results expected from them. The art of approximation is learned through practice. It requires extreme care and yet will not be acquired without a sense of adventure, that must always be tempered by cautious checks against rigorous theory and common sense. Models that result from approximations, based on the art of knowing which terms or components can be disregarded in an analysis, are also justifiable if they simplify the explanation and avoid the confusion that a rigorous theory may cause the first-time reader. This practice is adopted where possible in this book; for example, in the application of flowgraphs to networks and in the illustrations of error estimation in antenna near-field measurement. Menu-driven procedures and computer-type documentation allow the engineer and test technician alike to perform set operations equally effectively. But the former should also know how the equipment works, be familiar with any supporting theory, and therefore be capable of devising new experiments or even performing standard measurements that properly take account of the limitations of the instruments and their working environments. There is always the temptation to believe a computer printout, but good engineers do not accept the results of complicated multiple procedures without making simple approximate calculations to check their likelihood, and this is only possible when based on a thorough understanding of the instruments. This book seeks to correct such tendencies by encouraging approximations that are uncompromisingly set in the context of a thorough theoretical background.

The material is based on a course in the postgraduate programme of the Open University, devised in collaboration with, and sponsored by, The Plessey Company PLC, and I would like to take this opportunity to acknowledge the support of Mr. T. G. P. Rogers on behalf of the Directors. It was developed from an agreed syllabus and first taught to a group of in-service students during a period when I was on secondment as a consultant to the company, and has since been recorded on video tape for use in other companies. Plessey granted me access to test equipment at several sites, and many engineers generously gave their time and knowledge in this joint effort. I wish to record my thanks to Ray Pengelly, Jeff Buck and Jim Arnold of Plessey Research Ltd., Caswell; to John Fish and Geoff Purcell of Plessey Electronic Systems Ltd., Roke Manor; to Peter Bradsell and especially to Dave Jones of Plessey Radar Ltd., Cowes, whose unfailing support guaranteed our success. Among other people and companies that have given help are Will Foster of Marconi Instruments, St. Albans; Rey Rosenberg of Hewlett Packard Ltd., Boreham Wood; Phil Combes of Wiltron Ltd., Crowthorne; B. Fleming and K. I. Khoury of Flann Microwave Instruments, Ltd., Bodmin; and Eric Griffin, Frank Warner, Jezz Ide and Malcolm Sinclair of RSRE, Malvern. Finally, I wish to thank my students, Helen Stowe, Bob Graham, Jonathon Bluestone, Irfan Altaf, Alistair Doe and Derek Reeves, who, by kindly bearing with me during the first presentation of the course, made possible this book.

Chapter 1

Swept-frequency principles

1.1 Introduction

It would be difficult to imagine a modern microwave laboratory that did not
have one or more time-swept frequency sources. They are to be found in every
type of experiment, in standards laboratories, on research benches and in production test areas; and are especially suited to the rapid acquisition of results in
broad-band measurements. It is for this reason that we begin with the principles
of swept sources. Microwave components, such as filters, power splitters and
amplifiers, operate over a frequency range, referred to as the bandwidth,
though it is often true that performance outside that range can be equally
important, as for instance when frequency conversion causes in-band effects
due to out-of-band responses. Broad-band characterisation of components and
systems is therefore an essential element of modern design and testing. It was
always possible to make a series of spot-frequency measurements, over a required frequency band, but this involved many adjustments of source, detector
and other components of a test set, and became time consuming and costly in
engineering effort. As component bandwidths increased so did those of the test
sets, including sources and detectors, so that it became possible to sweep a
source through a desired frequency range and to get a continuous detected
output of measured quantities, such as transmission or reflection return loss.
But swept sources do have limitations, many of which, such as power levelling,
frequency stability or match, will be taken up in later Chapters. A complete
appraisal of the operation and performance of a swept spectrum analyser is not
possible at this stage. Receiver noise, detection methods, display formats etc. all
affect the output result to varying degrees, depending on the precise nature of
the measurement. In covering many of these aspects we will see that response
sensitivity depends on a number of factors, such as system noise level, bandwidth, averaging procedures, data rates and, in swept systems, on the sweep
rate. In order to avoid confusion, in this first Chapter we take the opportunity
to review the fundamentals of frequency resolution as we examine how the
detection sensitivity is affected by sweep rate.

The spectrum of a constant-amplitude swept-frequency source might at first sight appear to be a continuous band of frequencies of equal amplitude. In practice, sources are swept repetitively, with a rapid flyback to the start frequency between each sweep. The time discontinuities at the start and stop of the sweep cause the spectrum to spread beyond the swept range, and also result in the appearance of amplitude ripples within the range. These effects are similar to those at a diffraction edge in optics or from the edges of antenna apertures. There may be further amplitude and phase disturbances if the source wave is detected after reflection from, or transmission through, a component under test. A set of continuous filters, or a single filter after down conversion to an intermediate frequency, can resolve a swept-frequency spectrum into smaller frequency bands for spectral analysis of amplitude and phase. Spectral resolution increases with decreased filter bandwidth, but, because frequency varies with time in the sweep, the filters are both frequency and time dependent in a special way that leads to a fundamental limit on attainable minimum resolution bandwidth.

A variety of test-set arrangements is possible, as we shall see in later Chapters, but for the moment just two are illustrated in Fig. 1.1. A simple transmission measurement in Fig. 1.1A has a voltage-controlled oscillator (VCO) with constant output amplitude which gives a series of ramp frequency changes with time when driven by an equivalent ramp control voltage. A diode detector and

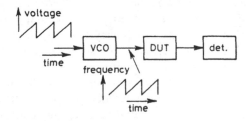

Fig. 1.1A   *Swept transmission measurement*

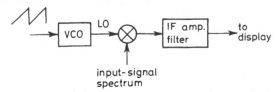

Fig. 1.1B   *Time display of spectral response at intermediate frequency*

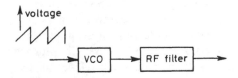

Fig. 1.1C   *RF filtering of swept frequency*

display are connected to the output of the device under test (DUT), to measure transmitted level as a function of frequency. In Fig. 1.1B we see how a signal spectrum can be investigated by mixing it with a swept source and observing the result after filtering at an intermediate frequency. When a component of the signal spectrum is separated in frequency from the instantaneous swept frequency by the intermediate frequency, an output as a function of frequency appears in the display if the ramp is used to drive the horizontal displacement. It would appear that the observable spectrum detail should depend only on the spectral response of the DUT in Fig. 1.1A or the IF-filter bandwidth in Fig. 1.1B, but this is not true when the square root of the sweep rate approaches the reciprocal of the filter time response. We can see this by considering the filtering to be at the swept-source output, as in Fig. 1.1C.

1.2 The swept spectrum

We can begin to analyse this situation by first considering the linear time dependence of the swept source with frequencies and times as defined in Fig. 1.2.[1] A linear scan from f_1 to $f_1 + S$ with period T gives an instantaneous frequency $f(t)$ at time t, where

$$f(t) = f_1 + \frac{S}{T} t \qquad (1.1)$$

in the range $nT < t < (n + 1)T$.

The instantaneous angular frequency is also given by the time rate of change of phase, or

$$2\pi f(t) = \frac{d\phi(t)}{dt} \qquad (1.2)$$

On substituting for $f(t)$ and integrating for the instantaneous phase we have

$$\phi(t) = 2\pi \left\{ f_1 t + \frac{S}{2T} t^2 \right\} \qquad (1.3)$$

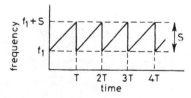

Fig. 1.2 *Repeated linear frequency/time sweeps*

and if V_0 is the amplitude of the output RF voltage, the instantaneous voltage becomes

$$V(t) = V_0 \exp j2\pi \left\{ f_1 t + \frac{S}{2T} t^2 \right\} \tag{1.4}$$

The spectrum of the swept output $S(f)$ is the Fourier transform of $V(t)$:

$$S(f) = \int_{-\infty}^{+\infty} V(t) \exp(-2\pi f t) \, dt$$

On substituting for $V(t)$ from eqn. 1.4,

$$S(f) = \int_{-\infty}^{+\infty} V_0 \exp -j \left\{ 2\pi (f - f_1)t - \pi \frac{S}{T} t^2 - \frac{\pi}{S/T} (f - f_1)^2 \right\}$$

$$\times \exp -j \frac{\pi}{S/T} (f - f_1)^2 \, dt$$

The function $V(t)$ is repetitive, with a line spectrum of Fourier terms separated by the pulse repetition frequency. The spectrum for a single cycle in the interval $0 < t < T$ gives the shape of the spectrum envelope as

$$S(f) = V_0 \int_0^T \exp -j\pi \left[\sqrt{S/T} t - \frac{f - f_1}{\sqrt{S/T}} \right]^2 \exp -j\pi \frac{(f - f_1)^2}{S/T} \, dt \tag{1.5}$$

This is a Fresnel equation which can be put in a standard form by writing

$$v^2 = 2 \left[\sqrt{S/T} t - \frac{f - f_1}{\sqrt{S/T}} \right]^2 \tag{1.6}$$

with

$$v = v_1 = \sqrt{\frac{2T}{S}} (f - f_1) \qquad \text{at } t = 0 \tag{1.7}$$

and

$$v = v_2 = \sqrt{\frac{2T}{S}} [f - (f_1 + S)] \qquad \text{at } t = T \tag{1.8}$$

Therefore

$$S(f) = V_0 \exp -j\pi \frac{(f - f_1)^2}{S/T} \int_{v_1}^{v_2} \exp j \frac{\pi}{2} v^2 \, dt$$

But

$$dv = \sqrt{\frac{2S}{T}} \, dt$$

Therefore

$$S(f) = \sqrt{\frac{T}{2S}}\, V_0 \exp\, -j\pi\, \frac{(f-f_1)^2}{S/T} \int_{v_1}^{v_2} \exp j\frac{\pi}{2} v^2\, dv \qquad (1.9)$$

Now

$$\int_{v_1}^{v_2} \exp j\frac{\pi}{2} v^2\, dv = \int_{0}^{v_2} \exp j\frac{\pi}{2} v^2\, dv - \int_{0}^{v_1} \exp j\frac{\pi}{2} v^2\, dv$$

The exponential expressions on the right-hand side can be expanded to give the following form:

$$\int_{0}^{v} \exp j\frac{\pi}{2} v^2\, dv = \int_{0}^{v} \cos\frac{\pi}{2} v^2\, dv + j \int_{0}^{v} \sin\frac{\pi}{2} v^2\, dv$$

$$= C(v) + jS(v) \qquad (1.10)$$

$C(v)$ and $S(v)$ are standard forms of the Fresnel integral whose values are tabulated[2] as a function of v.

The solution of eqn. 1.9 can now be written in terms of these Fresnel integrals as

$$S(f) = V_0 \sqrt{\frac{S}{T}}\, \{[C(v_1) - C(v_2)] + j[S(v_1) - S(v_2)]\} \exp\left[-j\pi\, \frac{T}{S}(f-f_1)^2\right]$$

$$(1.11)$$

Its generalised graphical form is shown in Fig. 1.3, in which the nominally equal amplitude band of frequencies from f_1 to $f_1 + S$ is distorted by Fresnel diffraction; showing a 6 dB fall from the mean peak level at the extremes of the sweep range, amplitude ripples in the normal pass band and spectral leakage beyond it. The ripples decay towards the centre frequency from each edge, the number

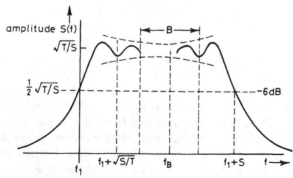

Fig. 1.3 *Fresnel spectrum for single linear ramp*

depending on the scan width. If a section of the swept band is passed through a rectangular filter of bandwidth B, the Fresnel pattern reappears with the -6 dB levels now at the filter edges and with a reduced number of ripples.[1]

Fresnel integrals are characterised by the square term v^2 in the integrand, and they occur in a number of mathematical models of physical reality. For instance, in optics, they are used to calculate the amplitude maxima and minima in a wave diffracted by a straight edge, or to find the radiation pattern of an illuminated hole in an otherwise opaque screen.[3] When the observation point is at a great distance from the hole, waves from all parts of its area can be considered to propagate along parallel paths, and the summation of each field contribution depends on a linear phase relation to the hole dimensions. As the hole and the observer become closer, the phase of each wave component is no longer linearly related to the hole dimensions, and the distance is defined as being in the Fresnel region when the dimensional dependence of phase is second order. Since phase is represented as an exponent, it is possible to express the summation in a Fresnel integral form with v in eqn. 1.10 as a dimensional variable in the aperture. Fig. 1.3 may now represent the amplitude of the radiation pattern in a region close to an aperture, with the horizontal axis re-defined as the displacement angle from the normal to that aperture. At greater distances, where the phase dependence is linear, ripples are not evident in the central peak region of the pattern, and only a single peak is seen. This is normally referred to as the Fraunhofer or far-field region. These ideas are also found in antenna theory, and a full discussion is given in Chapter 12 when we examine the behaviour of aperture antennas.

1.3 Minimum-resolution bandwidth

In Fig. 1.3 the first minimum in the ripple pattern occurs at $f_1 + \sqrt{S/T}$, where f_1 is the lower edge frequency of the filter or the start of the ramp when no filter is applied. It can be seen that, if B is reduced much below $\sqrt{S/T}$, there will be a single peak whose level will eventually decrease below $\sqrt{T/S}$ as B is still further reduced. The effect is illustrated in Fig. 1.4, where the time-filtered interval is shown to be BT/S. In Fig. 1.4b just two peaks occur in the filter pass band, whereas in Fig. 1.4c the bandwidth of the filter is too small for even the first peak fully to form. In consequence, the pattern -6 dB bandwidth exceeds the filter bandwidth. The condition $B = \sqrt{S/T}$ roughly corresponds to the minimum bandwidth obtainable by filtering. It is related to a gradual transition from Fresnel to Fraunhofer diffraction in the region of minimum bandwidth. As we have seen, the latter is distinguished from the former in having a single main peak; and as we will see in Chapter 12 the monotonic edges in Fig. 1.3 may also become a series of smaller amplitude sidelobes. We see that there is a progressive transition and that the criterion for change depends on the application. In this case the issue is loss of main peak level, causing increased pattern

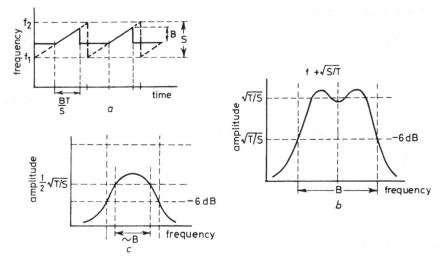

Fig. 1.4 *Filtered Fresnel spectrum of linear ramp*
Filter output is pulse train with PRF = 1/T

bandwidth; whereas in the transform theory of antenna pattern determination, side-lobe nulls are often the important test. Both are related to a quadratic phase deviation, due to a square term in the time interval in eqn. 1.3 in the former case, and to a similar spatial effect in the radiating aperture in the latter.

We can compare the criteria through the quadratic phase 'error' in the two cases. In Chapter 12 we see that the phase error between the centre and edge of a radiating aperture should be less than $\pi/8$ for Fraunhofer or far-field conditions. In swept-frequency filtering an optical criterion can be used, as we will now show.

Let v_1 and v_2 correspond to the edges of a rectangular filter sampling a swept frequency. Eqns. 1.7 and 1.8 show that

$$\Delta v = v_1 - v_2 = \sqrt{\frac{2T}{S}} \, B \tag{1.12}$$

The scan rate remains at S/T, but the frequency scan is only B. Δv is a chord on the Cornu spiral[3] corresponding to the filter bandwidth; and in optics a criterion for a transition to Fraunhofer diffraction, in the sense of the formation of a single main peak with no loss in mean level, occurs when $\Delta v \approx \sqrt{2}$. If we substitute this in eqn. 1.12 the minimum resolution bandwidth is

$$R_0 = B = \sqrt{S/T} \tag{1.13}$$

The equivalent time-aperture quadratic phase error can be found by reference to eqn. 1.3 in which

$$\phi_q = \pi \frac{S}{T} t^2$$

If T' is the scan time in the filter

$$T' = \frac{T}{S} B$$

and

$$\phi_q = \pi \frac{T}{S} B^2 \qquad (1.14)$$

Therefore, when B is the minimum resolution bandwidth,

$$\phi_q = \pi$$

which is eight times as large as the antenna quadratic phase condition in eqn. 12.54.

1.4 Gaussian filter

Real filters are not rectangular in shape. A Gaussian shaped filter is typical of cascaded amplifiers in IF circuits, so, as a comparison, we will find the minimum resolution bandwidth for a Gaussian filter with the characteristics shown in Fig. 1.5. If δ is the variance, the frequency response of a Gaussian is

$$H(\omega) = \exp\left[-\frac{1}{2}\left(\frac{\omega}{\delta}\right)^2 \right] \qquad (1.15)$$

where $\omega = 2\pi f$, the angular frequency.

At the -3 dB point $\omega = \pi B$ and

$$H(\omega) = 1/\sqrt{2}$$

giving

$$\delta = (\pi/\sqrt{\ln 2})B \qquad (1.16)$$

where B is the 3 dB bandwidth of the filter.

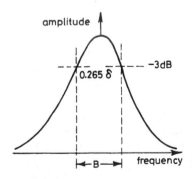

Fig. 1.5 *Gaussian frequency response*

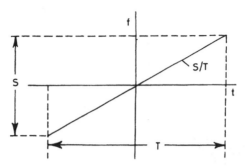

Fig. 1.6 *Single-frequency ramp*

To find the response to a linear sweep we can use a mathematical function which includes zero frequency with a sweep from $-T/2$ to $+T/2$ as in Fig. 1.6. Only the quadratic term appears in the phase to give the swept voltage as

$$V(t) = \exp\left(j\pi \frac{S}{T} t^2\right) \tag{1.17}$$

This has a Fourier transform[7]

$$S(\omega) = \tau\sqrt{2\pi} \exp -\frac{1}{2}(\tau\omega)^2 \tag{1.18}$$

where

$$\tau = \sqrt{\frac{j}{2\pi}\frac{T}{S}}$$

The product of eqns. 1.15 and 1.18 gives the frequency response of the swept filter, $Y(\omega)$.

$$Y(\omega) = S(\omega)H(\omega) = \tau\sqrt{2\pi} \exp\left[-\frac{1}{2}\left(\tau^2 + \frac{1}{\delta^2}\right)\omega^2\right] \tag{1.19}$$

The inverse Fourier transform of $Y(\omega)$ gives the time response $y(t)$ as

$$y(t) = \frac{1}{\left[1 + \left(\frac{2\pi}{\delta^2}\frac{S}{T}\right)^2\right]^{1/4}} \exp\left[-\frac{\delta^2 t^2/2}{1 + \left(\frac{T\delta^2}{S\,2\pi}\right)^2}\right] \tag{1.20}$$

$Y(\omega)$ and $y(t)$ are both Gaussian forms, since Fourier transformation of a Gaussian results in a new Gaussian.[5,6]

The method to find the filter bandwidth in eqn. 1.16 can be used to find the time bandwidth corresponding to the -3 dB points in eqn. 1.20. This is illustrated in Fig. 1.7 by

$$\Delta t' = 2\frac{\sqrt{\ln 2}}{\delta}\left[1 + \left(\frac{\delta^2 T}{2\pi S}\right)^2\right]^{1/2} \tag{1.21}$$

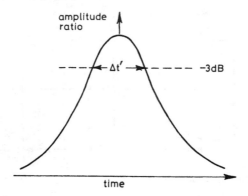

Fig. 1.7 *Gaussian time response*

Remembering that δ defines the filter bandwidth, we can see that, when the square root of the sweep rate is much less than the bandwidth, or

$$\sqrt{\frac{S}{T}} \ll \frac{\delta}{2\pi} \qquad (1.22)$$

eqn. 1.20 reduces to

$$y(t) = \exp\left[-\frac{1}{2}\left(\frac{2\pi}{\delta}\frac{S}{T}\right)^2 t^2\right] \qquad (1.23)$$

and eqn. 1.21 to

$$\Delta t = \frac{\sqrt{\ln 2}}{\pi}\delta\frac{T}{S} \qquad (1.24)$$

Substituting for δ from eqn. 1.16 gives

$$\Delta t = B\frac{T}{S} \qquad (1.25)$$

which is just the scan time in the filter.

If we define the resolution bandwidth of the swept filter as

$$R = \frac{S}{T}\Delta t' \qquad (1.26)$$

the ratio of this to the filter bandwidth is

$$\frac{R}{B} = \frac{\Delta t'}{\Delta t} = \left[1 + 0.195\left(\frac{S}{TB^2}\right)^2\right]^{1/2} \qquad (1.27)$$

in which eqn. 1.16 has again been used to substitute for δ. Thus for large filter bandwidth $R \approx B$, and decreasing B also decreases R. But if B decreases too much the second term in eqn. 1.27 quickly increases and $R > B$. There is a minimum resolution bandwidth R_0 which we can find by differentiating R with respect to B and setting

$$\frac{dR}{dB} = B - \frac{0 \cdot 195}{B^3}\left(\frac{S}{T}\right)^2 = 0$$

The solution for B is

$$B_0 = \sqrt{\frac{1}{2 \cdot 27}\frac{S}{T}} \qquad (1.28)$$

with $R_0 = \sqrt{2}B_0$ on substituting for B_0 in eqn. 1.27. It follows that the minimum resolution bandwidth is[4]

$$R_0 \approx \sqrt{\frac{S}{T}} \qquad (1.29)$$

This is comparable with the $-6\,$dB minimum resolution bandwidth of the rectangular filter.

Referring again to eqn. 1.20 we can define a response sensitivity as

$$\alpha = \frac{1}{\left[1 + \left(\frac{2\pi}{\delta^2}\frac{S}{T}\right)^2\right]^{1/4}} \qquad (1.30)$$

Using eqn. 1.16 to substitute for B gives

$$\alpha = \left[1 + 0 \cdot 195\left(\frac{S}{TB^2}\right)^2\right]^{-1/4} \qquad (1.31)$$

A plot of bandwidth ratio against a normalised sweep rate shows, in Fig. 1.8, that $R = B$ and $\alpha \approx 1$ when $B \geqslant R_0$, but that, for $B < R_0$, system performance deteriorates through increasing resolution bandwidth and decreasing sensitivity.

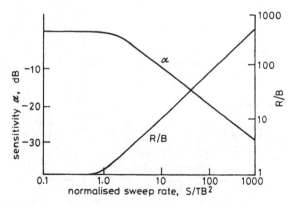

Fig. 1.8 *Sensitivity and resolution bandwidth against normalised sweep rate*

1.5 Spectrum analyser

It is possible to analyse the spectral structure of a band of frequencies by passing it through a set of contiguous filters. The resolution bandwidth is related to the filter bandwidths, and determines the fineness of detail in the measured amplitude as a function of frequency. In a swept spectrum analyser, the contiguous filters are replaced by a single filter at an intermediate frequency, and the band of frequencies is heterodyned into it by means of a swept local oscillator and a mixer. The filter output is a time-scanned record of the spectral components. These components can be displayed as a function of frequency to a resolution given by the filter bandwidth and the frequency scan rate.

A simplified block diagram of a swept-frequency spectrum analyser is shown in Fig. 1.9. The time base controls a linear sweep from 0 to T seconds and synchronises the display scan. The input signal (the one under examination) is mixed with the local-oscillator output to produce a component at the intermediate frequency which is selected by the IF filter. In the display part of Fig. 1.9, the IF outputs, for an input signal consisting of components at two frequencies f_1 and f_2, are shown appearing at sweep times t_1 and t_2. Normally the IF outputs would be detected before display and appear as positive vertical displacements on a time axis.

Fig. 1.10 illustrates the case when the local-oscillator frequency is always greater than the detected signal input by an amount equal to the intermediate frequency f_{IF}. The upper ramp line from f_0 to $f_0 + S$ represents the instantaneous frequency of the local oscillator. It repeats in time intervals T. Detectable signal frequencies always lie in a band of three lines, separated from, and below, the local-oscillator ramp by f_{IF}. The centre of these lines extends from $f_0 - f_{IF}$ to $(f_0 + S) - f_{IF}$; and the two outer lines indicate the resolution bandwidth at the intermediate frequency. Two signals are shown at frequencies f_1 and f_2. They fall within the resolution bandwidth at times t_1 and t_2 for a time interval

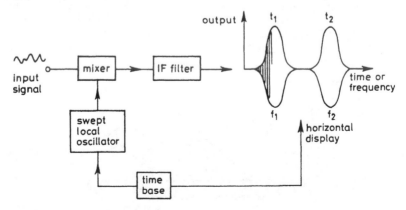

Fig. 1.9 *Elements of spectrum analyser*

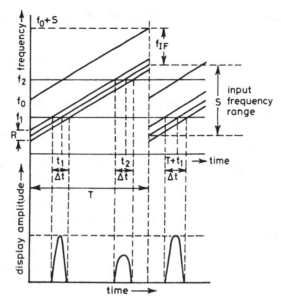

Fig. 1.10 *Time response and resolution bandwidth*
$f_0 = f + f_{IF}$; $f = f_1$ or f_2

Δt, during which they are able to generate an IF output to the display. We have already seen how the resolution bandwidth is determined by the IF filter and the frequency scan rate.

1.6 Mixer products

The non-linear transfer characteristic of a mixer can be approximated with increasing accuracy by progressively including higher-order terms of a polynomial expansion of the form

$$v_0(t) = av_i(t) + bv_i^2(t) + cv_i^3(t) + \cdots \tag{1.32}$$

in which $v_0(t)$ is the output voltage for input voltage $v_i(t)$. If $v_1(t)$ and $v_2(t)$ are the voltages at frequencies f_1 and f_2 and $V_0(t)$ is the instantaneous voltage of the swept local oscillator

$$v_i(t) = V_0(t) + v_1(t) + v_2(t) \tag{1.33}$$

A good mixer is a signal multiplier in which the second-order coefficient b in eqn. 1.32 is much larger than the others. In practice, neither a, c nor some of the higher-order ones are insignificant, with the result that the input frequencies and many higher-order products are present at the input to the IF filter. The only required output stems from the second-order term with coefficient b, and, for the moment, this is the one we will consider.

On substituting for $v_i(t)$ in eqn. 1.32 and retaining only second-order terms, we have

$$v_0(t) = b[V_0^2(t) + 2V_0(t)v_1(t) + 2V_0(t)v_2(t) + v_1^2(t) + 2v_1(t)v_2(t) + v_2^2(t)]$$

(1.34)

An intermediate frequency is formed from the sum or difference combinations found in the terms $2V_0(t)v_1(t)$ and $2V_0(t)v_2(t)$. The remaining terms in eqn. 1.34 generate a constant level, and other sums and differences of frequencies that can usually be eliminated by the IF filter.

There are more products generated by the other neglected terms of the polynomial expansion of $v_0(t)$. In general, the most difficult to eliminate by filtering are those deriving from the third coefficient, with frequencies $(2f_2 - f_1)$ and $(2f_1 - f_2)$ which are very close to f_1 and f_2. In Appendix 1 we see that third-order intermodulation distortion is an important factor in the specification of a spectrum analyser.

Assuming the spurious frequency components can be reduced to negligible levels, the output voltage of the mixer reduces to

$$v_0(t) = 2bV_0(t)v_1(t) + 2bV_0(t)v_2(t)$$

(1.35)

The spectrum of the product of two time-domain functions is the convolution of the individual spectra.[8] Thus, if $H(f)$, $X(f)$ and $Y(f)$ are the spectra of $V_0(t)$, $v_1(t)$ and $v_2(t)$, respectively, the spectrum of $v_0(t)$ is

$$Z(f) = \int_{-\infty}^{+\infty} H(p)X(f-p)\, dp - \int_{-\infty}^{+\infty} H(p)Y(f-p)\, dp$$

(1.36)

where p is a running parameter in the integration.

We will assume that the IF filter is rectangular and that $V_0(t)$ has a linear frequency/time ramp. The spectrum of $V_0(t)$ is therefore simplified to the Fresnel pattern in Fig. 1.11, and the convolution of it with the spectra of $v_1(t)$ and $v_2(t)$ is shown in Fig. 1.12.

The line spectra of $v_1(t)$ and $v_2(t)$ in Fig. 1.12a are convoluted with the local-oscillator spectrum in Fig. 1.12b to give four bands, one pair for sum and

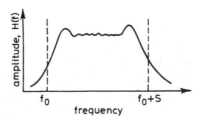

Fig. 1.11 *Ramp frequency response*

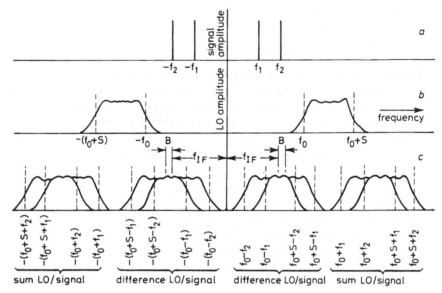

Fig. 1.12 *Mixer products*

a second pair for difference combinations. Spectral overlapping occurs in the IF passband shown at f_{IF} in Fig. 1.12c. Separation of the frequency components is, in fact, temporal, as can be seen by referring to Fig. 1.13, in which the time apertures of the two signals are separated by nearly a whole sweep period. f_1 appears in the IF filter near the start, and f_2 near the end, of the sweep period. The closer they are the greater is the degree of overlap and the shorter is the time separation. If Δt is the time resolution of the swept filter the signals are separated provided

$$|t_2 - t_1| \geqslant \Delta t \tag{1.37}$$

and in the limiting case of minimum resolution bandwidth, with $R = R_0$ in eqn.

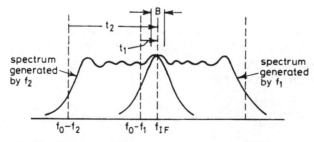

Fig. 1.13 *Time and frequency resolution*

1.26 and using eqn. 1.29 for R_0,

$$\Delta t = \Delta t' = T\frac{R_0}{S} = \frac{1}{R_0} \tag{1.38}$$

An alternative measure of the minimum separation between two frequencies f_1 and f_2 is the lowest obtainable beat frequency $\Delta f = f_2 - f_1$. But a sinusoid has a single frequency only if it exists, and could therefore be measured, for all time; and this is the significance of the infinite limits in the Fourier transform of a time function. In practice, such long measurements are impossible; or they would have no practical use if they were possible. The question is: how can we decide whether an observation time is long enough for a given purpose? The answer is: judgment based on reasonable assumptions. For instance, the frequency of a single sinusoid can be determined by observing it for at least one period. There is uncertainty about the measured result if only one period is observed, countered by the observer's reasonable assumption that the sinusoid exists. When this idea is applied to frequency measurement, the observation time must be at least equal to the reciprocal of the beat frequency, or

$$\Delta t \geqslant \frac{1}{\Delta f}$$

This mutual exclusiveness of frequency and time interval is one form of the uncertainty principle, usually written as

$$\Delta f \, \Delta t \geqslant 1$$

and gives from eqn. 1.38,

$$\Delta f \geqslant R_0 \tag{1.39}$$

The minimum observable separation is therefore limited by the resolution bandwidth of a spectrum analyser. As we will see in Appendix 1, spectral resolution is also affected by system noise, particularly residual frequency modulation and close-to-carrier phase noise.

1.7 Pulsed signals

To complete this preliminary discussion of spectrum analysis, we will briefly look at the measurement of pulsed signal spectra. The simplest pulse is rectangular in shape. If its pulse length is t_0 seconds the spectrum is a $(\sin X)/X$ function which we can write as

$$S(f) = \frac{\sin (\omega t_0)/2}{(\omega t_0)/2} \tag{1.40}$$

where ω is the angular frequency.

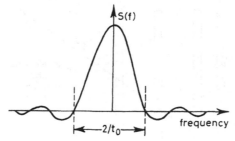

Fig. 1.14   *Spectrum of a single pulse*

The pattern of $S(f)$ is sketched in Fig. 1.14. Most of the pulse energy lies in the frequencies between the first nulls within the bandwidth $2/t_0$. If the pulse length t_0 is small compared to the observation time Δt, its spectrum is wider than the resolution bandwidth of the spectrum analyser and the spectral structure will be evident in the displayed result. As Δt approaches t_0, the main spectral lobe fills the filter bandwidth and the response appears as a CW signal. In practice, if

$$t_0 \leqslant 0{\cdot}1 \, \Delta t$$

the main features of the pulse spectral envelope are observable. But the uncertainty principle states that

$$R_0 \, \Delta t \geqslant 1$$

therefore

$$t_0 \leqslant \frac{0{\cdot}1}{R_0}$$

or $R_0 t_0 \leqslant 0{\cdot}1$, for observation of the pulse envelope. The uncertainty principle relates to the observation time Δt, which must be greater than t_0.

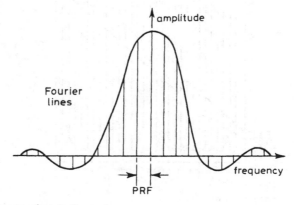

Fig. 1.15   *Spectrum of repeating pulse*

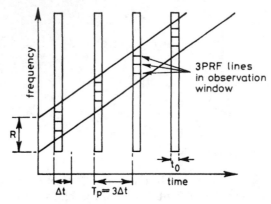

Fig. 1.16 *Fourier lines per pulse*
PRF $< R$, $T_p > \Delta t$, $T_p = 1/\text{PRF}$

Pulses must be repeated to make continuous measurements, but this causes a line spectrum to occur inside the envelope spectrum of a single pulse with line separation equal to the pulse repetition frequency. The effect is shown in Fig. 1.15. When the pulse repetition frequency is less than the resolution bandwidth there are several lines per resolution window. It is clear from Fig. 1.16 that, for three PRF Fourier lines per observation window, the observation time is greater than the pulse width and the spectrum is widely spread compared to the sweep range, even though three lines are integrated in the resolution filter.

A second example is shown in Fig. 1.17 in which the PRF is greater than the resolution bandwidth, labelled R, since in practice the analyser is not always set

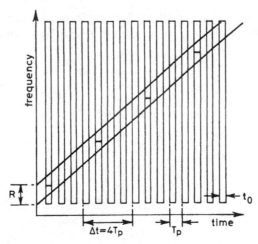

Fig. 1.17 *Fourier lines per pulse*
PRF $> R$, $T_p < \Delta t$, $T_p = 1/\text{PRF}$

to the minimum resolution. Line separations are greater than R and a line occurs only every fourth pulse. The limiting case for satisfactory observation is $PRF \leqslant R$. When the PRF is much more than this, base-line lifting can be seen in the display. It is caused because the response from the previous pulse has not sufficiently decayed when the current one occurs.[9]

Pulse measurements are made at lower sensitivity than for CW signals because the pulse width must be less than the time-observation window. It therefore appears for only a fraction of the total observation time, and the sensitivity in eqn. 1.31 is reduced in the ratio

$$\frac{\alpha_0}{\alpha} \approx \frac{t_0}{\Delta t} \tag{1.42}$$

But typically

$$t_0 \leqslant 0\cdot 1\,\Delta t \quad \text{and} \quad R\,\Delta t \geqslant 1$$

therefore

$$\alpha_0 \lesssim 0\cdot 1\alpha \quad \text{or} \quad \alpha_0 \lesssim t_0 R\alpha$$

In practice, the resolution bandwidth and sweep rate are adjusted for the best visual display.

1.8 Summary

Swept-frequency sources are used in broad-band and automatic measurements. We have been mainly concerned with the fundamental limitations set by sweep rate and bandwidth, in order to lay solid foundations on which to base our further investigations of modern swept-measurement systems. In consequence, and also because their great versatility precludes further discussion at this stage, this brief introduction to spectrum analysers is completed in Appendix 1 by looking at the specification of a particular analyser at a stage where we may assume sufficient familiarity with many of the uses of other specialised instruments to which analysers may also be applied.

1.9 References

1 MILNE, K.: 'The combination of pulse compression with frequency scanning for three-dimensional radars', *Radio & Electronic Eng.*, 1984, **28**, pp. 89–106
2 JAHNKE, E., and EMDE, F.: 'Tables of functions' (Dover Publications, 1945)
3 JENKINS, F. A. and WHITE, H. E.: 'Fundamentals of optics' (McGraw-Hill, 1951) p. 372
4 ENGELSON, M.: 'Modern spectrum analyser theory and applications' (Artech House, 1984)
5 PANTER, P. F.: 'Modulation, noise and spectral analysis' (McGraw-Hill, 1965) p. 47
6 ROSIE, A. M.: 'Information and communication theory' (Blackie, 1966) p. 33
7 'Spectrum analysis, pulsed RF'. Hewlett-Packard Application Note 150-2, Nov. 1971
8 HARMAN, W. W.: 'Principles of the statistical theory of communication' (McGraw-Hill, 1963)
9 ADAMS, S. F.: 'Microwave theory and applications' (Prentice-Hall, 1969) Chap. 5.3

1.10 Examples

1   Suppose that the gate time τ of an observation sample period encompasses only half the period of a sinusoid of frequency ω. Use the convolution form to find the spectrum of the product of the sample gate and half sinusoid. Is it possible from this result to measure the frequency of the sinusoid?

2   In adjusting a spectrum analyser to observe the spectrum of a repeating 2 μs rectangular RF pulse, the sweep rate was set at 10 GHz per second at the minimum resolution bandwidth. The sweep scan was set to display the main lobe and a pair of sidelobes, one on each side of the pulse carrier frequency. What is the pulse repetition frequency which causes about 10 Fourier lines to occur in the main lobe of the pulse spectrum? How many pulses occur between each observed spectral line? What would you expect to happen to the display if the pulse length is doubled?

3   Calculate the $-6\,\mathrm{dB}$ bandwidth of the spectrum obtained by filtering a linearly swept frequency with a rectangular filter of bandwidth

$$B = \sqrt{\frac{S}{2T}}$$

Chapter 2

Error Models

2.1 Introduction

Signals enter or leave a microwave component via its input or output ports. The component can be specified entirely in terms of the input reflections at the ports and the internal transmission ratios between them. There are other ways of completely specifying a component, e.g., as a network of impedance elements, but it is often difficult to give these elements a firm physical significance at microwave frequencies. Microwave field descriptions fit comfortably with wavelike signals and their interactions with the device through reflections and internal transmissions. Most measurement procedures reduce the device to a one- or two-port structure, usually by putting known terminations on all ports except those being measured. In most cases we will consider the device under test (DUT) to be a two-port network with the reflection and transmission properties indicated in Fig. 2.1.

A microwave test set is a network providing a source-wave input and detected-wave outputs from the DUT. For a two-port DUT, the test set has at least three ports, consisting of the source input and the reflection and transmission outputs. In general, therefore, a measurement network is a multi-port, in which the DUT is embedded, and through which the wanted signals must be measured. The observed reflection and transmission waves are dependent on both the DUT and the measuring network. Imperfections in the test set, arising from internal reflections and input mismatches, can cause large errors in the measured ratios unless steps are taken to calibrate their effects. By replacing the DUT with known standard devices, it is possible to calibrate the test set, at least to the levels of accuracy of the standards and any external instrumentation. An equivalent error network, or error model, of the test set can then be derived from the calibration measurements and subsequently used to correct the results from unknown DUTs.

In this Chapter we develop the fundamental principles underlying the error models of reflection and transmission test sets. The analysis relies on S-parameters and the ideas of flowgraphs. These are introduced by considering a simple

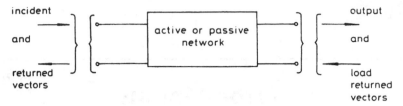

incident

and

returned
vectors

active or passive
network

output

and

load
returned
vectors

Fig. 2.1 *Two-port network*

reflectometer with a single mismatch error, in which we first sum the multiple reflections in a geometric series and then show how this can be done using the ideas of flowgraphs, but with little reduction in complexity or solving time. Finally we introduce the non-touching loop rule and demonstrate its simplicity and speed by solving the reflectometer problem and other simple examples. At this stage we are ready to apply the rule to vector and scalar error models.

2.2 *S*-parameters

S-parameters can be explained most easily by first considering a two-port network with incident and reflected waves at the input, and transmitted and load reflected waves at the output, as already shown in Fig. 2.1. If we define a_1, a_2 as waves entering, and b_1, b_2 as waves leaving, the linear two-port network in Fig. 2.2, we can write

$$b_1 = a_1 S_{11} + a_2 S_{12}$$
$$b_2 = a_1 S_{21} + a_2 S_{22}$$

(2.1)

These equations express the fact that the output wave at a port is a linear combination of the inputs at all the ports. Since the waves have amplitude and phase, the *S*-parameters are complex. *S*-parameters can be expressed as wave ratios by placing a matched load on each port in turn. Thus for a matched load on port 2, $a_2 = 0$ and

$$S_{11} = \frac{b_1}{a_1}, \qquad S_{21} = \frac{b_2}{a_1}$$

(2.2)

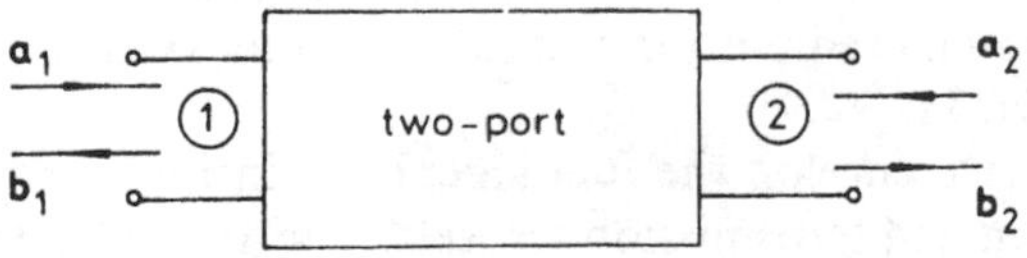

a_1

①   two-port   ②

a_2

b_1

b_2

Fig. 2.2 *Definition of input and output ports*

Similarly for a matched load on port 1, $a_1 = 0$ and

$$S_{22} = \frac{b_2}{a_2}, \qquad S_{12} = \frac{b_1}{a_2} \tag{2.3}$$

S_{11}, S_{22} are reflection coefficients at the ports and S_{21}, S_{12} are transmission coefficients through the network.[2,6]

The extension of eqns. 2.1 to N-ports is straightforward. Thus

$$b_1 = a_1 S_{11} + a_2 S_{12} + a_3 S_{13} + a_4 S_{14} + \cdots$$

$$b_2 = a_1 S_{21} + a_2 S_{22} + a_3 S_{23} + a_4 S_{24} + \cdots$$

$$b_3 = a_1 S_{31} + a_2 S_{32} + a_3 S_{33} + a_4 S_{34} + \cdots$$

$$b_4 = a_1 S_{41} + a_2 S_{42} + a_3 S_{43} + a_4 S_{44} + \cdots$$

$$\cdot \quad \cdot \quad \cdot \quad \cdot \quad \cdot \quad \cdot \quad \cdot \quad \cdot \quad \cdot \quad \cdot \quad \cdot \quad \cdot \quad \cdot$$

$$b_N = a_1 S_{N1} + a_2 S_{N2} + a_3 S_{N3} + a_4 S_{N3} + \cdots \tag{2.4}$$

2.3 Simple reflectometer

The following hypothetical example of a reflectometer measurement of the input reflection of a bandpass filter is to provide a means of comparing different methods of circuit analysis. The arrangement of such a test set is illustrated in Fig. 2.3. Output from the sweeper divides equally in the 3 dB splitter to form a reference and test channel. The test signal passes to the bandpass filter under test via a pad attenuator and directional coupler. The reflected signal at the filter input is sampled by the directional coupler, and, after suitable level adjustment for the coupling ratio, is divided in the ratio meter by the reference signal. The result is the measured complex reflection coefficient S_{11M} of the filter. The matched load at the filter output sets $a_2 = 0$, as in eqn. 2.2.

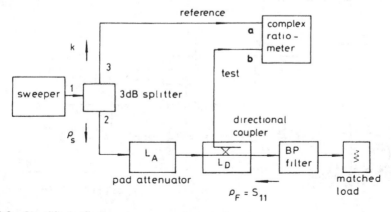

Fig. 2.3 *Simplified reflectometer*

The accuracy of this measurement depends on the quality of each component in the system. Reflections between them should be at a minimum, and the example would become very complicated should the reflection and transmission coefficients of every component be included. We will later show how the methods of signal flowgraphs can simplify the analysis, but at this stage we will use a straightforward reflection and propagation approach to demonstrate the complexity introduced by an error in only one of the components.

Suppose, then, that the 3 dB splitter is the only non-perfect component. All others transmit with linear phase delay and suffer no reflections at the inputs and outputs. Assume the splitter to split equally in the forward direction from the sweeper, but any reflected wave from the filter which enters port 2 is reflected with complex reflection coefficient ρ_S and transmitted to port 3 with coupling coefficient k. The reflection coefficient ρ_F of the filter is to be measured at the side output of the directional coupler whose main-line loss is L_D. A pad attenuator, with complex loss L_A, is inserted in the main line to reduce the error effects of multiple reflections between the filter and the imperfect 3 dB splitter. Fig. 2.4 shows how to collect the terms of the infinite series for the resulting waves in the reference and test channels. There are multiple reflections in the test channel between the splitter and filter, and with each pass the wave attenuates and phase-shifts along the line. A fraction k of each reflected wave couples to the reference channel and adds to the forward component $A/\sqrt{2}$, due to the 3 dB split.

The total reflected signal at the filter, found by summing the right-hand column in Fig. 2.4, is

$$b_1 = \rho_F L_A L_D \frac{A}{\sqrt{2}} + \rho_S \rho_F^2 L_A^3 L_D^3 \frac{A}{\sqrt{2}} + \rho_S \rho_F^3 L_A^5 L_D^5 \frac{A}{\sqrt{2}} + \cdots$$

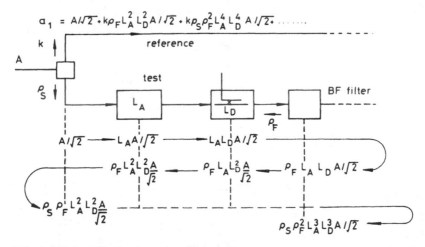

Fig. 2.4 *Multiple-reflection terms in reflectometer*

This geometric series with a common ratio less than unity can be summed as

$$b_1 = \rho_F L_A L_D \frac{A}{\sqrt{2}} \frac{1}{1 - \rho_S \rho_F L_A^2 L_D^2} \tag{2.5}$$

The reference channel wave is

$$a_1 = \frac{A}{\sqrt{2}} + k\rho_F L_A^2 L_D^2 \frac{A}{\sqrt{2}} + k\rho_S \rho_F^2 L_A^4 L_D^4 \frac{A}{\sqrt{2}} + \cdots$$

and sums as

$$a_1 = \frac{A}{\sqrt{2}} + k\rho_F L_A^2 L_D^2 \frac{A}{\sqrt{2}} \frac{1}{1 - \rho_S \rho_F L_A^2 L_D^2} \tag{2.6}$$

If the reference wave a_1 is multiplied by the product $L_A L_D$ to compensate for the padding and directional-coupler losses in the test channel, it becomes the best estimate of the forward wave at the bandpass filter. Since the reflection coefficient of the filter is the ratio of the forward to reflected waves at the filter input, the measured coefficient becomes

$$S_{11M} = \frac{b_1}{L_A L_D a_1} \tag{2.7}$$

In eqn. 2.7 the coupling ratio of b_1 to the side arm of the directional coupler would also have to be compensated for by making an equal adjustment to a_1, but since it cancels it will be ignored in all further discussion.

Substituting for b_1 and a_1 in eqn. 2.7 gives the following expression for S_{11M}:

$$S_{11M} = \frac{\rho_F}{1 + \rho_F L_A^2 L_D^2 (k - \rho_S)} \tag{2.8}$$

In the error-free case $S_{11M} = \rho_F$, so that the error ratio in eqn. 2.8 becomes

$$E = \frac{1}{1 + \rho_F L_A^2 L_D^2 (k - \rho_S)} \tag{2.9}$$

The error ratio gives the uncertainty in measurement due to the imperfections, k, ρ_S. The phases associated with each coefficient vary with frequency, each at a different rate because of the varying lengths of transmission line through which they interact. It is difficult to make estimates of the effects even for one frequency, though in vector error correction a computer can be employed. In scalar error correction it is usual to make worst-case estimates. We can see how this is done by first writing eqn. 2.9 in an amplitude and phase form as

$$E = \frac{1}{1 + |\rho_F| \exp j\phi_F |L_A|^2 \exp j2\phi_A |L_D|^2 \exp j2\phi_D (|\rho_S| \exp j\phi_S + |k| \exp j\phi_K)} \tag{2.10}$$

where ϕ_F, ϕ_A, ϕ_D, ϕ_S and ϕ_K are the phases. The largest or worst-case deviations from $E = 1$ occur when all the interfering effects contribute in phase. This is equivalent to taking only the amplitudes of ρ_F, ρ_A, ρ_D, ρ_S and ρ_K in eqn. 2.10 and choosing signs for the maximum and minimum values of $|E|$. We can show this by writing for the reciprocal of $|E|$,

$$|E|^{-1} = |1 + |\rho_F||L_A|^2|L_D|^2\{|\rho_S| \exp j\phi_1 + |k| \exp j\phi_2\}| \tag{2.11}$$

where

$$\phi_1 = \phi_F + 2\phi_A + 2\phi_D + \phi_S$$

and

$$\phi_2 = \phi_F + 2\phi_A + 2\phi_D + \phi_K$$

After some manipulation we have

$$|E|^{-1} = \{1 + 2|\rho_F||L_A|^2|L_D|^2[|\rho_S| \cos \phi_1 + |k| \cos \phi_2]$$
$$+ |\rho_F|^2|L_A|^4|L_D|^4[|\rho_S|^2 + 2|\rho_S||k| \cos (\phi_1 - \phi_2) + |k|^2]\}^{1/2}$$

This has a maximum when $\phi_1 = \phi_2 = 0$ and a minimum when $\phi_1 = \phi_2 = 180°$. Thus the maximum or minimum error ratio is given by selecting the appropriate sign in

$$|E| = \frac{1}{1 \pm |\rho_F||L_A|^2|L_D|^2(|\rho_S| + |k|)} \tag{2.12}$$

As an example, a filter in its stop band might have $|\rho_F|$ close to unity. If there is no padding attenuator, $|L_A| = 1$, and for a 20 dB coupler $|L_D| = 0.995$. A 3 dB splitter might have a 20 dB return and leakage loss to give $|\rho_S|$ and $|k|$ both equal to 0·1. The error-ratio limits are then

$$|E| = \frac{1}{1 \pm (0.995)^2(0.2)}$$

or

$$-1.57 \text{ dB} \quad \text{and} \quad +1.92 \text{ dB}$$

If a 6 dB pad is inserted, $L_A = 0.5$ and the error limits become -0.42 dB and $+0.44$ dB, clearly illustrating the beneficial effects of padding. Padding does, however, have the penalty of reduced dynamic range, since all signals are 6 dB nearer the noise level. Replacing the pad with an isolator reduces uncertainty with no loss in dynamic range, assuming it introduces no further internal reflections.

2.4 Flowgraphs

A network with more than one internal imperfection becomes very difficult to analyse by infinite summations of reflected waves. Matrix algebra, particularly with computer assistance, can be used to solve even the most complex circuits, provided all the discontinuities can be properly located, but it suffers the disadvantage of a decrease in physical insight with increasing complexity. The flowgraph method, to be described in this Section, can be equally difficult if all the terms are left in, but its close correspondence with the physical behaviour of the circuits allows simplification through well-founded approximations with no loss of physical insight. It is particularly helpful in understanding a complex network, even though a final analysis might resort to matrix algebra.

A flowgraph representation of a two-port network is shown in Fig. 2.5. The complex wave functions a_1, b_1, a_2, b_2 are the nodal points of signal flow along the paths designated by the arrows. Signals a_1 and a_2, directed into the two-port, cause output b_1, b_2. By following the arrows on the flow lines, eqns. 2.1 can be derived. Thus b_1 receives contributions $S_{11}a_1$ from a_1 and $S_{12}a_2$ from a_2 with a similar pair from a_1 and a_2 for b_2. The S's are called the path gains.

Flowgraph ideas can be extended to any number of ports. For instance, the 3 dB splitter of the previous Section can be represented as in Fig. 2.6. In this case, we have to consider only a $1/\sqrt{2}$ split with the leakage and reflection errors k, ρ_S. All other paths are assumed to have zero gains. Thus

$$S_{21} = S_{31} = \frac{1}{\sqrt{2}}, \qquad S_{32} = k, \qquad S_{22} = \rho_S$$

and the remaining paths can be deleted to give the simplified form in Fig. 2.7.

Networks are driven from signal sources which are often not matched to the input port; they are also connected at their outputs to other networks, detectors and a variety of terminations. A flowgraph representation is therefore required if we wish to incorporate these components into an analysis. Some examples are given in Fig. 2.8. A source has waves b out and a in, with reflection coefficient ρ_S. Detectors are connected to a meter with a transfer constant K, but they reflect with coefficient ρ_D. When a pair of two-ports are cascaded as in Fig. 2.9, the connected nodes $b_2 \rightarrow a_1'$ and $b_1' \rightarrow a_2$ must be correctly arrowed and given a gain of 1. The arrows then form a closed loop around b_2, a_1', b_1', b_2. The arrow labelled '1' indicates a zero electrical line length and no loss.

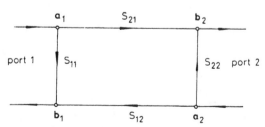

Fig. 2.5 *S-parameters for two-port*

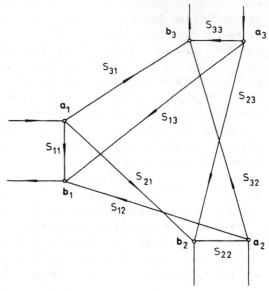

Fig. 2.6 *Flowgraph for splitter*
$S_{21} = S_{31} = 1/\sqrt{2}$
$S_{32} = k$
$S_{22} = \rho_s$

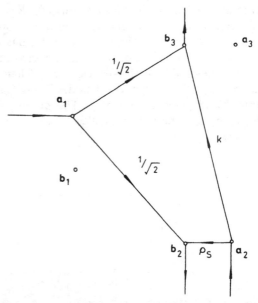

Fig. 2.7 *Reduced flowgraph for splitter*

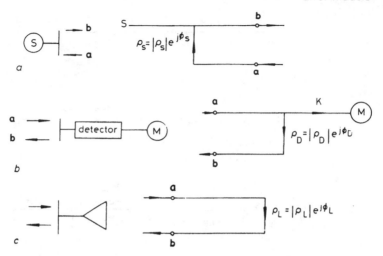

Fig. 2.8 *Flowgraph representations*
  a Signal source
  b Detector
  c Termination

Signal-flowgraph analysis as described by Kuhn[1] can be used to solve network problems whilst retaining a physical grasp of the procedure. It is based on four rules illustrated in Fig. 2.10. They are:

Rule 1:   The common node between two series branches can be eliminated by multiplying the path gains.
Rule 2:   Two branches joining two common nodes can be eliminated by adding the path gains.
Rule 3:   A branch which begins and ends on a single node can be eliminated by dividing the gain of every other branch entering the node by $(1 - \text{gain of loop})$.
Rule 4:   A node may be duplicated provided each separate flow path is retained.

These topographical rules can be used instead of the series summation to find the error ratio for the reflectometer measurement, by combining the flowgraph

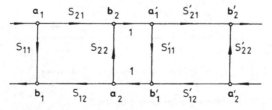

Fig. 2.9 *Flowgraph for cascaded two-ports*

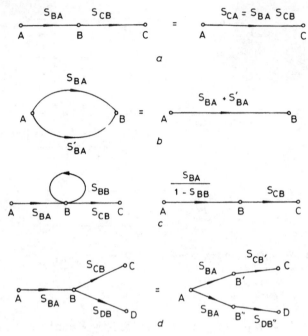

Fig. 2.10 *Topographical rules*
  a Rule 1: Branches in series
  b Rule 2: Branches in parallel
  c Rule 3: Self-loop on a single node
  d Rule 4: Node duplication

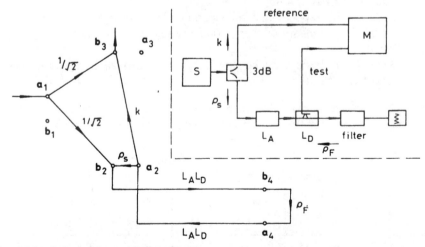

Fig. 2.11 *Reduced flowgraph for reflectometer*

of the 3 dB splitter with a flowgraph for the rest of the circuit as shown in Fig. 2.11. In the Figure the reference channel signal is b_3 and the reflected signal, which in fact would be coupled through the side arm of the directional coupler, is a_4. Measured reflection coefficient is the ratio a_4/b_3, suitably adjusted to equalise the losses $L_A L_D$ in the test channel (and also, of course, to adjust for the side-arm coupling ratio, which for convenience we have ignored).

It is easier to manipulate the flowgraph if it is re-arranged as in Fig. 2.12a. Figs. 2.12b—f show how the topographical rules are successively applied to reduce the flowgraph to just two paths: from a_1 to a_4 and b_3. Thus in Fig. 2.12b, if we wish to find the magnitude and phase at node a_4, there is an implicit path from it to the observer that must be preserved along with the path $L_A L_D$ from a_4 to a_2. We can apply rule 1 to the two series branches from b_2 to a_4 to get the single path gain $L_A L_D \rho_F$; and rule 4 to duplicate a_4, so that $a_4' = a_4$. Next we note, from the direction of the arrows, that there is a loop around b_2, a_4, a_2, b_2 with loop gain $L_A^2 L_D^2 \rho_S \rho_F$, found by applying rule 1; and that the loop can be detached if we duplicate node a_2 using rule 4 to get Fig. 2.12c. Only one branch, with gain $1/\sqrt{2}$, enters the node b_2. In Fig. 2.12d, using rule 3, this branch is divided by $1 - \rho_S \rho_F L_A^2 L_D^2$; and the series branches between nodes a_4 and b_3 are multiplied to eliminate the common node a_2 by a further application of rule 1. In Fig. 2.12e node b_2 is duplicated and a_4 is eliminated to find, from rule 1, two branches with common nodes, a_1 and b_3, that are eliminated by applying rule 2 in Fig. 2.12f.

The reference and test channel can now be written directly from Fig. 2.12f as

Reference

$$b_3 = \frac{a_1}{\sqrt{2}}\left[1 + \frac{k\rho_F L_A^2 L_D^2}{1 - \rho_S \rho_F L_A^2 L_D^2}\right] \tag{2.13}$$

Test

$$a_4 = a_4' = \frac{a_1}{\sqrt{2}}\left[\frac{\rho_F L_A L_D}{1 - \rho_S \rho_F L_A^2 L_D^2}\right] \tag{2.14}$$

After adjusting the reference channel with the test-channel loss factor, the measured reflection coefficient is

$$S_{11M} = \frac{a_4'}{L_A L_D b_3} \tag{2.15}$$

On substituting for a_4' and b_3 this becomes identical with eqn. 2.8.

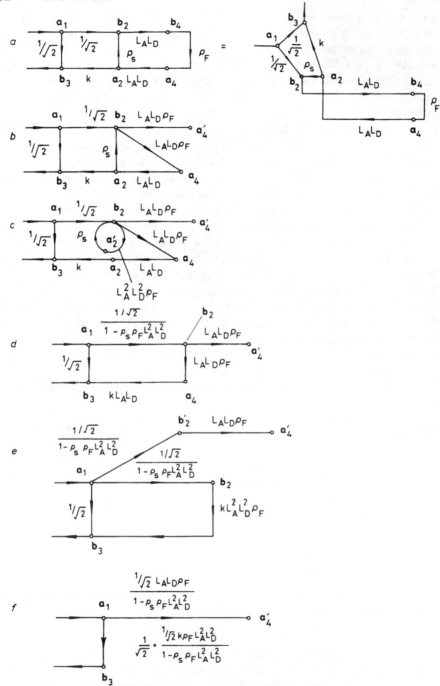

Fig. 2.12   *Application of topological rules to simple reflectometer*
     a Equivalent flowgraph   *c* Rules 1 and 4   *e* Rules 1 and 4
     b Rules 1 and 4   *d* Rules 1 and 3   *f* Rules 1 and 2

2.5 Non-touching loop rule

There is little apparent improvement in terms of complexity or length between topographic flowgraph methods and geometric series summations. The full power of flowgraphs is evident only when they are combined with the non-touching loop rule. It is then possible to find the transfer function between any two nodes of a network almost by inspection. The rule will not be proved, but stated and demonstrated by application to two examples. A full discussion of the topic is given in References 3—5.

If T is the ratio of an input to an output node, the non-touching loop rule states that

$$T = \frac{\sum\limits_{K} T_K \Delta_K}{\Delta} \tag{2.16}$$

where

T_K = path gain on the Kth forward path between the nodes

$\Delta = 1 - $ (sum of all individual loop gains) + (sum of loop gain products of all possible combinations of two non-touching loops) − (sum of the loop-gain products of all possible combinations of three non-touching loops) + · · ·

Δ_K = sum of all terms in Δ not touching the kth path

A path is a continuous succession of branches, either forming a loop by passing each node in the loop only once, or a forward path connecting the input to the output node. Path gain is the product of all the gains in the path. A loop originates and terminates on the same node. Loop gain is the product of the gains in the path. Non-touching loops are separate loops which do not touch at any node in the network. Combinations of two non-touching loops are called second-order loops, combinations of three non-touching loops, third order, etc.[2,5]

The non-touching loop rule can be applied to find the measured reflection coefficient in the reflectometer example by starting from Fig. 2.12a. There is only one path from a_1 to a_4, and there are no loops in the network that do not touch this path; so the only forward path gain is

$$T_1 = \frac{1}{\sqrt{2}} L_A L_D \rho_F = \sum_K T_K \Delta_K \tag{2.17}$$

There is only one loop in the network, so that

$$\Delta = 1 - \rho_S \rho_F L_A^2 L_D^2 \tag{2.18}$$

The ratio a_4 to a_1 is therefore

$$\frac{a_4}{a_1} = T = \frac{1}{\sqrt{2}} \left[\frac{\rho_F L_A L_D}{1 - \rho_S \rho_F L_A^2 L_D^2} \right] \tag{2.19}$$

which is identical with eqn. 2.14 derived by the topographical method. b_3 can be written down by inspection as

$$\frac{b_3}{a_1} = \frac{1/\sqrt{2}(1 - \rho_S\rho_F L_A^2 L_D^2) + 1/\sqrt{2}k\rho_S L_A^2 L_D^2}{1 - \rho_S\rho_F L_A^2 L_D^2} \tag{2.20}$$

which on cancelling terms is identical with eqn. 2.13, to give, on ratioing with eqn. 2.19, the same measured reflection coefficient as previously derived by geometric series summation and from the topographical rules.

Notice that, in applying the non-touching loop rule to find the ratio of b_3 to a_1, there are two paths. The first one is direct from a_1 to b_3 with a gain of $1/\sqrt{2}$. But it has a non-touching loop around b_2, b_4, a_4, a_2 and b_2 with a loop gain of $\rho_S\rho_F L_A^2 L_D^2$. The second path, with no non-touching loop, is via a_1, b_2, b_4, a_4, a_2, to b_3, and has a gain $(1/\sqrt{2})\rho_S L_A^2 L_D^2$. In the denominator there is just one loop.

2.6 Transmission measurement

A second example, of a transmission measurement test set, is illustrated in Fig. 2.13. As before, we make the somewhat unrealistic assumption that the errors are due only to reflection on the device side of the 3 dB splitter and at the input to the detector. Reference channel errors are ignored. The DUT might be an attenuator with transmission coefficient T and reflection coefficients ρ_1 and ρ_2. Measurement is made by noting the ratio-meter outputs, first with the DUT removed as in the flowgraph of Fig. 2.13a; and then with it set in place as in Fig. 2.13b. ρ_S and ρ_D are the reflection coefficients of the splitter and detector, respectively, E_S is the signal from the splitter and K is the detector conversion factor for meter reading M.

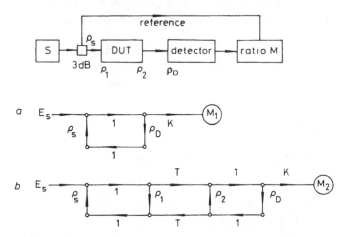

Fig. 2.13   *Transmission flowgraph*

In both flowgraphs there are no non-touching loops in the path from E_S to the meter. There is a first-order loop with gain $\rho_S\rho_D$ with no DUT in place, and three first-order loops plus a second-order loop, with gain $\rho_S\rho_1\rho_2\rho_D$ when the DUT is inserted. If M_1 and M_2 are the meter readings for the two cases, the input/output ratios are

$$\frac{M_1}{E_S} = \frac{K}{1 - \rho_S\rho_D} \tag{2.21}$$

$$\frac{M_2}{E_S} = \frac{KT}{1 - \rho_S\rho_1 - \rho_2\rho_D - \rho_S T^2\rho_D + \rho_S\rho_1\rho_2\rho_D} \tag{2.22}$$

The ratio of M_2 to M_1 is the measured transmission coefficient

$$T_M = \frac{(1 - \rho_S\rho_D)T}{1 - \rho_S\rho_1 - \rho_2\rho_D - \rho_S T^2\rho_D + \rho_S\rho_1\rho_2\rho_D} \tag{2.23}$$

In general, the higher-order loops make diminishing contributions with increasing order, because of the greater number of small terms appearing in the product.

2.7 Errors in reflection measurement

Reflection was traditionally measured from the standing wave induced in a line by an unmatched termination. By measuring the separation of the minima, the standing-wave ratio and the position of a chosen minimum when a known calibration load replaces the DUT, the DUT complex reflection coefficient can be found. Phase is not measured directly, but is derived from amplitude and dimensional measurements. In waveguides, illustrated in Fig. 2.14, the probe is inserted in a broad-wall slot from a sliding carriage.

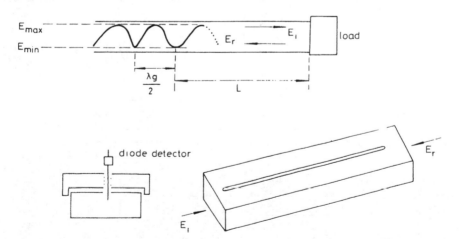

Fig. 2.14 *Standing waves on transmission line measured by slotted line*

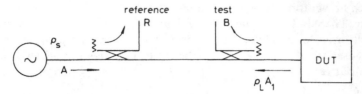

Fig. 2.15 *Two-coupler reflectometer*

Standing-wave indicators are not good broad-band devices and do not easily lend themselves to automation. Modern practice is to use directional couplers. In this way, the incident and reflected waves are separated from one another if the test-set components are perfect. In practice, component flaws impair signal flow, so that wave separation is not maintained and relative levels no longer depend solely on the device under test.

In Fig. 2.15 the input wave A is divided by a directional coupler to provide a reference flow path which is essentially a fraction of the forward wave. The main flow path continues to reflection at the DUT and a fraction of the reflection flow couples to the test wave B. Errors occur because of reflections at the couplings, the detectors, the load and the source. Most importantly, the directivity of the directional couplers causes direct coupling between the separated flow paths.

In broad-band instruments to 20 GHz the couplers are of coaxial or stripline construction, but the narrow-band two-hole waveguide directional coupler, outlined in Fig. 2.16, gives a simple understanding of the relationship between coupling and directivity factors. Reflection coefficients at the four input ports are labelled ρ_1, ρ_2, ρ_3, ρ_4. C is the coupling from ports 1 to 3 or 2 to 4 and D is the ratio between the waves at ports 4 and 3 for input wave at port 1. Coupling from port 1 to 4 is therefore CD. Reciprocal flows, which occur from any port

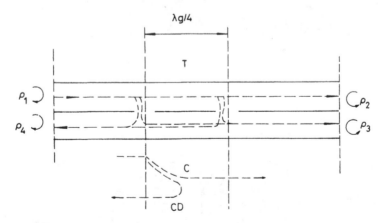

Fig. 2.16 *Waveguide two-hole directional coupler*

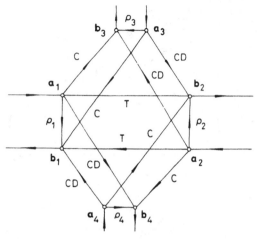

Fig. 2.17 *Flowgraph for directional coupler*

with similar couplings to the appropriately related outputs, lead to the flowgraph representation shown in Fig. 2.17. Coupling and absorption loss results in a transmission loss T between ports 1 and 2.

If we suppose that the coupler is used to take a reference sample of a forward wave at port 1, then port 4 can be connected to a matched load and the paths between it and the main line can be deleted from the flowgraph. We may further assume that the coupling between a_3 and b_1 will be small if the detector at port 3 is well matched to the line. The only significant coupling is $C_1 D_1$ from a_2 to b_3, because that represents a fraction of the reflected wave from the DUT, at port 3, which couples to the reference channel. It can be seen in Fig. 2.18 that,

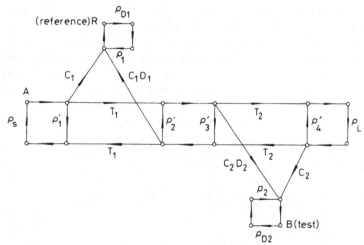

Fig. 2.18 *Flowgraph for two-coupler reflectometer*

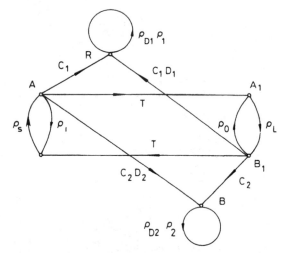

Fig. 2.19 *Reduced flowgraph for two-coupler reflectometer*

when two couplers are used in the reflectometer, there is a mixing of the forward and reflected waves at both the reference and test ports due to the directivity factors of the couplers. Most practical test sets have two couplers, one for reference, the other for the test channel, with the fourth ports feeding a matched load. We will perform an analysis on the flowgraph representation in Fig. 2.19, where a single directional coupler is employed. In fact, owing to the linear combinations involved, this form is equally correct, even when separate couplers are used, because the input and output reflection coefficients, ρ_i and ρ_0 can be determined for the new four-port resulting from terminating two ports of the original six-port network.

In Fig. 2.19 the input signal A couples with ratio C_1 to the reference channel, and via directivity coupling $C_2 D_2$ to the reference port. The incident wave A_1 is reflected as B_1, which couples via directivity $C_1 D_1$ to R and via C_2 to B. The loops at R and B ports are due to multiple reflections between the detector ρ_D and the input reflection at the ports. ρ_i and ρ_0 are the input and output reflection coefficients of the test set and ρ_S, ρ_L are the source and DUT reflection coefficients.

The non-touching loop rule can be applied to find the following ratios:

$$\frac{B}{A} = \frac{T\rho_L C_2(1 - \rho_{D1}\rho_1) + C_2 D_2(1 - \rho_{D1}\rho_1)(1 - \rho_0\rho_L)}{\Delta} \tag{2.24}$$

$$\frac{R}{A} = \frac{C_1(1 - \rho_0\rho_L)(1 - \rho_{D2}\rho_2) + T\rho_L C_1 D_1(1 - \rho_{D2}\rho_2)}{\Delta} \tag{2.25}$$

$$\frac{A_1}{A} = \frac{T(1 - \rho_{D1}\rho_1)(1 - \rho_{D2}\rho_2)}{\Delta} \tag{2.26}$$

where

$$\Delta = (1 - \rho_{D1}\rho_1)(1 - \rho_{D2}\rho_2)(1 - \rho_0\rho_L)(1 - \rho_S\rho_i) - T^2\rho_L\rho_S$$

+ higher-order terms.

The ratio R/A_1 shows the relationship between the sampled forward wave R and the actual incident wave on the load, A_1. Thus, from eqns. 2.25 and 2.26,

$$\frac{R}{A_1} = \frac{C_1(1 - \rho_0\rho_L)}{T(1 - \rho_{D1}\rho_1)} + \frac{T\rho_L C_1 D_1}{T(1 - \rho_{D1}\rho_1)} \tag{2.27}$$

This can be re-arranged to give

$$A_1(1 - \rho_0\rho_L) = R\frac{T}{C_1}(1 - \rho_{D1}\rho_1) - TD_1\rho_L A_1 \tag{2.28}$$

To show how the reflectometer behaviour depends on the load reflection co-efficient ρ_L, we write

$$A_1\rho_L = B_1 \tag{2.29}$$

to give, on substituting in eqn. 2.28,

$$A_1 = \frac{RT}{C_1}(1 - \rho_{D1}\rho_1) - (\rho_0 - TD_1)B_1 \tag{2.30}$$

In relating A_1 to R, eqn. 2.30 gives an effective source reflection $(\rho_0 - TD_1)$ and an effective transmission coefficient $T/C_1(1 - \rho_{D1}\rho_1)$. These are labelled ρ_E and T_E, respectively, on the equivalent flowgraph in Fig. 2.20. The incident wave on the DUT becomes

$$A_1 = RT_E + \rho_e\rho_L A_1$$

or

$$A_1 = \frac{RT_E}{1 - \rho_E\rho_L} \tag{2.31}$$

The significance of effective input reflection coefficient at the test port will become apparent in the error-network model of the next Section.

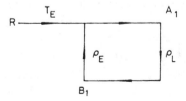

Fig. 2.20 *Equivalent flowgraph*

2.8 Vector calibration

When both phase and amplitude are measured the process is usually referred to as a vector measurement. System errors can be corrected by making a vector calibration of a number of standards with known reflection coefficients. The method uses an error model of the reflectometer based on the ratio of eqns. 2.24 and 2.25. Thus the measured ratio of reflected and transmitted waves is

$$\frac{B}{R} = \frac{T\rho_L C_2(1 - \rho_{D1}\rho_1) + C_2 D_2(1 - \rho_{D1}\rho_1)(1 - \rho_0\rho_L)}{C_1(1 - \rho_0\rho_L)(1 - \rho_{D2}\rho_2) + T\rho_L C_1 D_1(1 - \rho_{D2}\rho_2)} \tag{2.32}$$

This can be re-arranged as

$$\frac{B}{R} = \frac{C_2(1 - \rho_{D1}\rho_1)}{C_1(1 - \rho_{D2}\rho_2)} \left[D_2 + \frac{T(1 - D_1 D_2)\rho_L}{1 - (\rho_0 - TD_1)\rho_L} \right] \tag{2.33}$$

If we write

$$E_D = E_{11} = \frac{C_2(1 - \rho_{D1}\rho_1)}{C_1(1 - \rho_{D2}\rho_2)} D_2 \qquad\qquad \text{as a directivity error}$$

$$E_R = E_{21}E_{12} = \frac{C_2(1 - \rho_{D1}\rho_1)}{C_1(1 - \rho_{D2}\rho_2)} T(1 - D_1 D_2) \quad \text{as a transmission error}$$

and

$$E_S = E_{22} = \rho_0 - TD_1 \qquad\qquad\qquad \text{as an effective source error}$$

the result can be represented in a network with path gains equal to the error terms E_D, E_R and E_S. Fig. 2.21 shows the error network from which the measured reflection coefficient is found by the non-touching loop rule to be

$$S_{11M} = \frac{B}{R} = E_{11} + \frac{E_{21}E_{12}S_{11}}{1 - E_{22}S_{11}} = E_D + \frac{E_R\rho_L}{1 - E_S\rho_L} \tag{2.34}$$

The transmission terms $E_{21}E_{12}$ cannot be separated and are therefore shown combined as E_R in the lower path, though the upper path would be equally correct. The effective source reflection coefficient of eqn. 2.30 becomes the effective source error in the error model; and directivity leakage, via $C_2 D_2$, strongly influences the directivity error E_D.

Application of this error model to error correction through the use of standards is deferred for discussion in Chapter 5 on vector analysis. At this stage,

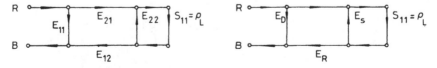

Fig. 2.21   *Error-term flowgraphs*

it is already clear that, if the magnitudes and phases of the reflection coefficients S_{11M}, for three known loads, placed successively at the test port, are measured, the resulting six simultaneous equations can be solved for E_D, E_R and E_S.

2.9 Scalar calibration

When only amplitude can be measured at the detection ports the process is referred to as a scalar measurement. The complex reflection ratio contains four quantities; i.e. two each of amplitude and phase. A simple reflectometer with just two fixed detection ports yields only the modulus of reflection coefficient. It is therefore in some respects less satisfactory than a standing-wave indicator, because in the latter the movement of a probe along a transmission line allows determination of both amplitude and phase. As we will see in Chapter 6, in six-port network analysers, a simple reflectometer may be modified to yield complex reflection coefficients, but only through the complication of adding two more detection ports.

In the simple reflectometer, Fig. 2.19 shows that the true reflection coefficient, ρ_L is the ratio B_1/A_1, whereas the measured ratio is B/R. The expression for B/R in eqn. 2.33, when expanded to second order, becomes

$$\frac{B}{R} = \frac{C_2(1 - \rho_{D1}\rho_1)}{C_1(1 - \rho_{D2}\rho_2)} [D_2 + T(1 - D_1 D_2)\rho_L + T(1 - D_1 D_2)(\rho_0 - TD_1)\rho_L^2]$$

(2.35)

If only amplitudes are measured, this becomes

$$\left|\frac{B}{R}\right|_{\rho_L} = \left|\frac{C_2(1 - \rho_{D1}\rho_1)}{C_1(1 - \rho_{D2}\rho_2)}\right| |D_2 + T(1 - D_1 D_2)\rho_L + T(1 - D_1 D_2)(\rho_0 - TD_1)\rho_L^2|$$

(2.36)

Calibration seeks to eliminate or at least reduce the effects of the error terms in eqn. 2.36 by substituting a short and/or open circuit for the DUT. For instance, when the DUT is replaced by a short circuit, $\rho_L = -1$ and the measured ratio is

$$\left|\frac{B}{R}\right|_{\rho_L = -1} = \left|\frac{C_2(1 - \rho_{D1}\rho_1)}{C_1(1 - \rho_{D2}\rho_2)}\right| |D_2 - T(1 - D_1 D_2) + T(1 - D_1 D_2)(\rho_0 - TD_1)|$$

(2.37)

The coupling- and detector-port reflections are eliminated by taking the ratio

$$\frac{|B|_{\rho_L}}{|B|_{\rho_L = -1}} = \frac{|B/R|_{\rho_L}}{|B/R|_{\rho_L = -1}} = \left|\frac{D_2 + T\rho_L + T(\rho_0 - TD_1)\rho_L^2}{D_2 - T + T(\rho_0 - TD_1)}\right|$$

(2.38)

where only the first-order error terms have been retained.

A similar ratio results if the short circuit is replaced with an open circuit at the test port. In this case, $\rho_L = 1$ and

$$\frac{|B|_{\rho_L}}{|B|_{\rho_L=1}} = \frac{|B/R|_{\rho_L}}{|B/R|_{\rho_L=1}} = \left| \frac{D_2 + T\rho_L + T(\rho_0 - TD_1)\rho_L^2}{D_2 + T + T(\rho_0 - TD_1)} \right| \tag{2.39}$$

If, on a decibel scale, the reference R is at 0 dB, then the reflected waves $|B|_{\rho_L}$, $|B|_{\rho_L=-1}$ and $|B|_{\rho_L=1}$ may vary with frequency, as shown in Fig. 2.22. Ripples, due to the changing phase relations of multiple reflections with the short circuit, tend to be in antiphase to those for an open circuit. This is due to the sign reversal of the transmission term in the denominator of eqns. 2.38 and 2.39. Since both short and open circuits do not affect the amplitude of a reflected wave, the average, dotted, line through the ripples gives an error-corrected reference, corresponding to the amplitude of the forward wave at the DUT. Thus, in Fig. 2.22, the measured reflection coefficient in decibels is the distance between the dotted line and the measured reflection $|B|_{\rho_L}$, when the DUT is at the test port. The ripples are caused by the phasing of reflections and couplings between imperfect components of the reflectometer. Only those terms having rapid phase variations with frequency can be eliminated by the averaging process, leaving a residual error over the frequency range.

It is not necessary to have a reference channel, since the measured reflection from a short or open circuit gives the forward wave at the DUT. This is illustrated by the test set and flowgraph of Fig. 2.23. Application of the non-touching loop rule gives

$$B = A \frac{T\rho_L C_2 + C_2 D_2 (1 - \rho_0 \rho_L)}{1 - \rho_S \rho_i - \rho_{D2} \rho_2 - T^2 \rho_S \rho_L - \rho_0 \rho_L + \rho_S \rho_i \rho_0 \rho_L + \rho_S \rho_i \rho_{D1} \rho_{D2} + \rho_0 \rho_L \rho_{D1} \rho_{D2}}$$

   $+ \text{ higher-order terms}$

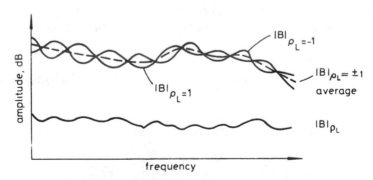

Fig. 2.22 *Open- and short-circuit calibration curve*

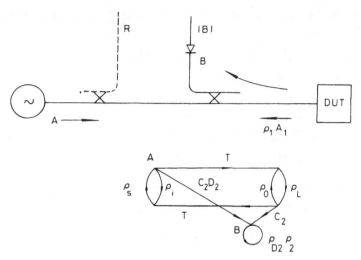

Fig. 2.23 *Single-coupler reflectometer*

Ignoring the higher-order terms and re-arranging, we have

$$B \approx AC_2 \frac{T\rho_L - D_2(1 - \rho_0\rho_L)}{(1 - \rho_S\rho_i)(1 - \rho_{D2}\rho_2)\left[1 - \rho_0\rho_L - \dfrac{T^2\rho_S\rho_L}{(1 - \rho_S\rho_i)(1 - \rho_{D2}\rho_2)}\right]}$$

or

$$|B| \approx |A| \left| \frac{C_2}{\tau} \frac{D_2 + (T - D_2\rho_0)\rho_L}{[1 - (\rho_0 - T^2\rho_S/\tau)\rho_L]} \right| \tag{2.40}$$

with

$$\tau = (1 - \rho_S\rho_i)(1 - \rho_{D2}\rho_2)$$

Ignoring products with D_2 and expanding to second order in ρ_L, we have

$$|B| \approx |A| \left| \frac{C_2}{\tau} [D_2 + T\rho_L + T(\rho_0 - T^2\rho_S/\tau)\rho_L^2] \right| \tag{2.41}$$

Comparison with eqn. 2.35 shows that the effective source error is $(\rho_0 - T^2\rho_S/\tau)$. Thus the reference channel eliminates the effects of source reflection and replaces it with a term depending on the directivity of the reference coupler.

Calibration of the single-coupler reflectometer in Fig. 2.23 results in the following ratios for the measured reflection coefficient:

$$|S_{11M}| = \frac{|B|}{|B|_{\rho_L = \pm 1}} = \left| \frac{D_2 + T\rho_L + T(\rho_0 - T^2\rho_S/\tau)\rho_L^2}{D_2 \pm T + T(\rho_0 - T^2\rho_S/\tau)} \right| \tag{2.42}$$

An approximate worst-case estimate of the errors can be made by expanding eqn. 2.42 as follows:

$$\frac{|B|}{|B|_{\rho_L = \pm 1}} \approx \left| \frac{-D_2/T - \rho_L - (\rho_0 - T^2\rho_S/\tau)\rho_L^2}{1 \pm D_2/T - (\rho_0 - T^2\rho_S/\tau)} \right|$$

$$\approx \left| \frac{1}{1 \pm D_2/T} \left[\frac{-D_2/T - \rho_L - (\rho_0 - T^2\rho_S/\tau)\rho_L^2}{1 - (\rho_0 - T^2\rho_S/\tau)} \right] \right|$$

$$\approx |(1 \mp D_2/T)[-D_2/T - \rho_L(1 + \rho_0 - T^2\rho_S/\tau) - \rho_L^2(\rho_0 - T^2\rho_S/\tau)]|$$

$$\approx |-D_2/T - \rho_L(1 + \rho_0 - T^2\rho_S/\tau \pm D_2/T) - \rho_L^2(\rho_0 - T^2\rho_S/\tau)|$$

Worst-case error is found by the addition of magnitudes to give

$$|S_{11M}|_{max} = |D_2/T| + |\rho_L| + |\rho_L|(|\rho_0| + |T^2\rho_S/\tau| + |D_2/T|)$$
$$+ |\rho_L|^2(|\rho_0| + |T^2\rho_S/\tau|)$$

But

$$\Delta\rho_{max} = |S_{11M}|_{max} - |\rho_L|$$

or

$$\Delta\rho_{max} = A + B|\rho_L| + C|\rho_L|^2 \qquad (2.43)$$

with

$$A = \left|\frac{D_2}{T}\right| \qquad \text{directivity error}$$

$$C = |\rho_0| + \left|\frac{T^2\rho_S}{\tau}\right| \qquad \text{effective source match}$$

$$B = A + C \qquad \text{calibration error}$$

A similar worst-case estimate with a reference channel is identical to eqn. 2.43, but with

$$C = |\rho_0| + |TD_1| \qquad (2.44)$$

A reference channel therefore improves accuracy provided the quality of the directional coupler is such that

$$|TD_1| < \left|\frac{T^2\rho_S}{\tau}\right|$$

2.10 Source levelling

If $|\rho_0| = 0$, a reference channel reduces the effective source reflection to the directivity of the forward-wave sampling coupler. A similar reduction can be

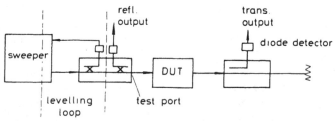

Fig. 2.24 *Levelling-circuit components*

achieved by applying feedback to maintain a constant output amplitude from the source. Fig. 2.24 shows an arrangement in which a sample of the forward wave is detected in a negative feedback loop to control the output level of the source. The device under test is shown in a reflection/transmission test set, with three directional couplers, for levelling, reflection and transmission coupling, respectively. The couplers and diode detectors introduce frequency-dependent errors which can be reduced by careful matching of their frequency responses. The levelling directional coupler and detector then cancel the variations of the other two through negative feedback on the sweeper output level. Whilst this has undoubted advantages, a reference channel has the same effect but without the trouble and expense of selecting matching components, though there may be other penalties in the external ratioing instrumentation. In a levelling circuit, forward level depends on the sweeper output and any source-reflected components in the main channel. Since the reflected component is also sampled in the levelling coupler its effects are reduced, thus giving an improvement in source match as well as frequency response.

To see how this happens we have, in Fig. 2.25, a forward wave a_i from a source with reflection coefficient ρ_S passing through a PIN diode attenuator and directional coupler, with directivity D coupling factor C and transmission coefficient T. At a given frequency the attenuation in the PIN attenuator is A and the wave emerges with amplitude

$$|a_L| = \sqrt{A}\,|a_i| \tag{2.45}$$

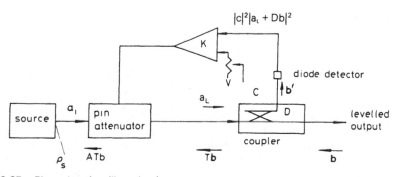

Fig. 2.25 *Flows in a levelling circuit*

A sample, Ca_L, is coupled into a diode detector, whilst the main-line forward wave continues towards the load, or DUT, where it reflects with complex wave amplitude b. A fraction CD of this also couples into the diode detector and the rest continues towards the source, where, after reflection, it contributes to a_i, the wave at the input to the attenuator. The driving voltage at the feedback amplifier is proportional to $|Ca_L + CDb|^2$ assuming a square-law diode with output voltage proportional to power input. For a PIN diode we may assume a logarithmic relation between attenuation and driving current; thus

$$20 \log \frac{|a_L|}{|a_i|} = 10 \log A = K|C|^2 |a_L + Db|^2 \tag{2.46}$$

where K is the loop conversion factor.

When frequency changes, or a different DUT is connected to the test port, a_i, a_L and b are affected, but the change in a_L is reduced by feedback to the PIN attenuator to a degree dependent on the loop gain, and on interference from the directivity-coupled reflected wave b. If δa_i, δa_L are the changes in a_i and a_L and A' is the new attenuation, we have

$$|a_L + \delta a_L| = \sqrt{A'}|(a_i + \delta a_i)| \tag{2.47}$$

But the new output from the diode is $[|C|^2 |a_L + \delta a_L + Db + D\delta b|]^2$, since both the forward and reflected waves in Fig. 2.26 have changed in the main line and δa_L has been affected by changes in Db as well as those in a_i. Therefore

$$10 \log A' = K|C|^2 |a_L + \delta a_L + Db + D\delta b|^2$$

and on combining with eqn. 2.46,

$$10 \log \frac{A'}{A} = K|C|^2 |\,|a_L + \delta a_L + Db + D\delta b|^2 - |a_L + Db|^2| \tag{2.48}$$

which, in worst-case form, and to second order, becomes

$$|\delta a_L| + |D\delta b| = \frac{10 \log (A/A')}{2K|C|^2 |a_L|}$$

and

$$|\delta a_L| + |D\delta b| \to 0, \quad \text{if } 2K|C|^2 |a_L| \gg 10 \log (A/A') \tag{2.49}$$

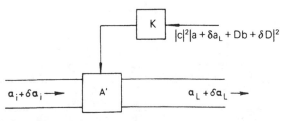

Fig. 2.26 *Levelling-attenuator control*

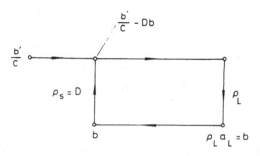

Fig. 2.27 *Equivalent source reflection*

The action of a high-gain feedback loop is therefore to maintain constant the level at the output of the diode. In the absence of a directivity contribution, this also keeps a_L constant. It is indeed the case, from eqn. 2.49, that the perturbations $|\delta a_L|$ and $|D\delta b|$ are reduced to negligible proportions if the loop gain $K|C|^2$ is large, but the constancy of the drive level from the detector diode means that the wave b' from the side arm of the coupler is also constant. Therefore

$$b' = Ca_L + CDb$$

or

$$a_L = \frac{b'}{C} - Db \tag{2.50}$$

In eqn. 2.50, b'/C is the forward wave from a leveller with zero directivity and b is the reflected wave in the line from the load. The flowgraph in Fig. 2.27 shows how, for high loop gain, the source appears to have an effective reflection coefficient equal to the coupler directivity.

2.11 Vector-transmission calibration

A vector-transmission measurement relies on comparison of a transmitted wave with the phase and amplitude of a reference wave. A combined reflection and transmission test set is shown in Fig. 2.28. Its equivalent flowgraph representation, also in the Figure, combines the reflection-error model of Fig. 2.21 with an approximate version of the flow paths in a transmission-sampling directional coupler. The forward wave is sampled as F via coupling C_3. Possible errors are reduced, as in earlier examples, to the directivity path C_3D_3, and the multiple reflections between the detector and the transmission port mismatch. An application of the non-touching loop rule gives the ratio of the sampled transmitted

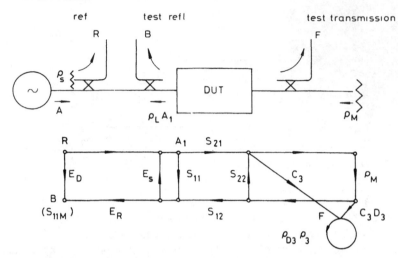

Fig. 2.28   *Reflection/transmission test set and flowgraph*

wave to the forward wave, A_1, at the input to the DUT as

$$\frac{F}{A_1} = \frac{S_{21}(C_3 D_3 \rho_M + C_3)}{\begin{array}{c}1 - \rho_M S_{22} - \rho_{D3}\rho_3 - E_S S_{11} - E_S E_{21} E_{12}\rho_M + E_S S_{11}\rho_{D3}\rho_3 \\ + \rho_M S_{22}\rho_{D3}\rho_3 + E_S S_{11} S_{22}\rho_M\end{array}}$$

where ρ_M is the reflection from a matched termination in the main line and S_{11}, S_{22}, S_{21}, S_{12} are the required scattering parameters of the device under test.

The dominant terms are $\rho_M S_{22}$, $E_S S_{11}$; and $E_S S_{21} S_{12}\rho_M \to 0$ in a unilateral device. Therefore

$$\frac{F}{A_1} \approx \frac{S_{21}(C_3 D_3 \rho_M + C_3)}{1 - E_S S_{11} - \rho_M S_{22}}$$

or

$$\frac{F}{A_1} \approx \frac{S_{21} E_T}{1 - E_S S_{11} - E_L S_{22}} \tag{2.51}$$

where

$$E_T \approx C_3, \quad \text{since } C_3 \gg C_3 D_3 \rho_M$$

and

$$E_L = \rho_M$$

The reflection/transmission-error model then reduces to a five-term network as shown in Fig. 2.29, and requires that five known standards be measured in place of the DUT in order to calibrate the test set.

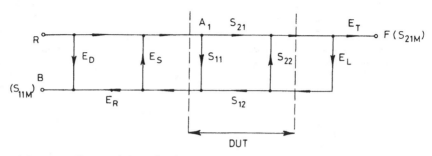

Fig. 2.29 *Error flowgraph for reflection/transmission test set*

2.12 Scalar-transmission calibration

A reference in a scalar measurement is provided by a 'through' calibration in which the DUT is removed and the detector is connected directly to the source, as shown in Fig. 2.30. With the waves and paths as defined in the Figure, we have for the ratio of the forward wave at the detector to the wave at the source,

$$\frac{|A_1|}{|A|} = \left|\frac{T_F}{1 - \rho_S\rho_D T_F}\right| \tag{2.52}$$

When the DUT is inserted in the flowgraph of Fig. 2.31, we have as the new ratio for the transmitted wave, A'_1,

$$\frac{|A'_1|}{|A|} = \left|\frac{S_{21}T_F}{(1 - \rho_S S_{11})(1 - \rho_D S_{22}T_F) - \rho_S\rho_D S_{12}S_{21}T_F}\right| \tag{2.53}$$

The ratio of eqns. 2.52 and 2.53 gives the measured transmission coefficient in the presence of the component errors of the transmission test set as

$$|S_{12M}| = \frac{|A'_1|}{|A_1|} = \left|\frac{(1 - \rho_S\rho_D T_F)S_{21}}{(1 - \rho_S S_{11})(1 - \rho_D S_{22}T_F) - \rho_S\rho_D S_{12}S_{21}T_R}\right| \tag{2.54}$$

Over a swept frequency range phasing ripples can be related to multiple reflection loops in the test set, and we will show in Chapter 6 how these are enhanced for error correction by means of air lines.

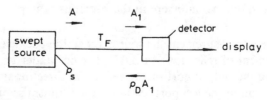

Fig. 2.30 *Through measurement*

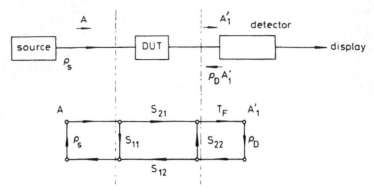

Fig. 2.31   *Transmission measurement*

2.13 References

1 KUHN, N.: 'Simplified signal flowgraph analysis', *Microwave J.*, Nov. 1963, pp. 59–66.
2 ADAM, S. F.: 'Microwave theory and applications' (Prentice-Hall, 1969)
3 MASON, S. J.: 'Feedback theory—Some properties of signal-flow graphs', *Proc. IRE*, 1953, **41**, pp. 1144–1156
4 MASON, S. J.: 'Feedback theory—Further properties of signal-flow graphs', *Proc. IRE*, 1956, **44**, pp. 920–926
5 ANDERSON, R. W.: '*S*-parameter techniques for faster, more accurate network design'. Hewlett Packard Application Note, 95-1
6 LIAO, S. Y.: 'Microwave devices and circuits' (Prentice-Hall, 1980) Chap. 4

2.14 Examples

1   The flowgraph expressions in Fig. 2.32 are incomplete. What are the missing terms? Would you expect them to be always significant?

2   A substitution method for measuring the scattering parameters of a DUT is illustrated in Fig. 2.33. The reference-coupler directivity is D and its forward coupling is C. First a measurement is made with the DUT removed and then repeated with it in place. Using the reflection and coupling values shown, calculate the measured transmission coefficient of the device under test.

3   By making reasonable assumptions about the system reflections, estimate the range of error in the measurement in the previous example. Take the reference directivity as 35 dB.

4   In a reflectometer measurement the source reflection coefficient is $|\rho_S| = 0 \cdot 1$. A scalar measurement of reflection at a DUT (device under test) can be made by first calibrating the output port of the reflection directional coupler with a short and open circuit at the test port. Assuming no transmission loss, estimate the error in the measured reflection coefficient if the test port input reflection

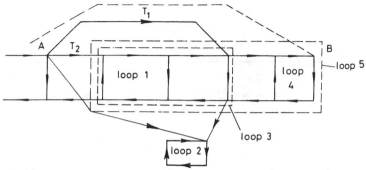

Fig. 2.32 *Flowgraph expressions for example 1*
Paths *A—B* and non-touching loops are
T_1 (loops 1 and 2) with $\Delta_k = 1 - G_1 - G_2$
T_2 (loop 1) with $\Delta_k = 1 - G_2$
1st-order loops: 1, 2, 3, 4, 5
2nd-order loops: 1×2, 1×4, 2×4, 2×3, 4×3, 5×2
3rd-order loops: $1 \times 2 \times 4$, $2 \times 3 \times 4$
$\Delta = 1 - (G_1 + G_2 + G_3 + G_4 + G_5) + (G_{12} + G_{14} + G_{24} + G_{23} + G_{43} + G_{52}) - G_{124} + G_{234}$
G is loop gain

$|\rho_0| = 0\cdot1$, the directivity of the coupler is -35 dB and the detector is matched. If the measurement is repeated with a reference channel and the effective source match at the test port is $0\cdot2$, what is the directivity of the reference coupler?

5 The output from a sweeper varies about its mean level by $\pm 2\cdot0$ dB, when feeding a matched load. After levelling the variation is $\pm 0\cdot5$ dB. What is the transducer coefficient if the levelled power output is 10 mW with a 20 dB levelling coupler? What is the effective source reflection coefficient if the levelling coupler directivity is -30 dB?

6 In a scalar transmission measurement the reflection coefficients of the source and detector are $0\cdot1$ and the transmission loss in the measurement system is $0\cdot25$ dB. If the DUT is an attenuator with a 20 dB loss and input/output return loss of -16 dB, calculate the transmission measurement uncertainty limits.

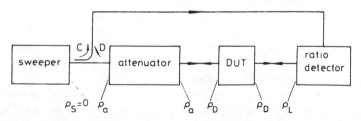

Fig. 2.33 *Substitution method for measuring scattering parameters of a DUT (example 2)*

Chapter 3

The Smith chart

3.1 Generalised wave functions

The Smith chart remains one of the most highly developed aids to microwave circuit design. In its common form it consists of loci of constant normalised resistance and reactance, plotted on a polar diagram representing the reflection coefficient at a fixed plane in a transmission system. A load reflection coefficient is related to voltages and/or currents measured at selected positions in an implicit transmission line; but in a microwave circuit, the perceived quantity flowing in the system is the power associated with an electromagnetic wave. Detection, e.g. in a standing-wave indicator, can give voltage or current outputs proportional to the wave amplitude at the detection point. Phase must be inferred from the relative positions of the detection probe and the reflecting load. Measured quantities can be manipulated on the Smith chart using an easily understood visual presentation, but computers may perform the same function with matrix algebra, e.g. on measured S-parameters, thus apparently obviating the use of the Smith chart. That this is not so is evident from the fact that even computerised measurements are displayed in Smith-chart form, and that extensive software is available to assist the engineer in manipulating the display in real time.

Underlying this modern use of a traditional method is the massive investment in design procedures already achieved before the advent of computers. Computing power has served to increase the application of the invested experience attached to the Smith chart, even though network analysers, with reflection and transmission test sets, provide results in measured S-parameter form that are well adapted to computerised calculations on matrices.

The Smith chart is conveniently derived from traditional transmission-line theory, with voltages and currents explicitly determined by the lumped or distributed circuit components. We therefore relate the S-parameters to component impedances and admittances through the currents and voltages on an equivalent transmission line, by first considering the simple network of Fig. 3.1.

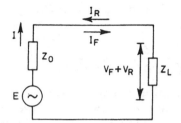

Fig. 3.1 *Forward- and reflection-voltage waves in impedance circuit*

The current on the load impedance is

$$I = \frac{E}{Z_L + Z_0} \qquad (3.1)$$

and the power delivered to Z_L is

$$P_L = \text{Re } Z_L |I|^2$$

$$= \frac{R_L |E|^2}{(R_L + R_0)^2 + (X_L + X_0)^2}$$

$$P_L = \frac{|E|^2}{4R_0 + (R_L - R_0)^2/R_L + (X_L + X_0)^2/R_L} \qquad (3.2)$$

where R_L, R_0 are the real and X_L, X_0 are the imaginary parts of Z_L and Z_0.

Maximum power is transferred to the load when its impedance is the complex conjugate of the source impedance. This follows because $R_L = R_0$ and $X_L = -X_0$ when the numerator in eqn. 3.2 is a minimum and

$$P_{max} = \frac{|E|^2}{4R_0} \qquad (3.3)$$

When the load is not matched to the source we may write the voltage V_L as

$$V_L = V_F + V_R = \frac{EZ_L}{Z_0 + Z_L} \qquad (3.4)$$

V_F is interpreted as the voltage in a forward wave with amplitude $E/2$. It is therefore the voltage across Z_L when $Z_L = Z_0^*$. V_R is a reflected voltage at Z_L whose magnitude is the difference between the unmatched and matched voltages. Therefore

$$V_R = \frac{EZ_L}{Z_0 + Z_L} - \frac{E}{2} = \frac{E}{2}\frac{Z_L - Z_0}{Z_0 + Z_L} \qquad (3.5)$$

The reflected voltage also appears across Z_0, the source impedance, to produce a current

$$-\frac{E}{2Z_0}\frac{Z_L - Z_0}{Z_L + Z_0}$$

The current in Z_0 due to the forward wave is $E/2Z_0$, so that the total current in the source impedance is

$$I = \frac{E}{2Z_0} - \frac{E}{2Z_0}\frac{Z_L - Z_0}{Z_L + Z_0} = \frac{E}{Z_L + Z_0}$$

which is the result expected in a circuit with total impedance $Z_L + Z_0$.

Separation of signal paths into forward and reverse wave flows provides an analogy with the wave amplitudes, *a* and *b*, in the scattering matrix or flowgraph. There is, however, no reason why *a* or *b* should be voltage amplitudes, as can be seen by considering Fig. 3.2. In this admittance circuit, the forward, matched load current is

$$I_F = \frac{I}{2}$$

and the unmatched current is

$$I_L = \frac{IY_0}{Y_L + Y_0}$$

giving a reflected current

$$I_R = \frac{I}{2}\frac{Y_0 - Y_L}{Y_0 + Y_L} \qquad (3.6)$$

Eqns. 3.5 and 3.6 give the same reflection coefficient

$$S_{11} = \rho = \frac{V_R}{V_F} = \frac{I_R}{I_F} = \frac{Z_L - Z_0}{Z_L + Z_0} = \frac{Y_0 - Y_L}{Y_0 + Y_L} \qquad (3.7)$$

Wave amplitudes are not constrained to be either voltage or current, and we are free to choose what they should be. Since the perceived flow is power, an aesthetically practical way to define a wave in a transmission system with a real Z_0 is[1,2]

$$a = \frac{V^+}{\sqrt{Z_0}} \qquad (3.8)$$

$$b = \frac{V^-}{\sqrt{Z_0}} \qquad (3.9)$$

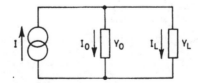

Fig. 3.2   *Current waves in admittance circuit*

where V^+ and V^- are the forward and reverse voltages. Phase is retained in the wave functions whilst power flows follow as

$$\text{Forward power flow} = \frac{1}{2}\,aa^* = \frac{1}{2}|a|^2 = \frac{1}{2}\frac{(V^+)^2}{}Z_0 \tag{3.10}$$

$$\text{Reverse power flow} = \frac{1}{2}\,bb^* = \frac{1}{2}|b|^2 = \frac{1}{2}\frac{(V^-)^2}{Z_0} \tag{3.11}$$

The relationship between forward and reverse currents and voltages in the waves is illustrated in Fig. 3.3. For clarity, the transmission line is split in the Figure at position z, where the total voltage is

$$V = V^+ + V^- \tag{3.12}$$

and the current is

$$I = I^+ - I^- \tag{3.13}$$

In terms of wave amplitudes

$$V = \sqrt{Z_0}(a + b) \tag{3.14}$$

$$I = \frac{1}{\sqrt{Z_0}}(a - b) \tag{3.15}$$

The equation for current was found by applying $I^\pm = V^\pm/Z_0$.

Eqns. 3.14 and 3.15 define the voltage and current on a transmission line in terms of a particular form of the generalised wave functions a and b. In conjunction with the transmission-line equations they will enable us to relate the S-parameters to the resistive and reactive components of the Smith chart.

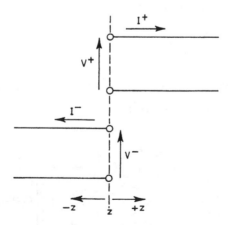

Fig. 3.3 *Forward and reverse voltages and currents in a transmission line*

3.2 Transmission-line equations

An equivalent circuit of a transmission line is shown in Fig. 3.4. It has series resistance R and inductance L per unit length; and parallel conductance G and capacitance C, also per unit length. These apparently lumped quantities are distributed on transmission lines such as coaxial cable or parallel line pairs. They are also distributed on stripline and in waveguide, but, in these cases, voltages and currents are less precisely definable at the load or source connections. However electric/magnetic field ratios give impedances that may, through eqns. 3.14 and 3.15, serve to relate the transmission equations to wave propagation in the transmission medium.

Let V be the voltage across the line at a point, distance z from an origin taken at the source. The drop in voltage in a short length δz is

$$\delta V = -(R + j\omega L)I\delta z$$

where ω is the angular frequency.

Therefore in the limit

$$\frac{dV}{dz} = -(R + j\omega L)I \tag{3.16}$$

Similarly the loss of current in a short length δz is

$$\delta I = -(G + j\omega C)V\delta z$$

and again

$$\frac{dI}{dz} = -(G + j\omega C)V \tag{3.17}$$

Differentiating the first equation and substituting in the second, we get

$$\frac{d^2 V}{dz^2} = (R + j\omega L)(G + j\omega C) = \gamma^2 V \tag{3.18}$$

with

$$\gamma = \sqrt{(R + j\omega L)(G + j\omega C)} \tag{3.19}$$

The second-order differential eqn. 3.18 has the general solution

$$V = A \exp(-\gamma z) + B \exp(\gamma z) \tag{3.20}$$

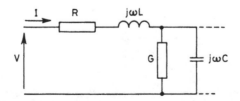

Fig. 3.4 *Distributed lumped-components model of transmission line*

If we compare this with eqn. 3.14, we have

$$a = \frac{A}{\sqrt{Z_0}} \exp(-\gamma z) \tag{3.21}$$

$$b = \frac{B}{\sqrt{Z_0}} \exp(\gamma z) \tag{3.22}$$

Eqn. 3.15 then becomes

$$I = \frac{A}{Z_0} \exp(-\gamma z) - \frac{B}{Z_0} \exp(\gamma z) \tag{3.23}$$

a and *b* are wave functions with propagation constant γ. If

$$\gamma = \alpha + j\beta \tag{3.24}$$

α is the attenuation constant in nepers per unit length in the line due to R and G terms in eqn. 3.19, and β is the phase shift per unit length, dependent primarily on L and C. For example, if $R = G = 0$,

$$\gamma = j\omega \sqrt{LC} = j\beta \tag{3.25}$$

and the phase change in a wavelength λ is

$$\beta = 2\pi f\lambda \sqrt{LC} \tag{3.26}$$

But in a sinusoidal function the phase change in a wavelength is 2π.
   Therefore

$$f\lambda = \frac{1}{\sqrt{LC}} \tag{3.27}$$

and $1/\sqrt{LC}$ is the propagation, or phase, velocity of the wave. Also, from eqns. 3.26 and 3.27,

$$\beta = \frac{2\pi}{\lambda} \tag{3.28}$$

and

$$\frac{\omega}{\beta} = \frac{1}{\sqrt{LC}} = v_P, \quad \text{the phase velocity} \tag{3.29}$$

If the load is at distance l from the source, its impedance follows from the ratio of eqns. 3.20 and 3.23, and is, if $\alpha = 0$

$$Z_L = \frac{V_l}{I_l} = Z_0 \frac{a+b}{a-b} = Z_0 \frac{A \exp(-j\beta l) + B \exp(j\beta l)}{A \exp(-j\beta l) - B \exp(j\beta l)}$$

$$\frac{Z_L}{Z_0} = \frac{1+b/a}{1-b/a} = \frac{1+\rho_L}{1-\rho_L} \tag{3.30}$$

with

$$\frac{b}{a} = \frac{B}{A} \exp (2j\beta l) = \rho_L$$

the ratio of the forward and reverse wave at the load. On re-arranging eqn. 3.30, the reflection coefficient is

$$\rho_L = \frac{Z_L - Z_0}{Z_L + Z_0} \tag{3.31}$$

and is similar to eqn. 3.7 for a simple impedance network.

The complex ratio B/A refers to the reflection coefficient as measured at the source. We may consider the rest of the line and its termination to be the load at any point, distance $z = l'$ from the source. Since the line length from the source is expressed in the phase term $j2\beta l'$, in a polar plot of the magnitude and phase of $(Z_L - Z_0)/(Z_L + Z_0)$, the modulus of ρ_L is

$$|\rho_L| = \frac{|B|}{|A|}$$

and the phase change causes this to rotate anticlockwise as we move towards the load. This is illustrated in Fig. 3.5.

If we now define the reflection phase at the load terminating a line at distance l from the source as ϕ, we will find it convenient to refer to z' as the distance from the load towards the source. At z' the reflection coefficient is

$$\rho_{z'} = \rho_L \exp (-j2\beta z')$$

with

$$\rho_L = |\rho_L| \exp j\phi$$

at the load.

At intervals of $2\beta z' = 2\pi$, i.e. towards the source, the reflection coefficient returns to its value at the load. Alternatively a single clockwise rotation of $|\rho_L|$

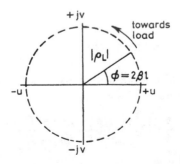

Fig. 3.5 *Polar diagram*

in Fig. 3.5 results in z' changing by $\Delta z'$ with

$$\beta \, \Delta z' = \pi$$

or

$$\Delta z' = \frac{\lambda}{2}$$

The reflected voltage in the line at l' from the source corresponding to z' from the load is

$$B \exp(j\beta l') = \rho_L \exp(-j2\beta z')A \exp(-j\beta l')$$

On adding the forward wave, the total voltage at l' is

$$V = A \exp(-j\beta l') + \rho_L \exp(-j2\beta z')A \exp(-j\beta l')$$

$$= A \exp(-j\beta l')[1 + \rho_L \exp(-j2\beta z')] \tag{3.32}$$

With $l' = l - z'$ and $\rho_L = |\rho_L| \exp(j\phi)$, this becomes

$$V = A \exp(-j\beta l')[1 + |\rho_L| \exp j(\phi - 2\beta z')] \tag{3.33}$$

From which we have the modulus of the line voltage as

$$|V| = |A|\{1 + 2|\rho_L| \cos(\phi - 2\beta z') + |\rho_L|^2\}^{1/2} \tag{3.34}$$

with maximum and minimum values

$$|V| = |A|[1 \pm |\rho_L|]$$

The ratio of $|V|_{max}$ to $|V|_{min}$ is the voltage standing-wave ratio

$$S = \frac{|V|_{max}}{|V|_{min}} = \frac{1 + |\rho_L|}{1 - |\rho_L|} \tag{3.35}$$

But

$$|\rho_L| = \frac{|b|}{|a|}$$

giving the standing-wave ratio in wave amplitude form as

$$S = \frac{|a| + |b|}{|a| - |b|} \tag{3.36}$$

or

$$|\rho_L| = \frac{S - 1}{S + 1} \tag{3.37}$$

Any $|\rho_L|$ on the polar plot in Fig. 3.5 can therefore also be expressed as a standing-wave ratio according to eqn. 3.35. Constant standing-wave ratios (SWR) or constant $|\rho_L|$ are circles centred on the origin. It follows from eqn 3.30 that the normalised complex impedance Z_L/Z_0 can also be plotted on a polar plot, since, for every value of ρ_L, there is a unique normalised impedance Z_L/Z_0, found by re-arranging eqn. 3.31 as

$$\frac{Z_L}{Z_0} = \frac{1+\rho_L}{1-\rho_L}$$

Other quantities depending on $|\rho_L|$ are also useful to indicate the radial distance to a point on the polar plot. Most common are SWR expressed as 20 log S, return loss as 20 log $|\rho_L|$ and the mismatch loss as 10 log $(1 - |\rho_L|^2)$. Mismatch loss is the fractional power absorption in the load.

3.3 Derivation of the Smith chart

In the polar plot of Fig. 3.5, the reflection coefficient is complex with components[3,4]

$$\rho_L = u + jv \tag{3.38}$$

u and v have values between 0 and ± 1. The modulus $\sqrt{u^2 + v^2}$ equals $|\rho_L|$ and the phase of ρ_L is $\tan^{-1}(u/v)$. Let

$$\frac{Z_L}{Z_0} = r + jx = \frac{R}{Z_0} + \frac{jX}{Z_0} = \frac{1+\rho_L}{1-\rho_L}$$

Therefore

$$r + jx = \frac{1 + u + jv}{1 - u - jv}$$

Rationalising

$$r + jx = \frac{1 - u^2 - v^2 + j2v}{(1-u)^2 + v^2} \tag{3.39}$$

Equating real parts

$$r = \frac{1 - u^2 - v^2}{(1-u)^2 + v^2}$$

or

$$u^2 - \frac{2ru}{1+r} + \frac{r}{1+r} = \frac{1}{1+r} - v^2$$

Subtracting $r/(1+r)^2$ from both sides,

$$u^2 - \frac{2ru}{1+r} + \frac{r^2}{(1+r)^2} = \frac{1}{(1+r)^2} - v^2$$

or

$$\left(u - \frac{r}{1+r}\right)^2 + v^2 = \frac{1}{(1+r)^2} \tag{3.40}$$

This equation of a circle has a radius $1/(1+r)$ and centre at the point $u = r/(1+r), v = 0$. Each r in the range 0 to ∞ describes a circle touching the $u = 1, v = 0$ point with centre on the u-axis. For example, if $r = 1$, the centre is at $u = 0.5$. The largest circle, for $r = 0$, has a centre at the origin.

Circles of constant reactance are found on equating the imaginary parts of eqn. 3.39:

$$x = \frac{2v}{(1-u)^2 + v^2}$$

or

$$u^2 - 2u + 1 + v^2 - \frac{2v}{x} = 0$$

If we add $1/x^2$ to both sides, we have

$$(u - 1)^2 + \left(v - \frac{1}{x}\right)^2 = \frac{1}{x^2} \tag{3.41}$$

The centre of this circle is at $u = 1, jv = \pm j/x$ and its radius is $1/x$. As an example, if $x = \pm 1$, the centres are at $v = \pm 1$ and $u = 1$ with radius 1. Centres therefore lie on a line $u = 1$ for all values of x. When the polar plot is composed of circles of constant normalised resistance and reactance it is called an impedance Smith chart.

Fig. 3.6 will allow us to identify the main features of an impedance Smith chart. The outer circle is marked in phase angle of voltage reflection coefficient. That the upper semicircle is the region of inductive reactance can be shown by noting that a short-circuit load produces a phase reversal between the forward and reflected wave, or a reflection phase of $\pm 180°$. The impedance at a distance z' from a short circuit is, from eqn. 3.30,

$$\frac{Z_{z'}}{Z_0} = \frac{1 + \rho_L \exp(-j2\beta z')}{1 - \rho_L \exp(-j2\beta z')}, \quad \text{with} \quad \rho_L = -1$$

Therefore

$$\frac{Z_{z'}}{Z_0} = \frac{1 - \exp(-j2\beta z')}{1 + \exp(-j2\beta z')}$$

$$= j \tan \beta z' \tag{3.42}$$

For the first quarter of a wavelength from the $\pm 180°$ position towards the generator, or source, $\tan(\beta z')$ is positive, indicating inductive reactances in the upper semicircle. Similarly, movement from the $\pm 180°$ position towards

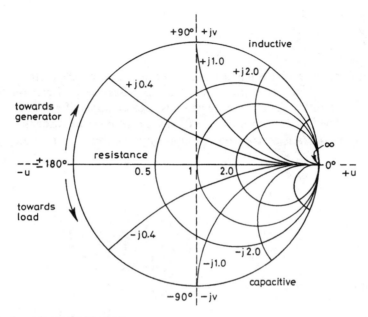

Fig. 3.6 *Impedance Smith chart*

the load is into the lower capacitive-reactance semicircular region. The dividing *u*-axis is purely resistive.

An admittance form of the Smith chart could be found by equating conductances and susceptances in a similar way to eqn. 3.39, but it is instructive to use eqn. 3.30 to define $Z_{z'}/Z_0$ at a position z' from the load as

$$\frac{Z_{z'}}{Z_0} = \frac{1 + \rho_L \exp(-j2\beta z')}{1 - \rho_L \exp(-j2\beta z')} \tag{3.43}$$

with

$$\rho_L = \frac{Z_L - Z_0}{Z_L + Z_0}$$

to give

$$\frac{Z_{z'}}{Z_0} = \frac{Z_L/Z_0 \cos(\beta z') + j \sin(\beta z')}{Z_L/Z_0 j \sin(\beta z') + \cos(\beta z')} \tag{3.44}$$

At a quarter-wave distance from the load, $z' = \lambda/4$, $\beta z' = \pi/2$ and eqn. 3.44 becomes

$$\frac{Z_{z'}}{Z_0} = \frac{Z_0}{Z_L} \tag{3.45}$$

The quarter-wavelength line is an impedance transformer, converting the terminating impedance to its reciprocal or admittance value. This property is

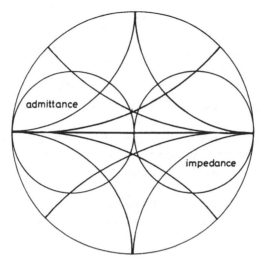

Fig. 3.7 *Admittance/impedance Smith-chart overlays*

useful in changing an impedance chart to an admittance Smith chart, simply by rotating it through 180° after plotting the corresponding impedance. An overlaid impedance/admittance chart as outlined in Fig. 3.7 is of great value in design procedures involving a mixture of series and parallel network components.

3.4 Standing waves and vector analysis

Vector analysis involves knowledge of phase and amplitude of measured quantities. It is sometimes possible to derive phase from scalar, or amplitude only, measurements, but extra information, such as some physical length in the test set, is also required. Standing waves, generated by mismatches in a transmission line, provide a system whose physical dimensions can yield the phase information. The amplitude of the standing wave, usually required at its maximum and minimum values, can be detected with a lightly coupled pick-up probe capable of movement along the line. On twin wire the probe is a small loop coupled to the magnetic field of the wire-guided waves. In a coaxial line or rectangular waveguide, it consists of a short wire inserted into a longitudinal slot cut in the outer wall of the former or the broad wall of the latter. If the probe is connected to a diode detector, operating in its square-law region, output current is proportional to the square of the wave amplitude at the probe position. The schematic arrangement of a waveguide standing-wave indicator in Fig. 3.8 does not show the sliding carriage and vernier scale necessary to stabilise the probe and indicate its position. Probe positioning and depth of insertion are a matter of careful design,[7] but we will not pursue this further because of the decreasing

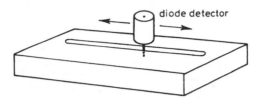

Fig. 3.8 *Slotted-line standing-wave indicator*

importance of standing-wave indicators in broad-band complex reflection-coefficient measurements, now almost exclusively made by reflectometry. It has been introduced here to show how a vector measurement of impedance is possible using a square-law detector.

Consider first a length of transmission line with a short-circuit load. The voltage at distance z' from the load is found by putting $|\rho_L| \exp(j\phi) = -1$ in eqn. 3.32 to give

$$V = 2A \exp(-j\beta l)j \sin(\beta z') \tag{3.46}$$

The resulting voltage, shown as a function of z' in Fig. 3.9, has peak amplitude twice the amplitude of the forward wave $A \exp(-j\beta l)$, with peaks and nulls (anti-nodes and nodes) at fixed points along the line, separated by half-wavelength intervals. The time dependency, $\exp(j\omega t)$, not shown in eqn. 3.46, might be observed by placing a small probe at the peak of an anti-node where the level oscillates over a range $\pm 2A$. Of course, the detector will 'see' these variations only if its response time is short enough or the frequency is sufficiently low. A probe placed at a node would register no response. Such waves, called standing waves, result from interference between a forward and reverse wave of equal amplitude.

For terminations other than a short or open circuit, there is only partial reflection of the forward wave. Eqn. 3.33 gives the line voltage, which we now assume to be square-law detected as

$$|V|^2 = |A|^2\{1 + 2|\rho_L| \cos(\phi - 2\beta z') + |\rho_L|^2\} \tag{3.47}$$

The short-circuit case can be recovered by setting $|\rho_L| = 1$ and $\phi = 180°$, to give

$$|V|^2 = 4|A|^2 \sin^2(\beta z') \tag{3.48}$$

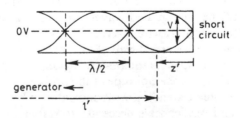

Fig. 3.9 *Standing waves from short circuit*

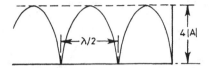

Fig. 3.10 *Square-law detection of standing-wave voltage*

The detected output, illustrated in Fig. 3.10, has a standing-wave ratio of infinity, since $|V|_{min} = 0$. In the method to be described, the probe is moved to find the position and separation of at least two nulls to find the wavelength in the transmission line.

When a short circuit is replaced by an unknown load Z_L, the standing-wave ratio may decrease and the pattern nulls shift, depending on the phase and amplitude of the new reflected wave. These effects can be shown by re-arranging eqn. 3.47 as

$$|V|^2 = |A|^2 \left\{ (1 - |\rho_L|)^2 + 4|\rho_L| \cos^2 \left(\frac{\phi}{2} - \beta z' \right) \right\} \tag{3.49}$$

There is a constant level $|A|^2(1 - |\rho_L|)^2$ and an oscillating level $4|A|^2|\rho_L| \cos^2 (\phi/2 - \beta z')$. These are again illustrated as a function of z' in Fig. 3.11.

In the Figure, the impedance at P is identical to the load impedance due to cyclic repetition of the pattern at half-wavelength intervals. At the peaks and nulls of a standing wave, the phase term in eqn. 3.33 is equal to ± 1. Both current and voltage are real, giving purely resistive impedance at these points. The length of the radial vector (or phasor) on the Smith chart for the load reflection is given by the standing-wave ratio, found from the maximum and minimum, measured as square values at the probe as

$$S^2 = \frac{|V|^2_{max}}{|V|^2_{min}} = \frac{(1 + |\rho_L|)^2}{(1 - |\rho_L|)^2} \tag{3.50}$$

At Q in the line this vector points to 180° and movement along the line to P causes clockwise rotation by $\phi'/2$. The normalised resistance and reactance of the load can be read directly from P on the Smith chart of Fig. 3.12. $\phi'/2$ is the

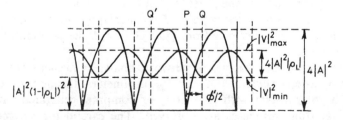

Fig. 3.11 *Null shift due to load change*

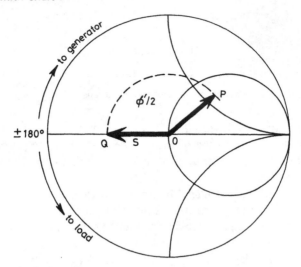

Fig. 3.12   *Smith chart plot*

change in $\beta z'$ due to movement QP towards the generator. It can be determined by noting that in eqn. 3.49 the first minimum from the load occurs when

$$\frac{\phi}{2} - \beta z' = -\frac{\pi}{2}$$

or

$$\beta z' = \frac{\pi}{2} + \frac{\phi}{2}$$

and this is the value of $\beta z'$ at Q in Fig. 3.11. Since P is the null position for a short circuit the line phase is π, so that the change in $\beta z'$ from Q to P is

$$\frac{\phi'}{2} = \pi - \beta z' = \frac{\pi}{2} - \frac{\phi}{2} \tag{3.51}$$

Since β is determined by the wavelength measurement with a short-circuit termination, eqn. 3.51 may be used to find the reflection phase ϕ at the load.

In practice, the distance Q—P is converted to a fraction of a wavelength and the vector OQ is rotated on the chart by that fraction towards the generator. If the minimum at Q' is the starting point, P is reached by movement towards the load, but since the total distance from Q to Q' is $\lambda/2$, the position of P on the chart is unchanged.

In Chapter 2 we saw that vector network analysers measure the reflection coefficient ρ_L as error-corrected S_{11} or S_{22} parameters. In one form of output they have a polar display with an overlay or computer-generated Smith chart on which real-time responses are observed. The careful measurements that ensure reliable results with a standing-wave indicator are replaced by error-correction procedures on a test set. A six- or twelve-term error correction, to be described in Chapter 5, is comparable in accuracy with the best manual measurements on a standing-wave indicator. The great advantage of the former is its much reduced time scale and its easy adaption to automation.

3.5 Application of the Smith chart

There are many applications of the Smith chart,[7] and to illustrate its use in design we give just two examples: (*a*) a double-stub tuner and (*b*) optimising the gain of a unilateral transistor amplifier.

3.5.1 Double-stub tuner

A short-circuited length of transmission line has an admittance

$$\frac{Y}{Y_0} = j \cot (\beta z) \tag{3.52}$$

Two 'stubbed' lines form a matching device with a wide tuning range if inserted into the main transmission line. In Fig. 3.13 the coaxial cable feeds an antenna with input impedance $R_A + jX_A$. Two matching stubs, consisting of short-circuited lengths of coaxial lines, are connected in parallel to the feeder at two places separated by distance l_2, and the distance to the antenna terminals is l_1. The antenna admittance can be plotted on the Smith chart by first finding its impedance point at A in Fig. 3.14 and then the diametrically opposite point at B. l_1 to the first stub tuner takes B towards the generator to C, where the parallel addition of the stub reactance takes the phasor to point D. Distance l_2 moves D on a circle of radius OD to point E where the parallel reactance of the second stub reduces the reflection to zero by moving the impedance to the match point 0, for a normalised impedance of unity. The required reactances at the stubs are obtained by adjusting their short-circuiting pistons. Solutions are not unique, since l_1 and l_2 can be chosen independently, but having set these lengths, only one setting of the stubs is capable of matching the load at a particular frequency. As frequency increases, the electrical length of l_2 approaches $\lambda/2$. This places a tuning-range limitation on the method.

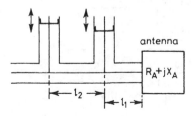

Fig. 3.13 *Double-stub tuner*

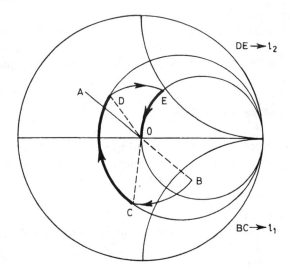

Fig. 3.14 *Smith chart plot for double-stub tuner*

3.5.2 Optimised gain of microwave transistor amplifier

A two-port transistor amplifier can be characterised through measurement of its S-parameters in a vector network analyser. To show how the Smith-chart display is used in gain optimisation, we begin with a 'black-box' two-port model of the amplifier[5] and its equivalent flowgraph as shown in Fig. 3.15. At microwave frequencies the amplifier input and output are connected by transmission lines, usually stripline, to the source and load. If Z_0 is the line impedance, the source and load reflection coefficients are

$$\rho_S = \frac{Z_S - Z_0}{Z_S + Z_0} \tag{3.53}$$

$$\rho_L = \frac{Z_L - Z_0}{Z_L + Z_0} \tag{3.54}$$

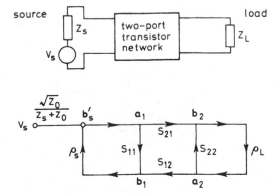

Fig. 3.15 *Two-port transistor amplifier with flowgraph*

The input node b'_s is the wave into a Z_0 matched transmission line. If V_S is the open-circuit voltage output at the source, the voltage across a Z_0 matched transmission line connected to the source is

$$V'_S = V_S \frac{Z_0}{Z_S + Z_0} \tag{3.55}$$

where Z_S is the source impedance. But from eqn. 3.9,

$$b'_s = \frac{V'_S}{\sqrt{Z_0}}$$

or

$$b'_s = V_S \frac{\sqrt{Z_0}}{Z_S + Z_0} \tag{3.56}$$

The power gain G of an amplifier is defined as the ratio of the output power P_L delivered to the load Z_L, to the input power P_{SA} available from the source:

$$G = \frac{P_L}{P_{SA}} \tag{3.57}$$

The power delivered to the load is the incident power minus the power reflected from the load:

$$P_L = \frac{1}{2}|b_2|^2 - \frac{1}{2}|a_2|^2$$

or

$$P_L = \frac{1}{2}|b_2|^2(1 - |\rho_L|^2) \tag{3.58}$$

To find the power available from the source we first connect it directly to the load, as in Fig. 3.16, when

$$P_L = \frac{1}{2}\frac{|b'_s|^2}{|1 - \rho_S\rho_L|^2}(1 - |\rho_L|^2) \tag{3.59}$$

This has a maximum value

$$P_{SA} = \frac{|b'_s|^2}{2(1 - |\rho_S|^2)} \tag{3.60}$$

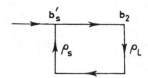

Fig. 3.16 *Through connection*

when the load reflection coefficient is the complex conjugate of the source reflection, or

$$\rho_S = \rho_L^*$$ (3.61)

The ratio of eqns. 3.58 to 3.60 gives the power gain as

$$G = \frac{|b_2|^2}{|b_S'|^2}(1 - |\rho_S|^2)(1 - |\rho_L|^2)$$ (3.62)

The ratio $|b_2|^2/|b_S'|^2$ can be found by applying the non-touching loop rule to the flowgraph in Fig. 3.15 to give

$$\frac{b_2}{b_S'} = \frac{S_{21}}{1 - S_{11}\rho_S - S_{22}\rho_L - S_{21}\rho_L S_{12}\rho_S + S_{11}\rho_S S_{22}\rho_L}$$ (3.63)

and

$$G = \frac{|S_{21}|^2(1 - |\rho_S|^2)(1 - |\rho_L|^2)}{|(1 - S_{11}\rho_S)(1 - S_{22}\rho_L) - S_{21}S_{12}\rho_S\rho_L|^2}$$

In a unilateral device $|S_{12}|^2 \ll |S_{21}|^2$, so that

$$G = \frac{|S_{21}|^2(1 - |\rho_S|^2)(1 - |\rho_L|^2)}{|1 - S_{11}\rho_S|^2|1 - S_{22}\rho_L|^2}$$ (3.64)

Transistor-amplifier design depends on many other factors than those considered here.[5] We wish only to demonstrate the use of overlapping Smith charts by concentrating on the input and output matching requirements for maximum gain. This condition is met when the input and output are conjugately matched to source and load, respectively. Thus in the flowgraph

$$\rho_S = S_{11}^*; \qquad \rho_L = S_{22}^*$$

where S_{11}^* and S_{22}^* are the complex conjugates of S_{11} and S_{22}. The gain now becomes

$$G_{max} = \frac{|S_{21}|^2}{(1 - |S_{11}|^2)(1 - |S_{22}|^2)}$$ (3.65)

Suppose the input and output line impedances are 50 Ω and that we match these to S_{11} and S_{22}, respectively, by means of the series/parallel components Z_1, Z_2/ Y_1, Y_2 in Fig. 3.17. The reflection coefficient looking towards Y_1, Z_1 is S_{11}^*; and

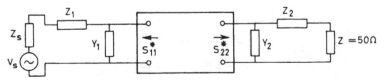

Fig. 3.17 *Matching elements*

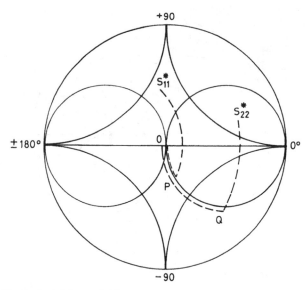

Fig. 3.18 *Smith chart plot for matching*

towards Y_2, Z_2 it is S_{22}^*, if we choose the values correctly. One method of selecting matching impedances depends on overlaying impedance and admittance charts as illustrated in Fig. 3.18. S_{11}^* and S_{22}^* are plotted on them, either from their complex components or their amplitudes and phases, and suitable matching impedances/admittances are selected to reduce them to the centre of the charts, where mutual reflection is zero. For instance, in the Figure, the dotted line S_{11}^* to P is on the admittance chart and corresponds to L_1 in Fig. 3.19, whereas P to O is on the impedance chart and corresponds to C_1. A similar argument reduces S_{22}^* to O via $S_{22}^*Q(L_2)$ and $QO(C_2)$. This can be checked in reverse by beginning at the centre of the chart for a generator matched into the network and progressing to the conjugate position S_{11}^* at the input to the active two-port. The method is suitable only for narrow-band small-signal active devices, but can still provide insight even with large-signal two-ports.

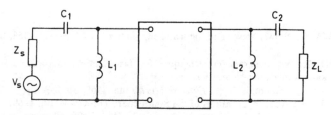

Fig. 3.19 *Matching elements*

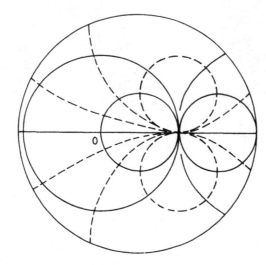

Fig. 3.20 *Negative Smith chart*

3.6 Negative resistance Smith chart

Oscillators, such as tunnel diodes, often have negative input impedances, resulting in reflection coefficients greater than unity. The Smith chart can be expanded beyond the $|\rho_L| = 1$ radius to allow for this. Such a circle, with $|\rho_L|$ expanded to 3·16, is illustrated in Fig. 3.20, where the central circular area is the normal positive-resistance Smith chart. Regions beyond that are for $|\rho_L| > 1$. An alternative negative-resistance version, allowing for reflection coefficients from one to infinity, is derived by writing[6]

$$\frac{1}{\rho_L} = u' + jv' = \frac{1}{u + jv}$$

for eqn. 3.38 and proceeding with an analysis similar to that for the positive chart. It turns out to be identical to the positive one, but as though viewed from the back of the paper. This form is very convenient for computerised displays, where the reverse transformation can be achieved easily.

3.7 References

1 KURAKAWA, K.: 'Power waves and the scattering matrix', *IEEE Trans.*, 1968, **MTT-16**, pp. 194–202

2 ANDERSON, R. W.: 'S-parameter techniques for faster, more accurate network design'. Hewlett–Packard Application Note, 95-1, Feb. 1967

3 HICKSON, R. A.: 'The Smith chart', *Wireless World*, Jan. 1960, pp. 2–9

4 SMITH, P. H.: 'Electronic applications of the Smith chart' (McGraw-Hill, 1969)

5 LIAO, S. Y.: 'Microwave devices and circuits' (Prentice-Hall, 1980), Chaps. 4 and 6

6 SMITH, P. H.: 'A new negative resistance Smith chart', *Microwave J.*, June 1965, pp. 83–92

7 ADAM, S. F.: 'Microwave theory and applications' (Prentice-Hall, 1969) Chap. 2

3.8 Examples

1 It was found that a certain impedance terminating a waveguide produced a VSWR of 7 and that the first minimum was 1/8 of a guide wavelength from it. Determine the normalised resistance and reactance of this terminating impedance. State the value of return loss and mismatch loss.

2 Discuss the advantages of having a transmission line matched. A rectangular waveguide is terminated in a power-measuring device that has a normalised impedance of $1 + j0.2$. Calculate the fraction of the incident power that is absorbed by the device, and the resultant standing wave ratio in decibels.

3 A coaxial connector, connected to a matched load by a length of air line, has a maximum return loss of 20 dB and the line is approximately 5λ at 3.0 GHz. It is found that a small adjustment of the line length gives zero reflection from the connector. What is the return loss due to the connector alone? Estimate the return loss when the frequency changes to 2.925 GHz.

Chapter 4

Signal generation

4.1 Signals and sources

Signals generated for measurement purposes depend on the test requirements of a specification or design development. They may themselves sometimes involve a separate design, particularly for high-power testing or when special signal processing is necessary. There is, however, a range of laboratory signal sources intended for general use, and it is these we are to address in this Chapter.

There is some confusion over terminology in signal generation that we may resolve for this discussion by referring to Fig. 4.1. Source, generator and oscillator are often loosely interchangeable terms, but in this Chapter, sources are to be the oscillator circuits at the centre of a signal generator or sweeper, both of which may be synthesised, if frequency accuracy is essential. We will define components as follows.

4.1.1 Signal generators
These are calibrated transmitters used for testing receivers, giving simulated signals necessary for design production and maintenance tests.[1] Among the facilities available are calibrated frequency, output level (from 0·01 pW to megawatts) and variable modulation (including AM, FM, phase, pulse etc.). The integrity of the signal depends on minimising harmonics, non-harmonic spurious outputs, microphonics, phase noise and electromagnetic interference (EMI) leakage. Signal generators may be divided into non-sweepers and sweepers.

4.1.2 Non-sweepers
Usually referred to as simple signal generators, they provide all or some of the facilities outlined above. Synthesised generators have frequency accuracies of the order of 1 in 10^9, achieved by phase locking the source oscillator to a temperature-controlled crystal oscillator. Signal generators with no phase locking are typically accurate to $\pm 1\%$, largely due to the temperature dependence of cavity oscillators, though accuracies of ± 10 MHz in 18 GHz are obtainable

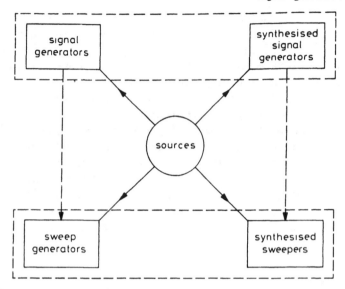

Fig. 4.1 *Sources, sweepers and generators*

by microprocessor frequency correction using PROM data on the characteristics of the oscillator.

4.1.3 Sweepers
There are swept signal generators which, unsynthesised, give $\pm 1\%$ accuracy, or 1 in 10^9, when synthesised.[2] In the latter case sweeping is performed by switching between closely spaced crystal-controlled fixed frequencies. A technique called 'lock and roll', in which phase lock is applied at the beginning, but is lost during the rest of the sweep, gives accuracies comparable to a PROM-corrected signal generator. Output power ranges are typically $-120\,\mathrm{dBm}$ to $20\,\mathrm{dBm}$, unlevelled, but reduced to a maximum of 2—10 mW when levelled.

4.1.4 Sources
There may be one or more sources at the centre of a signal generator. Some different types are listed below:

DC voltage	Type	Tuning
High	Klystron	Cavity/reflector
	Backward-wave oscillator	Slow-wave-structure/beam-current
	Gunn diode	
Low	Bi-polar	YIG/varactor in cavity
	GaAs FET	
	IMPATT	

Detailed descriptions of these devices are given in References 3, 4 and 6. Some broad general principles are that cavities exhibit low single-sideband and spurious noise. They also have a low broad-band noise floor, but suffer from warm-up drift. Backward-wave oscillators are broad band compared to klystrons, but are noisier with a shorter life. Both require high voltages. Modern semiconductor sources are readily synthesised, use low voltages and give power outputs comparable to those from low-power klystrons. They are broad band and more easily switched or swept in frequency. In consequence, they are rapidly replacing traditional high-voltage sources in the low- to medium-power generators found in most microwave test sets. For this reason we restrict our discussion of signal generators to a full description of a synthesised generator with a YIG tuned oscillator as its source. Once we have an understanding of the synthesised version, it is a simple matter to extend the discussion to lock-and-roll sweepers, fixed-frequency generators and unsynthesised sweepers.

4.2 YIG tuned oscillator

In this oscillator an active element provides a negative resistance in a YIG resonator consisting of a sphere of yttrium–iron–garnet in a DC magnetic field. Active elements may be a bipolar transistor, in the range 1—20 GHz or a GaAs FET from 7 to 40 GHz. Both are stable, but bipolars have 10—15 dB lower phase noise than GaAs FETs and a wider tuning range in spite of a lower maximum frequency. Each has a low gain per stage and an efficiency of $\sim 30\%$. Alternatively, the active element can be a Gunn or IMPATT diode, with a frequency range from 1 to > 100 GHz. They are difficult to stabilise and have lower efficiencies at 20 and $\sim 10\%$, respectively. We will confine our discussion to bipolars or GaAs FETs and show how the oscillator components, listed below, are combined to produce a stable yet flexible source.

The advantages of YIG tuned oscillators are:

● Low phase noise
● Wide operating frequency range (two octaves)
● Excellent tuning linearity
● Reliability

The main components in a YIG oscillator are:

● YIG resonator
● Variable DC magnetic field for tuning
● FM coil for tuning
● Negative dynamic resistance
● Output buffer and amplifier

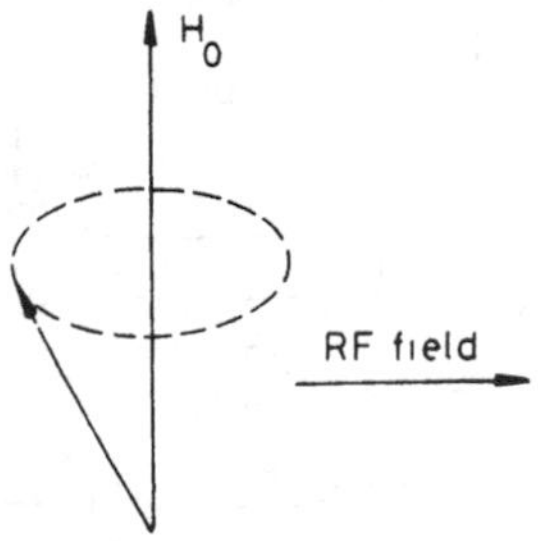

H_0

RF field

Fig. 4.2 *Precession in RF field*

Yttrium–iron–garnets are a family of ferrites doped to resonate at microwave frequencies when immersed in a suitable magnetic field. Elemental dipoles of the material, originating in the precession of electron spin about the DC magnetic field, are aligned in the direction of that field, with precession frequency at the natural magnetic resonance of the garnet. When an RF field is imposed perpendicularly to the DC magnetic field at the resonance frequency, an absorption or reflection occurs depending on the configuration of the circuits external to the garnet. In Fig. 4.2, electron spins precess about H_0 at the gyromagnetic frequency given by

$$f_0 = \gamma H_0$$

where

$$\gamma = 2{\cdot}8 \text{ MHz s}^{-1} \text{ Oe}$$

Since f_0 is linearly related to the magnetic field, the oscillator can be tuned by varying H_0. The resonating garnet is made in the shape of a sphere with diameter in a range 0·2—2 mm and a tolerance of $\pm 5\,\mu$m. Fig. 4.3 shows how a

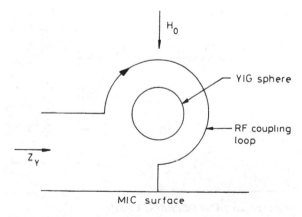

Fig. 4.3 *Coupling to YIG sphere*

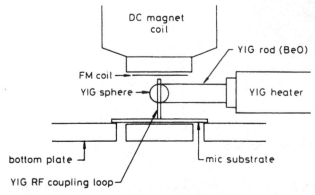

Fig. 4.4 *YIG-oscillator construction (After OSBRINK, 1983)*

coupling loop is placed around the YIG sphere to give orthogonal RF and DC
magnetic fields. A schematic of the total assembled structure is given in Fig. 4.4,
showing the RF coupling loop directly connected to an MIC substrate on which
are placed the active devices. Resistance to temperature variation is achieved by
heating the sphere and its supporting rod of non-magnetic beryllium oxide with
the rod rotated to position the sphere for minimum temperature sensitivity.
Heating to 105°C $\pm$ 5°C gives approximately a sevenfold reduction in changes
due to temperature variation. In the equivalent circuit of Fig. 4.5, R_L and L_L
represent coupling-loop loss and inductance; C_Y, L_Y give the resonant condition
and R_Y the Q loss of the YIG sphere. Resonances due to unwanted magnetic
modes of the spheres are not included in the equivalent circuit. They can cause
spurious outputs as well as detuning the oscillator.

For oscillation, the real part of the resonator impedance is cancelled by the
negative-resistance circuits on the MIC. Z_C, the input impedance of either a
bipolar or GaAs FET in Fig. 4.6, is in parallel with the resonator Z_Y. If at

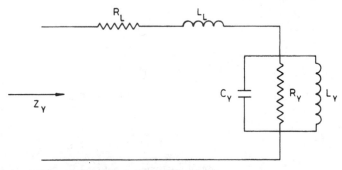

Fig. 4.5 *YIG oscillator circuit (After OSBRINK, 1983)*

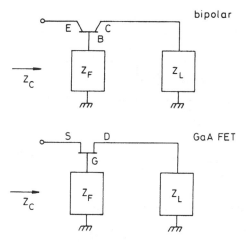

Fig. 4.6 *Bipolar and GaAs FET active elements (After OSBRINK, 1983)*

resonance Z_C has a negative real part, the reflection coefficient of the transistor is greater than unity since[5]

$$\rho_C = \frac{Z_C - Z_0}{Z_C + Z_0}$$

In general, reflection coefficient ρ_C equals the reciprocal of the resonator output reflection coefficient, and, for large signals, stable oscillations occur when the magnitude of the real parts of Z_Y and Z_C are equal and reactances are conjugately matched. Variation of DC magnetic field determines frequency setting to a high degree of linearity ($\sim 0.08\%$), whilst frequency modulation is accomplished by passing a modulating current though an FM coil, consisting of several turns of wire mounted on the magnet pole face.

4.3 Synthesised sweeper

We are now in a position to describe a complete synthesised sweeper with a YIG oscillator source. Output must be at the selected frequency, stabilised at the required power, and conditioned by levelling, amplifying or modulating as illustrated in Fig. 4.7.

4.3.1 Frequency

Accurate frequency can be made available at microwave frequencies by phase locking a source to a harmonic of a temperature-controlled low-frequency quartz oscillator. A non-linear circuit, based on a step-recovery diode,[7] at the output of the oscillator serves to generate harmonics, and, in a phase-lock loop,

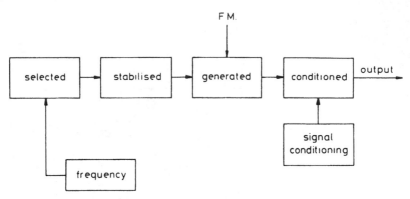

Fig. 4.7 *Signal selection, generation and conditioning*

one of them is compared with and stabilises the microwave source. This reference oscillator, typically at 10 MHz, is accurate to 1 part in 10^9 per day or 0·01 Hz average frequency variation. Short-term (10^{-5} to 1 s) phase jitter, discussed in more detail in Chapter 10, causes short-term frequency errors, typically one or two orders greater than long-term average errors, depending on the quality of the reference standard. Harmonic generation is effectively frequency multiplication, so that the transferred accuracy of the reference is unchanged as a percentage of the final frequency, and the 0·01 Hz variation at 10 MHz becomes 6 Hz at 6 GHz. Resolution, on the other hand, is a matter of the number of steps in the frequency range, with the smallest step size limited to the percentage accuracy. However, if we start from 10 MHz, the maximum number of harmonic steps to 6 GHz is 600 and the resolution is only 10 MHz. We could consider increasing the resolution by dividing the reference source frequency before generating harmonics, but this is limited by falling levels with increasing harmonic number. For a typical practical resolution of 10 kHz, division by 10^3 is necessary and, at 6 GHz the harmonic number is 60 000. But only about the first 100 harmonics are suitable for most applications in a step-recovery diode, because the power-conversion ratio is proportional to the reciprocal of the harmonic number. The problem is overcome by using a combination of harmonic generation and frequency conversion.

There are many different ways of combining these two techniques, most of them based on phase-lock loops to stabilise a variable oscillator against a fixed-frequency reference, and mixer circuits for frequency up- or down-conversion. In Fig. 4.8 we show how the phase-lock loop locks the phase of a 100 MHz voltage-controlled crystal oscillator (VCXO) to a temperature-compensated 10 MHz quartz reference by a method called indirect synthesis. In the loop the VCXO frequency is divided and compared with the reference in a phase-sensitive detector that gives an output only when the phases differ. After filtering to remove unwanted components, this output is fed, as negative feedback,

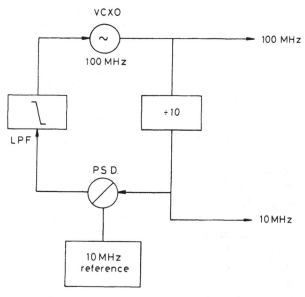

Fig. 4.8 *Indirect Reference*

to the VCXO, so that its frequency is controlled for minimum phase difference. In this way the frequency of the reference is effectively multiplied by 10 and the overall noise performance is improved. Within the phase-lock bandwidth the predominating noise is phase noise close to the 10 MHz reference carrier, whereas outside that range most noise is typical of the VCXO. Fig. 4.9 shows that an unlocked VCXO has higher noise than the quartz crystal close to the carrier, but since it is otherwise less, a careful choice of phase-lock bandwidth minimises overall noise. These ideas are covered in more detail in Chapter 10.

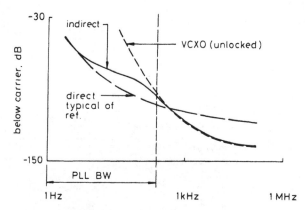

Fig. 4.9 *Carrier noise in indirect reference*

In transferring a low-frequency reference to a high microwave frequency we have to avoid generating very high harmonics in the step-recovery diode. This is only possible if the reference frequency can be raised sufficiently before driving the diode. It is easier to show how this can be done if we take a specific example and use the frequencies shown in Figs. 4.10 and 4.11. We begin at the microwave stage of frequency generation where a YIG oscillator, tunable from 2 to 6·2 GHz is to be locked to an accurate frequency source to a setting resolution of 10 kHz. In Fig. 4.10 the YIG oscillator can be coarse tuned to within ±35 MHz of any frequency in its range by adjustment of its DC magnetic field through the field amplifier. Its ouput is mixed with a harmonic of a derived reference source tunable from 198 to 220 MHz in steps of 0·5 MHz. The sampler is a special kind of mixer that incorporates a YIG pin switch filter that, under computer control, can select any single harmonic in the required frequency range. We will meet this device again in Chapter 10, and see in more detail how it is used in frequency counters. Since the YIG oscillator can be coarse tuned to within ±35 MHz there should be at least one harmonic in each 70 MHz interval. The output IF from the sampler then equals the separation of the YIG frequency from the selected harmonic. But the intermediate frequency is determined by the output phase-lock loop, incorporating the phase-sensitive

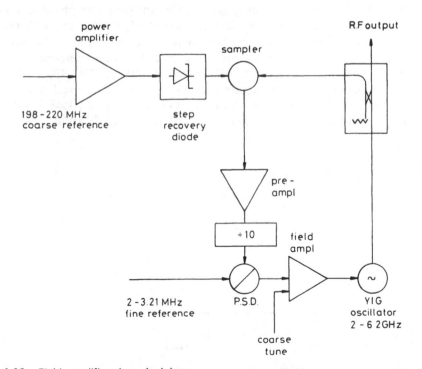

Fig. 4.10 *Field-amplifier phase-lock loop*

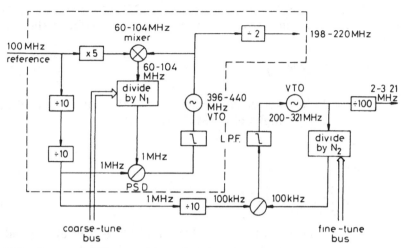

Fig. 4.11 *Course- and fine-tune phase-lock loops*

detector (PSD) and field amplifier and driven by a second fine-tuning reference, switchable in steps of 1 kHz, from 2 to 3·21 MHz. Since the intermediate frequency is divided by 10 before phase comparison, it can be set to a resolution of 10 kHz in the range 20—32·1 MHz by preselecting the fine tuning reference, again usually under computer control. The final frequency is therefore selected in two stages—by coarse tuning to a selected harmonic and fine tuning to the fine tune reference. Although capture is possible within ±35 MHz, the intermediate frequency range is only from 20 to 32·1 or 12·1 MHz. Harmonic spacing must therefore be closer than 24·2 MHz to ensure continuous tuning over the whole range. Only one beat frequency, either above or below the fine-tune reference, will be less than 12·1 MHz. Its value in the IF amplifier determines how far the YIG frequency is from the chosen harmonic, and its position *vis-à-vis* the reference gives the direction from it.

To find the average harmonic line density we consider the centre frequency, 209 MHz, of the coarse reference and find that, for instance, at 4 GHz the mean harmonic number closest to $4 \times 10^9/209 \times 10^6$ is 19, and since the reference is in 0·5 MHz steps, the line separation is $19 \times 0·5 = 8·5$ MHz, which is within the capture range of the fine-tuning intermediate frequency. This can be formalised into the following algorithms.

If f_C is the coarse-reference baseband frequency and f_{RF} the required microwave frequency, the harmonic number in the step-recovery diode is

$$m = \frac{\text{lower integer}}{\text{value}} \left[\frac{f_{RF}}{f_C} \right] \tag{4.1}$$

The frequency difference between the harmonic and f_{RF} is

$$\Delta f_{RF} = f_{RF} - m f_C \tag{4.2}$$

Baseband frequency can be shifted in steps of 0·5 MHz to compensate, but the RF shift, Δf_{RF}, is m times greater than an equivalent baseband shift of Δf_C. Therefore

$$\Delta f_C = \frac{\Delta f_{RF}}{m} = \frac{f_{RF}}{m} - f_C \tag{4.3}$$

If f_C is taken as the lower baseband frequency, 198 MHz, the baseband can be shifted by Δf_C, in N_1 multiples of 0·5 MHz, or

$$\Delta f_C = 0 \cdot 5 N_1$$

and from eqn. 4.3 the integer N_1 is

$$N_1 = \text{int} \left[2 \left(\frac{f_{RF}}{m} - 198 \right) \right] \tag{4.4}$$

and m is given by eqn. 4.1 with $f_C = 198$ MHz and all frequencies in megahertz. The total baseband shift of 22 MHz translates into 22m MHz in the RF band. Even at 2 GHz, $m = 10$ from eqn. 4.1, allowing a harmonic shift of 220 MHz, which just covers the range between two harmonics. After shifting, the new baseband frequency becomes $(198 + 0 \cdot 5 N_1)$ MHz and the minimum shift of a harmonic line in the RF band becomes

$$\frac{1}{2} \frac{f_{RF}}{f_C} = \frac{f_{RF}}{2(198 + 0 \cdot 5 N_1)} = \frac{f_{RF}}{396 + N_1} \tag{4.5}$$

From eqn. 4.1 and eqn. 4.4 $N_1 = 4$ at 2 GHz and 6·2 GHz with harmonic separation 5 and 15·5 MHz, respectively. Therefore line separation in each case remains within the IF range.

Finally we see in Fig. 4.11 how coarse tuning to 0·5 MHz steps in the range 198—220 MHz is provided by mixing a 396—440 MHz voltage-tuned oscillator (VTO) with a × 5 100 MHz reference, phase locking at 1 MHz and dividing the output frequency by two. Fine tuning is obtained by dividing the reference by 1000 before phase locking in the range 200—321 MHz and dividing by 100 to give 1 kHz steps. Resolution is determined by a dividing integer N_2, with 1211 1 kHz levels from 2 to 3·21 MHz, requiring 11 bits. Coarse tuning via N_1 has 45 0·5 MHz levels or 6 bits. A total of 17 bits, or 128 K levels, therefore specifies the final selection to a resolution of 10 kHz. This compares with a multiplication factor of 62 000 (16 bits in binary) for a 10 kHz step in 6·2 GHz.

4.3.2 Signal conditioning

Conditioning, taking place at the output of the YIG oscillator, is mostly concerned with amplitude control, since frequency modulation is by means of the FM coil. There is, however, some frequency control in Fig. 4.12 through a YIG-tuned multiplier (YTM) to extend the range to 18·6 GHz. A power amplifier following the PIN modulator provides a high-level signal to the YTM, in which a step-recovery diode produces harmonics, one of which is selected by a

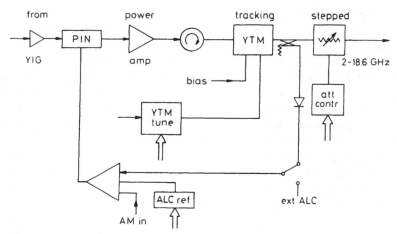

Fig. 4.12 *Output-level conditioning of YIG oscillator*

YIG tracking filter, not shown in the Figure. Tracking of input YIG and YTM depends on very linear magnetic-tuning structures in each device, and determines the re-setting accuracy of a selected harmonic to ± 20 MHz at 18 GHz. Frequency ranges from 2 to 6·2 GHz, 6·2 to 12·4 GHz and 12·4 to 18·6 GHz, with diminishing setting resolutions, are derived from the first three harmonics of the YTM. This falling resolution, and the complexity of tracking filters, has resulted in an increasing use of fundamental oscillators to cover each frequency range.

Automatic levelling takes place in a feedback loop with directional coupler, detector, amplifier and PIN diode attenuator. Amplifier gain has to be correct for stable negative feedback, but the levelled power is not dependent on a diode detection law, since it is effectively held constant against the level selection ALC reference inserted in the amplifier. We have already seen how effective source match depends on coupler directivity. Amplitude modulation may be added into the ALC circuit at the loop-amplifier input.

Finally, a calibrated step attenuator, often associated with a power level read-out, is placed at the output to set and measure output power. Power read-out is usually limited to the accuracy of square-law-detector diodes.

4.4 Comparison with unsynthesised generators

Unsynthesised generators are source-oscillator dependent since they have no phase locking. In sweep mode they are faster because the sweep is continuous rather than a series of phase-locked fixed points. Full synthesised accuracy is available only during fixed-frequency operation, with a switching time between frequencies of ~ 50 ms, thus restricting synthesised sweep widths in both resolution and bandwith. A narrow sweep of 100 Hz has 100 frequency steps for 1 Hz resolution and the sweep time is 5 s. Taking 100 steps over a 600 MHz bandwidth reduces resolution to 6 MHz. A good compromise is lock-and-roll mode with phase lock only at the beginning of each sweep.

Unsynthesised generators are often upgraded by means of synchronisation to external references. Similarly external normalisers can correct for amplitude versus frequency variations. Cost trade-offs are usually in favour of internally controlled instruments when considering a new generator, but may sometimes point to add-on components for older instruments.

Very rough comparisons of performance are given in the following table, but they should be treated with considerable caution, since only a careful reading of specifications, and preferably with a use in mind, can give a real feel for the relative performance of different instruments:

	Synthesiser	*Sweep generator*
Frequency range	10 MHz to 40 GHz	Similar
Stability	1×10^{-9} per day	± 10 MHz
Resolution	1 to 4 Hz	± 10 MHz
Swept resolution	0·1% of sweep width	± 30 MHz
Switching time	$\leqslant 20$ ms to within 10 Hz	Analogue
Output power	$+ 10$ dBm	Similar
Harmonics (Relative to carrier)	< -35 dBc	< -20 dBc
Non-harmonic spurious	-60 dBc	< -50 dBc

High resolution does not guarantee high accuracy and it is important to distinguish between them. For instance, the fundamental limit to swept resolution is the minimum resolution bandwidth of Chapter 1:

$$R = \sqrt{\frac{S}{T}}$$

If we take a sweep scan $S = 50$ MHz for time scan $T = 500$ ms., the minimum resolution is 10 kHz for a simple sweep generator. But accuracy depends on long-term stability in an unlocked YIG oscillator. This is 1 part in 10^3; so that at 10 GHz the accuracy is $10^{-3} \times 10$ GHz $= 10$ MHz. For a synthesised sweeper this becomes $10^{-9} \times 10$ GHz $= 10$ Hz, but resolution is a matter of switching time and the number of spot frequencies. If the switching time is 25 ms, so that in 500 ms there are 20 spot frequencies over a 50 MHz scan, the resolution is 2·5 MHz. On the other hand, in sweep mode, the synthesiser might have a resolution of 0·1% of sweep scan or 50 kHz.

These calculations demonstrate the need to be clear about an application, before deciding on the best type of instrumentation, especially when the most expensive solution could lead to a greater error, either of resolution or accuracy.

4.5 References

1 MINCK, J. L.: 'Signal simulations, signal generators and signal integrity,' *Microwave J.*, April 1983, pp. 75–83
2 MORRIS, M. L.: 'Sweep generators vie for synthesiser applications', *Microwave J.*, April 1983, pp. 135–142
3 LIAO, S. Y.: 'Microwave devices and circuits' (Prentice-Hall, 1980)
4 BADEN-FULLER, A. J.: 'Microwaves' (Pergamon, 1979)
5 OSBRINK, N. K.: 'YIG-tuned oscillator fundamentals' (MSDH, 1983) pp. 207–225
6 EDWARDS, T. C.: 'Introduction to microwave electronics' (Arnold, 1984)
7 ADAM, S. F.: 'Microwave theory and applications' (Prentice-Hall, 1969) p. 149

4.6 Examples

1   The frequency range of a YIG oscillator is from 12·4 to 18 GHz. Using phase-lock loops similar to those outlined in Section 4.3.1, specify the harmonic number from the step-recovery diode and the divider values, N_1 and N_2. You may assume 2 MHz steps into the multiplier, and a final frequency resolution of 1 kHz, and that the YIG capture range is unchanged.

2   What sources would you choose for (*a*) CW, (*b*) pulse measurement in a frequency range 26·5—100 GHz at power levels from 1 to 10 mW? If the power required is increased to several hundred watts and frequency modulation is necessary, what would be your choice? If necessary, consult manufacturer's specifications.

3   Frequencies can be extended beyond 6·2 GHz in the example quoted in Section 4.3.1 by means of a tracking YIG filter multiplier. Some signal generators have three separate oscillators to cover the range 2—6·2 GHz, 6·2—12.4 GHz and 12.4—18 GHz. What are the relative advantages of the two methods?

Chapter 5

Vector analysers

5.1 Vector measurement systems

Vector analysers measure amplitude and phase of reflected or transmitted signals with respect to an input reference, either directly or as complex ratios. There are many different types of analyser systems, distinguished by their error-correction capabilities, output display, degree of automation etc. A complete vector network system usually has an analyser and test set at its centre, to which is added a display unit, a suitable sweep oscillator, a controlling computer and further items as required by the measurement in progress. For instance, a computer is not always necessary for single measurements; displays are often simple meters etc. This discussion of vector network analysers takes place around a particular example, the Hewlett–Packard HP8510. By choosing an advanced commercially available instrument we are able to subsume the techniques of many other examples in a single description, digressing where necessary to include some important aspects not otherwise covered in the thematic context. By concentrating on modern instrumentation we are ignoring many interesting methods not currently in use. An alternative historical presentation can be found in References 1 and 2.

5.2 The HP8510

The major elements of a measurement system based on the HP8510 are shown[6] in Fig. 5.1. They can be broken into three parts: the source, which is a programmable synthesised sweeper or sweep oscillator; the test set, consisting of a splitter and bridge network for measuring forward and reverse S-parameters of a two-port device under test; and the analyser itself, which in this case includes a control microprocessor, a number of peripherals and access to an external data bus (HP–IB) for remote or automatic operation.

Sweeper scan range and frequency are selected via the analyser's system bus. In each sweep, measurements can be made at 51, 101, 201 or 401 discrete

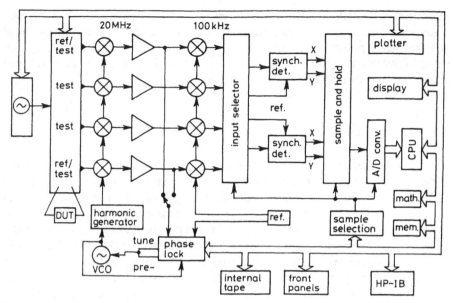

Fig. 5.1 *HP8510 block diagram (After BRAUN, 1984)*

frequencies. The test set is designed to avoid the need to reverse connections to the DUT when reversed signal flow is required. One reference and the two test channels are used to measure transmission and reflection for one signal flow through the DUT. The other reference is similarly used in the reverse direction. Conversion to first IF is made by locking a harmonic of the local oscillator VCO at 20 MHz above the selected reference channel. Tuning of the phase-locked loop is by digital division to ensure precise repetition from sweep to sweep. When the main loop, via the sampling mixer, is unlocked, pretune selects a harmonic by digital control of a second phase-lock loop. The harmonic is chosen to be within the IF bandwidth of the main loop, so that the VCO becomes a signal-tracking oscillator. A common crystal reference serves both phase-lock loops, as well as local oscillator for the second down conversion to 100 kHz. Auto-ranging is achieved by ratioing the test channels with the reference channel and controlling amplifier gains in the input selector to optimise sensitivity for all signal levels. Common-mode variations are also minimised by the same process. A digitally controlled synchronous detection precedes a sample-and-hold multiplier, that in turn drives an analogue/digital convertor. From there the readings enter a central processing unit prior to storage or display. Each reading takes about 40 μs for an analogue/digital conversion, and four readings are made for each data point.

Many of the standard components of the HP8510 are found as optional additions in earlier systems. Complete integration of a 16 bit microprocessor with the analyser and sweeper allows use of a more complex error-correction algorithm, whilst also reducing measurement time compared to other systems. Thus forward and reverse scattering parameters of a network are measured at 401 frequency points in 800 ms, giving virtual real-time operation during network adjustments. Dynamic accuracies of ± 0.5 dB and $\pm 0.3°$ in amplitude and phase are also claimed.

5.3 Microwave test sets

The S-parameter test set as part of an HP8510 network-analyser system incorporates both transmission and reflection measurements as shown in Fig. 5.2. If we begin with forward testing of a DUT, the input switch connects the RF via the splitter to reference channel a_1 and via a step attenuator in the incident channel to the device under test. A bias TEE, also inserted in the incident (and transmission) channel, is to allow DC biasing on active DUTs. A bridge circuit, similar to a directional coupler, but providing 6 dB or more isolation between any two forward ports, separates the reflected signal into test channel b_1. The transmitted signal is separated from the main channel by a second bridge and directed to test channel b_2. In reverse, signal flow is in the opposite direction through the device with channel a_2 as reference and b_2 and b_1 as reflection and transmission outputs, respectively.

The first IF conversion to 20 MHz is in a sampling mixer with a step-recovery diode as described in Chapter 4. Harmonics are generated from a local-oscillator VCO tuned over a 60—160 MHz range, one of which is locked over

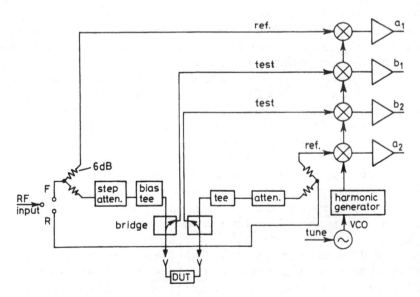

Fig. 5.2 *HP8510 test set (After BRAUN, 1984)*

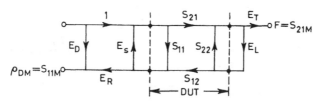

Fig. 5.3 *Error flowgraph for transmission*

an octave range to the incoming signal. Automatic retuning of the local oscillator allows a change of harmonic-number selection for each octave band in the range 45 MHz—26·5 GHz.

Disturbances due to reflections in the test set are kept to a minimum by the 6 dB loss in each arm of the splitters and the effective 12 dB loss for passes through the bridges. Thus, in error-free conditions, reflection measurements give S_{11} and S_{22}, and transmission measurements S_{21} and S_{12}. In the presence of errors, we have already seen in Chapter 2 how an error model for a one-way transmission and reflection measurement can provide insight into error correction by means of known calibration standards. This simple model, repeated in Fig. 5.3, has two deficiencies in that it does not take account of crosstalk between reference and test channels in the mixers, switches and IF amplifiers; and that it also requires the DUT to be reversed in the test set to measure S_{12} and S_{22}. Connector disturbance during reconnection can change some error terms between a forward and reverse run, leading to uncorrectable residual errors. Fig. 5.4 shows how the addition of an extra flow path takes account of transmission leakage crosstalk and Fig. 5.5 extends the model to correspond with a non-reversing or switched test set. The 12-term error model of Fig. 5.5 allows for different directivity errors (E_{DF}, E_{DR}) in the two bridges, crosstalk in both directions (E_{XF}, E_{XR}), different effective source errors (E_{SF}, E_{SR}) and transmission losses (E_{TF}, E_{TR}). By careful attention to the balancing of each pair of components it may sometimes by possible to obviate the need to determine every term. However, for devices with large reflectivities and non-reciprocal components the full model is necessary.

A calibration procedure, with known standards substituting for the DUT, is applied to determine the model error terms. Standards for reflection calibration

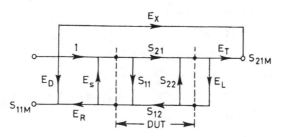

Fig. 5.4 *Six-term error model*

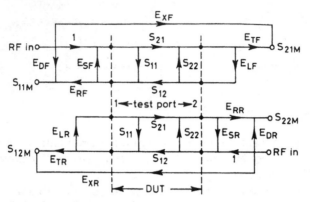

Fig. 5.5 *Twelve-term error model*

are commonly a matched load, a short and an open circuit. They are connected
and measured as S_{11} at the two ports, and have equivalent flowgraphs as shown
in Fig. 5.6. Their accuracy depends on careful design and maintenance, and on
their comparison with national standards at laboratories such as the National
Bureau of Standards in the USA or the National Physical Laboratory in the
UK. Errors in 'perfect' loads cause small reflections that can be evaluated by
placing the load at the end of an accurate sliding air line. Any reflection from
the load adds to the directivity error of the test port E_D. By sliding the load on
the air line the relative phases of both errors are changed to give the polar plot
of Fig. 5.7, when the small reflection δE due to the load adds vectorially to the
constant phase vector E_D of the directivity. Since varying air-line length rotates
δE about P, a true value for E_D is found by measuring $E_D + \delta E$ at several points
around the error circle in order to find its centre.

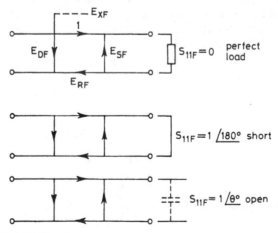

Fig. 5.6 *Calibration loads*

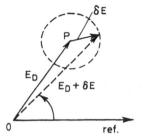

Fig. 5.7 *Sliding-load vector diagram*

Short-circuit standards are limited by connector connection repeatability and open circuits in coaxial line by a small fringing capacitance, that can be accurately modelled as a phase angle θ in Fig. 5.6. In waveguide an offset open circuit can be simulated by a $\lambda/4$ length of shorted waveguide. Connectors are an error source partly influenced by careful operator training. Manufacturers usually give instructions on how to handle connector changes, including cleaning methods to maintain good contacting surfaces. A good 7 mm connector has an S_{11} of -90 to -100 dB with repeatability over a 15 dB range, but adapter connections involving transitions between connector types can seriously degrade performance to less than -30 dB.

To calibrate a test set, each of the calibration terminations is measured at all frequencies in the required scan range. For the sliding load several (4—6) positions of the termination are taken at each frequency, with the operator making connections and adjustments on request from the computer. In the error model of Fig. 5.6, the measured reflection coefficient is

$$S_{11M} = E_{DF} + \frac{S_{11F}E_{RF}}{1 - E_{SF}S_{11F}} \tag{5.1}$$

Three known S_{11F} for forward and S_{22R} for reverse signal flow give three equations for E_{DF}, E_{RF}, E_{SF} and a further three for E_{DR} E_{RR}, E_{SR} with

$$S_{22M} = E_{DR} + \frac{S_{22R}E_{RR}}{1 - E_{SR}E_{22R}} \tag{5.2}$$

For crosstalk, E_{XF} can be determined by making a transmission measurement at port 2 with no connection to either port. Load match and transmission tracking errors, E_{LF} and E_{TF}, follow from measurement of reflection and transmission with a through connection of port 1 to port 2. These cases are represented by the flowgraphs of Figs. 5.8a and b, respectively. E_{XF} is therefore found directly, but E_{SF} and E_{LF} require two further equations easily derived from the developed flowgraph in Fig. 5.9, by writing

$$S_{11M} = E_{DF} + \frac{E_{RF}E_{LF}}{1 - E_{SF}E_{LF}} \tag{5.3}$$

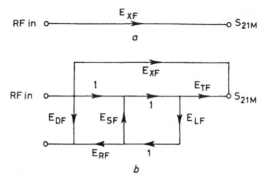

Fig. 5.8 *No-connection and through measurements*

$$S_{21M} = E_{XF} + \frac{E_{TF}}{1 - E_{SF}E_{LF}} \tag{5.4}$$

where S_{11M} and S_{21M} are the measured reflection and transmission coefficients in a forward transmission with a through connection. Transmission-tracking and load-reflection errors follow from the previous equations as

$$E_{TF} = \frac{1}{E_{RF}} \left[\frac{S_{11M} - E_{XF}}{E_{SF}S_{11M} - E_{SF}E_{DF} + E_{RF}} \right] \tag{5.5}$$

$$E_{LF} = \frac{S_{11M} - E_{DF}}{E_{SF}S_{11M} - E_{SF}E_{DF} + E_{RF}} \tag{5.6}$$

Six calibration measurements are necessary, leading to a 6-term error procedure, in analysers restricted to forward transmission only, whereas 12 error terms must be found by repeating the 6-term calibration in reverse connection in reversing test sets. After calibration, normal measurement of the DUT S-parameters can proceed, and their raw values corrected by means of an explicit expression, stated but not proved in eqns. 5.7—5.10, in terms of the measured

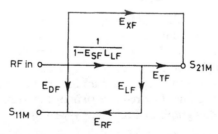

Fig. 5.9 *Topographical reduction*

S-parameters and errors found during calibration[3]:

$$S_{11} = \frac{(1 + DE_{SR})A - E_{LF}BC}{(1 + AE_{SF})(1 + DE_{SR}) - CBE_{LR}E_{LF}} \tag{5.7}$$

$$S_{12} = \frac{[1 + A(E_{SF} - E_{LR})]C}{(1 + AE_{SF})(1 + DE_{SR}) - CBE_{LR}E_{LF}} \tag{5.8}$$

$$S_{21} = \frac{[1 + D(E_{SR} - E_{LF})]B}{(1 + AE_{SF})(1 + DE_{SR}) - CBE_{LR}E_{LF}} \tag{5.9}$$

$$S_{22} = \frac{(1 + AE_{SF})D - E_{LR}BC}{(1 + AE_{SF})(1 + DE_{SR}) - CBE_{LR}E_{LF}} \tag{5.10}$$

with

$$A = (S_{11M} - E_{DF})/E_{RF} \qquad B = (S_{21M} - E_{XF})/E_{TF}$$

$$C = (S_{12M} - E_{XR})/E_{TR} \qquad D = (S_{22M} - E_{DR})/E_{RR}$$

It is clear that all four parameters must first be measured and substituted in the expressions, before a corrected version of any one of them can be calculated.

An algorithm[3] of the experimental procedure, in Fig. 5.10, begins with calibration of the test set, either for four or two parameters, depending on whether DUT reversal is necessary. Crosstalk, not shown, should also be measured, so that the flow chart becomes universal in the sense that by repeating the six coefficients derived for the 'one path' calibration, or by retaining all 12 coefficients for a 'two-path', either case can be calculated.

An example of an earlier non-switching test set, this time for a transmission measurement, is given in Fig. 5.11, to illustrate how operator involvement depends on the degree of automation. The harmonic convertor contains a sampling mixer with step-recovery diode, and gives an output at 20 MHz to the analyser. Electrical or manual adjustment of the reference-line length and DUT reversal are operator tasks, as are the considerable re-arrangements necessary to convert the test set to a reflection form illustrated in Fig. 5.12. Consequent cable movement, connector wear and component changes lead to increased error, in contrast to the static arrangement in the HP8510. Automatic performance of these tasks reduces measurement time, increases repeatability and accuracy, but also de-skills the operator, with increased equipment costs offset against cheaper labour.

For waveguide measurements, the two test sets can be realised by a waveguide network of the type illustrated[7] in Fig. 5.13. Waveguide size depends on frequency, giving very large layouts at low frequencies. Accuracy is maintained by splitting the sweeper input between two nearly identical waveguide paths and by selecting couplers in matched pairs. Fixed and sliding short circuits

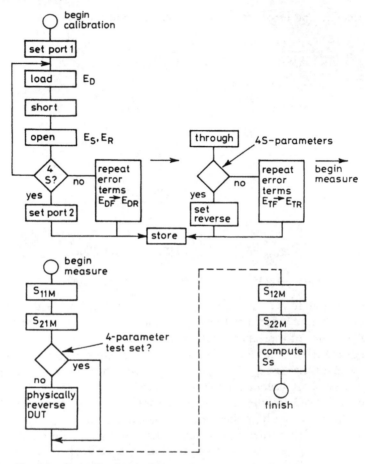

Fig. 5.10 *Algorithm for calibration and test*

reflect equal signals into the two channels via 10 dB directional couplers to allow transmission tests. Added waveguide in the reference channel cancels the DUT phase path. A reflection test is made by replacing the fixed short circuit with the DUT, and placing through connections in reference and test channels.

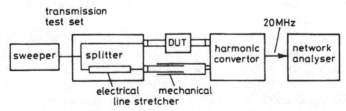

Fig. 5.11 *Coaxial transmission test set*

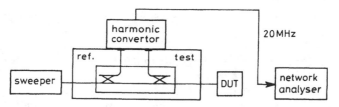

Fig. 5.12   *Coaxial reflection test set*

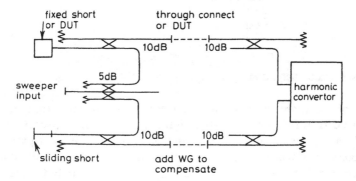

Fig. 5.13   *Waveguide reflection/transmission test set*

In both cases line stretching is possible in the reference channel by movement of the sliding short circuit, and error correction can be applied exactly as in the coaxial case.

5.4 Signal detection

The HP8510, in common with many other network analysers, is based on harmonic mixing with local-oscillator retuning and tracking to ensure accurate phase comparison and a dynamic range from 80 to 100 dB over a wide frequency band from 45 MHz to 26·5 GHz. Network analysers with fundamental mixing, illustrated[5] in Fig. 5.14, have wider dynamic ranges in excess of 100 dB, but are limited by a down-conversion process to an upper frequency not much above 2 GHz. In the example shown, beating between a swept and two fixed sources causes the local oscillator to track the split RF signal in a range 0·5—1300 MHz by a fixed amount equal to the intermediate frequency. Narrow IF bandwidth (10 kHz) and lack of spurious and harmonic responses, that are a feature of harmonic mixing, give sensitivities of −110 dBm. The two IF outputs are often synchronously detected before being processed for display.

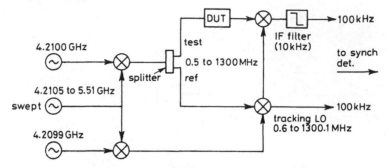

Fig. 5.14 *Fundamental mixing*

5.5 Synchronous detection

Output of simultaneous reflection and transmission data from the two synchronous detectors in the HP8510 is digitised for computer processing, whereas in earlier analysers results remain in analogue form prior to display. There are two mixers in a synchronous detector, each connected to both the test and reference signals but with a quadrature phase shift in the test signal as shown in Fig. 5.15. If $R \cos \omega t$ is the reference and $T \cos \omega t$ is the test signal, output from mixer A is

$$X = R \cos (\omega t) \, T \cos (\omega t + \phi) \tag{5.11}$$

and from B

$$Y = R \cos (\omega t + 90) \, T \cos (\omega t + \phi) \tag{5.12}$$

These become, after low-pass filtering,

$$X = M \cos \phi \tag{5.13}$$

$$Y = M \sin \phi \tag{5.14}$$

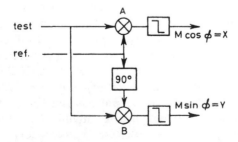

Fig. 5.15 *Synchronous detection*

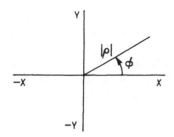

Fig. 5.16 *Polar display*

$\rho \propto \sqrt{(X^2 + Y^2)}$

$\phi = \tan^{-1}(Y/X)$

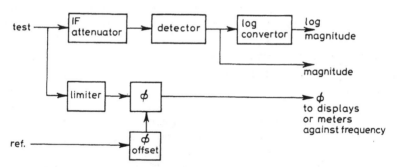

Fig. 5.17 *Magnitude and phase display*

where M is proportional to the test amplitude since, in the product $M = RT$, common-mode variations have been cancelled. Orthogonal display of X and Y produces a polar plot of the complex transmission ratio through the DUT with amplitude proportional to $\sqrt{X^2 + Y^2}$ and phase angle $\phi = \tan^{-1}(Y/X)$, as illustrated in Fig. 5.16. Polar plots are often displayed with a Smith chart, either in overlay or computer-generated form.

An alternative display of amplitude and phase on meters or an oscillosope is sometimes preferred, and Fig. 5.17 in block-diagram form shows how this can be realised in analogue circuits, with frequency proportional to the horizontal scan of a dynamic vertical display of amplitude or phase, or with all three indicated statically on meters.

5.6 Computer control

In the HP8510 complex signal components enter a sample-and-hold output multiplier prior to A/D conversion. All data processing and displays are performed by computer to provide a wide choice of display formats (log, dB,

VSWR etc.). It also controls operation of RF source, phase locking, auto-ranging, detectors and convertors: and detailed monitoring of gain and offset errors allows continuous subtraction of their effects during a self-calibration sequence. We have already discussed the computer's role in vector error correction of input data and, in Chapter 11, we will see how, by fast-Fourier transformation, it can facilitate transfer of data between time and frequency domains.

5.7 Accuracy

Calibration corrects the errors in the test set, but there are residual display inaccuracies due to a number of unavoidable factors, such as sweeper resolution, trace thickness or, in computerised systems, data-processing time. For example, a 500 ms scan of 50 MHz with typical 2 ms calculation time per frequency allows 250 points and limits resolution to 0·2 MHz. Residual errors that cannot be corrected by calibration have their origins in the repeatability of connections and switches, noise in detectors and harmonic skip in samplers, though the latter has been virtually eliminated in modern network analysers by digital tuning in phase-locked frequency tracking. A comparison between an older HP8409A, with a separate harmonic convertor, and the HP8510 network analyser highlights many design improvements. An error budget for the former[5] quotes RMS magnitude error of 0·2—0·1 dB and phase error of about 2°, compared to ±0·05 dB and ±0·3° at 10 dB above the reference level[4] for the HP8510.

Modern vector network analysers, with direct phase comparison, are virtually unchallenged in the frequency range to 40 GHz, but we will see that at higher frequencies they may be in competition with multiport devices depending only on amplitude measurements.

5.8 References

1 FITZPATRICK, J.: 'A history of automatic microwave network analysers', *Microwave J.*, April 1982, pp. 43–56
2 WARNER, F. L.: 'Microwave network analysers' *in* BAILEY, A. E. (Ed.): 'Microwave measurements' (Peter Peregrinus, 1985) p. 170
3 FITZPATRICK, J.: 'Error models for systems measurements', *Microwave J.*, May 1978, pp. 63–66
4 DONECKER, B.: 'Accuracy predictions for a new generation network analyser', *Microwave J.*, June 1984, pp. 127–141
5 'Vector measurements of high frequency networks'. Hewlett–Packard Seminar Handbook, 5952-9270
6 BRAUN, C. R.: 'Powerful network analyser refines microwave measurements', *MSN*, Jan. 1984, pp. 56–64
7 'Microwave network applications'. Hewlett–Packard Application Note 117-1, June 1970

5.9 Examples

1 Derive the equivalent three-term error model for the reflectometer discussed in Chapter 2 by finding expressions for E_0, E_R and E_S.

2 The reflectometer in the previous question is calibrated using imperfect standards. The short-circuit reflection coefficient is $-(1 + \varepsilon_S) \exp(j\delta_S)$ and the open circuit is $(1 + \varepsilon_0) \exp(j\delta_0)$. If the true reflection coefficient of the DUT is $0.5 \angle 0°$, estimate the worst-case measured values if $|\varepsilon_S|, |\varepsilon_0| \approx 0.1$, $L_A^2 L_D^2 (k - \rho_S) \approx 0.1$ and the phases are unknown.

3 What is the uncertainty in a transmission measurement expressed in terms of the six-term error model for a forward-transmission test set?

Chapter 6

Scalar analysers

6.1 Scalar analysis

When measurements are restricted to the amplitudes of wave functions in a test set, they are called scalar, and if the analyser is not required to generate phase information, it is also scalar. Since amplitude is the square root of power in a wave, expensive and complex synchronous detection against a frequency-stable reference may be replaced by a simple square-law diode detector with a single RF signal, though we will later find it is sometimes advantageous to derive a second (reference or ratioing) channel when the source is unlevelled or has a high reflection coefficient. Dynamic range is reduced to little better than 60 dB because of the limitations of detector diodes, but there are many devices, e.g., attenuators or filters, with operating ranges less than this and for which phase is sometimes of secondary importance. Simplicity and lower cost are always factors in the choice between vector and scalar analysis. When phase information is extracted from scalar results, by extending the number of measurement ports or by repeated measurements with different critical lengths in a test set, the analyser is truly vector and there is an increase in complexity and cost. We have already met the commonest form in the standing-wave indicator, where several amplitude measurements along a slotted line yield the complex reflection ratio at the load port. In the next Chapter we will generalise these ideas to the level of a six-port, but at this stage the discussion is restricted to simple transmission and reflection test sets from which only amplitude ratios are available.

6.2 Components of scalar analysers

As for vector analysers, measured parameters of a device under test include the effects of imperfections in components of the test set.[4] There are scalar error-correction procedures based on calibration with known standards, but in the absence of phase knowledge, the error-network models of vector analysis cannot be fully implemented. All components such as connectors, splitters,

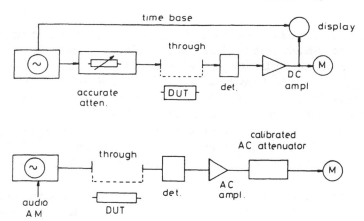

Fig. 6.1 *Scalar transmission test sets*

bridges etc. should be of the highest quality compatible with test requirements and costs.

The transmission test rigs in Fig. 6.1, distinguished by the nature of the test signal, have in common that the measured transmission coefficient is the ratio of two readings taken first with the DUT removed for a through or direct connection of source and detector and then with it inserted between them. Readings appear on meters or display screens, and may also be fed directly onto a computer. When the test signal is in CW, detection gives a DC output that is fed into a DC amplifier, and the transmission loss or gain is the difference in the settings of an RF attenuator, adjusted to keep the output constant with and without the DUT in place. Measurement accuracy depends on the RF attenuator, because the detector diode level remains constant and cannot be affected by poor square-law behaviour. Drift in DC amplifiers is potentially a serious error source, that is now reduced by microprocessor monitoring and control. An audio-modulated signal, obtained by square-wave switching the source, is detected as a chopped DC level, thus obviating the need for a DC amplifier and facilitating the use of an accurate audio attenuator. The method suffers from crystal-detector square-law error and the possibility of inaccuracy in measuring DUTs with an amplitude-response assymmetry. On the other hand, modulation-system bandwidths are generally narrower, and therefore reduce harmonic and spurious components in the source. Again, internal noise in a DUT is not modulated by an externally modulated input, thus improving its apparent noise level, particularly in a synchronous audio detector.

We are now in a position to look at each component separately in the order listed below:

- Source
- RF/audio attenuators
- Connectors
- Diode detectors
- Detection amplifiers
- Display

6.2.1 Source

Power level, frequency stability and match are the important characteristics of sources, already discussed in some detail in Chapters 2 and 4. Source match, improved by levelling or reference ratioing, affects uncertainty error due to multiple reflections between the DUT and source output port. It is often possible to replace the directional coupler in a levelling or reference circuit with a much cheaper stripline splitter. For instance in Fig. 6.2, the two parallel arms, each connected through a series 50 Ω pad to a 50 Ω load, combine to a 50 Ω input with a tracking ratio between the ports of the order of 0·25 dB and feedthrough (k in the reflectometer of Chapter 2) between 0·1 and 0·3.

Harmonic, sub-harmonic and spurious output from sweeper sources cause raised levels in the skirts of, say, a filter passband. When, in Fig. 6.3, the sweeper is at f_1, harmonics might be near f_0, to give the raised shoulders at the -30 dB level. Sub-harmonics and spurious outputs cause higher-frequency shoulders in a similar way. Harmonic shoulders at the level shown are typical of synthesised sweepers with YIG multipliers, whereas levels nearer -50 to -60 dB are found for fundamental oscillators consisting of three or four GaAs FETs to cover a range from 0·5 to 40 GHz.

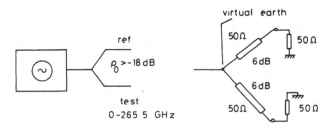

Fig. 6.2   *3 dB splitter*

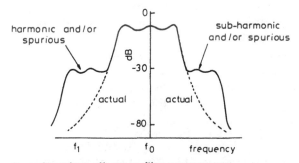

Fig. 6.3   *Harmonic and spurious effects on filter measurement*

6.2.2 Attenuators

Examples of attenuators in Fig. 6.4 are high-quality laboratory instruments, that may also serve as secondary standards. Resistive films in the rotary-vane attenuator (RVA) totally absorb any wave with an electric field parallel to the vane. Fixed vanes at input and output ensure no cross-polarised components exist following a transition from rectangular to circular guide, and the central circular rotating section propagates a TE_{11} mode. Attenuation is related to rotation angle θ of the central vane according to

$$\text{Attenuation in dB} = 40 \log (\sec \theta) + \alpha_0 \qquad (6.1)$$

where α_0 is the residual loss in the attenuator.

In the second example, a piston attenuator, the length of a cut-off circular waveguide in the form of a piston (not shown in the Figure) is altered to change the attenuation of an evanescent mode with attenuation constant

$$\alpha = \frac{2\pi}{\lambda_C} \sqrt{1 - \left(\frac{\lambda_0}{\lambda_C}\right)^2} \quad \text{neper per unit length} \qquad (6.2)$$

where λ_C is the dominant mode cut-off and λ_0 is the free-space wavelength. Pick-up loops couple energy to the mode, and the sliding section of the piston moves against a scale marked linearly in decibels.

Fixed attenuators, available in waveguide and coaxial lines, can be cascaded to higher values as required, but uncertainty error increases with the number of connectors. A coaxial-TEE attenuator, illustrated in Fig. 6.5, has the advantage of the wide frequency range typical of TEM devices.

At audio frequencies, resistive (RVD) or inductive (IVD) networks can be made to divide very accurately a voltage in a number of fixed ratios, and are therefore suitable as primary attenuation standards. A relative comparison of representative curves of the maximum errors of some different attenuators is given in Fig. 6.6, and fuller details are given in Reference 1.

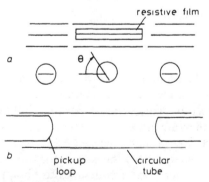

Fig. 6.4 *Rotary (a) and piston (b) attenuators*

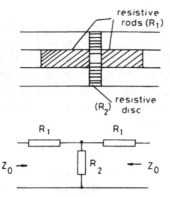

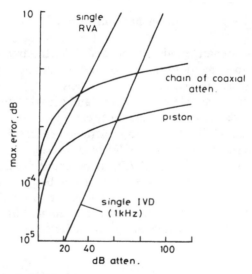

Fig. 6.5   *Coaxial TEE attenuator*

Fig. 6.6   *Errors in standard attenuators*

6.2.3 Connectors

Connectors are often the limiting factor in accurate level setting, and for this reason they are difficult to measure. Their imperfections, modelled by a small reactance at the connection plane, are comparable to those of the high-quality loads against which they are usually assessed. It is possible to estimate the magnitude of the reactance through repeated measurement of a connector with different matched loads spaced from the connector through different lengths of line and over a range of increasing frequencies to obtain the increase in error spread with jX shown in Fig. 6.7. S_{11M} for the connector–load combination is measured with full error correction on a vector network analyser. From the

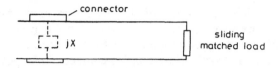

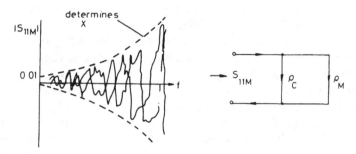

Fig. 6.7   *Connector mismatch*

flowgraph $|S_{11M}|^2 = |\rho_c + \rho_M|^2 \approx X^2/(X^2 + 4Z_0^2)$. Since fluctuations in X and Z_0 due to repeated measurement errors, predominantly serve to decrease $|S_{11M}|^2$ from that due to the connector alone, the dotted lines in Fig. 6.7 are the best estimate of $|X|^2$.

The following standing-wave ratios and return losses for some mated-pair connectors are taken from the 1986 Hewlett–Packard catalogue.[6]

	SWR	*Return loss (dB)*	
Type N	1·08	28·3	
APC 3·5	1·01 + 0·004f	34·2	
APC 7	1·003 + 0·002f	40·1	10 GHz
BNC	1·3—1·5	15·6	

These figures rise rapidly with mixed systems of connectors and adapters.

6.2.4 Diode detectors
Important specification parameters in diode detectors for scalar analysers are:

- Dynamic range
- Power accuracy
- Frequency response
- Input VSWR

- Tangential sensitivity
- Temperature stability
- Square-law range

Metal semiconductor Schottky barrier diodes are the most common detectors in scalar analysers. Their construction is described in Chapter 8, and at this stage we note their wide dynamic range (-55 to $+15$ dBm in a 1 kHz band-width), their low noise at 15 dB below a point-contact diode, power accuracy $\sim \pm 0.1$ dB, wide frequency range from 0.01 to 20 GHz and input VSWR ~ 1.5. Point-contact diodes are less stable and more unreliable than Schottky diodes, but their much lower junction capacitance makes them the only suitable detector at high frequencies up to 200 GHz.

Diode tangential sensitivity[2] is usually measured by finding the input level of a pulse, which on detection shows a notch with lower noise edge just tangential to the upper noise edge of the quiescent detected level. It varies from -50 to -55 dBm in a 10 MHz bandwidth for a Schottky barrier diode.

Typical temperature response of a detector is shown in Fig. 6.8, where, over a 50°C range, ouput changes by 1.2 dB for a constant RF input. In a tempera-ture-compensation circuit a piecewise linear approximation to the response is stored, e.g. as three linear regions, and used in a hardware or software correc-tion procedure.

To find the limits to square-law operation we begin with the diode character-istic curve of Fig. 6.9, in which the voltage and current are related as

$$i = I_S[\exp(v/D) - 1] \qquad (6.3)$$

where I_S is the reverse saturation current and $D = nkT/e \approx 25$—50 mV. k is Boltzmann's constant, T the diode temperature and e the electronic charge. $n = 1.2$ for a Schottky and 2 for a point-contact diode. On expanding eqn. 6.3,

$$i \approx I_S\left[\frac{v}{D} + \frac{v^2}{2D^2} + \cdots\right] \qquad (6.4)$$

which may be approximated as

$$i \approx I_S\left[\frac{v}{D} + \frac{v^2}{2D^2}\right] \quad \text{if } v \ll 50 \text{ mV}$$

For very small voltages there is a linear relation between the voltage across the diode and the current through it, but as a detector a diode should be well

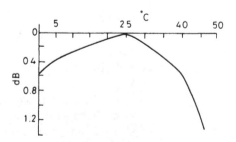

Fig. 6.8 *Piecewise approximation*

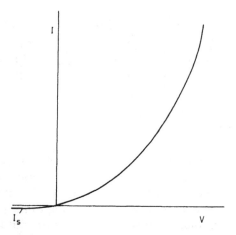

Fig. 6.9 *Diode characteristic curve*

matched to the RF transmission line and normally provides a rectified output current into a resistive load that is strongly dependent on the square term in the expansion. Fig. 6.10 shows a detector circuit consisting of a 50 Ω terminated line, the diode and an RC filter with detected output V_0 across the resistance R. If V_{RF} is an input RF voltage, the voltage across the diode is

$$v = V_{RF} - V_0 = V_S \cos \omega t - V_0 \tag{6.5}$$

On substituting for v in eqn. 6.4 and taking the average over one cycle, the average current in R is

$$\bar{i} = \frac{I_S}{2\pi} \int_0^{2\pi} \left[\frac{V_S \cos \omega t - V_0}{D} + \frac{V_S \cos \omega t - V_0{}^2}{2D^2} \right] d\omega t \tag{6.6}$$

with eqn. 6.4 expanded to second order. Integrated over one cycle $\cos \omega t$ is zero, leaving eqn. 6.6 as

$$\bar{i} = \frac{I_S}{2\pi} \int_0^{2\pi} \left[-\frac{V_0}{D} + \frac{V_S^2 \cos^2 \omega t}{2D^2} - \frac{V_S V_0}{D^2} \cos \omega t + \frac{V_0^2}{2D^2} \right] d\omega t$$

$$\bar{i} = I_S \left[-\frac{V_0}{D} + \frac{V_S^2}{4D^2} + \frac{V_0^2}{2D^2} \right] \tag{6.7}$$

But

$$V_0 = \bar{i}R$$

and since $V_0^2/2D^2 < V_0/D$, we have

$$V_0 = \frac{I_S R}{4[D^2 - (I_S R)/2]} V_S^2 \tag{6.8}$$

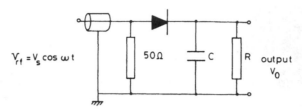

Fig. 6.10 *Detector circuit*

Eqn. 6.8 shows that the output voltage of the detector is proportional to the RF-power input level.[3] This square-law behaviour holds so long as the diode remains in conduction over the RF cycle. For large input signals, the detector circuit behaves as a peak detector with C charging to the peak voltage on each forward half cycle. The diode is now a linear detector with $V_0 \propto V_S$. The square-law range is thus bounded by noise at its lower limit and the onset of linear operation at its upper limit.

   We have so far discussed three kinds of detection: harmonic and fundamental mixing in vector analysers and Schottky diodes in scalar analysers. They are roughly compared on cost and operating scales[8] in Fig. 6.11.

6.2.5 Detection amplifiers

In modern scalar analysers detected output from the diode is fed to microprocessor-controlled DC amplifiers as sketched in Fig. 6.12. Detectors A and B might be in a reference and test, or transmission and reflection channel, and there is often a third reference channel. Thermocouples monitor the detector temperatures to provide compensating and offset shaping in the following circuits, and the inputs are multiplied and corrected for common-mode drift and offset tracking before passing via a low-pass filter to a log shaper (also

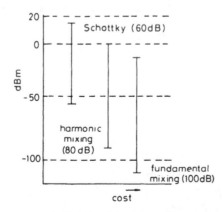

Fig. 6.11 *Comparison of scalar and vector dynamic ranges*

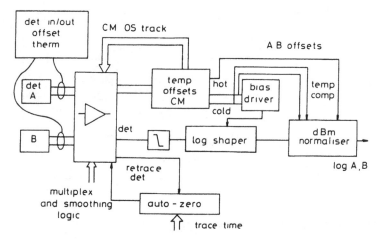

Fig. 6.12 *Scalar analyser signal processing*

continuously adjusted for temperature drift) and a dBm normaliser to give a log output accurate to ± 0.5 dBm. During retrace, each detector output is short-circuited and the zero adjusted automatically.

Many of the disadvantages of drift and temperature stability are therefore overcome by digitising the signals immediately after detection in each channel. Further signal processing, including square-law correction, often by means of stored look-up tables against level and temperature changes, is then software driven with some degree of operator control through menu selection.[12,13]

6.2.6 Display

An oscilloscope display of amplitude, dBm, VSWR etc. against frequency can also be printed in hard copy. Computer control allows internal processing before display and provides storage of previous test set-ups with menu-driven operator instructions.

6.3 Automated scalar network analyser

There are many commercially available scalar network analysers from companies such as Hewlett–Packard, Pacific Measurements, Marconi etc. There will be no attempt to make comparisons, and no recommendations are intended, when we choose the Wiltron 5600 scalar analyser as our case study of automated measurements. It has been chosen because its use of a bridge circuit in error correction gives an opportunity to study bridge behaviour and ripple error-correction techniques. A diagram of the system is given in Fig. 6.13.

A synthesised sweeper supplies RF to a standing-wave-ratio (SWR) bridge with built in reflection detector at channel *B* to the analyser, whilst channel *A*

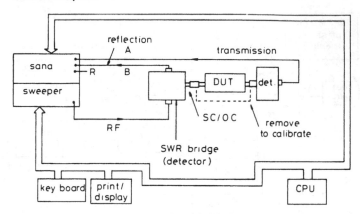

Fig. 6.13   *Scalar reflection/transmission test set*

is fed from a second detector for transmission measurements. All three channels have automatic internal calibration and stabilised DC amplifiers with dynamic range from $+16$ dBm to -50 dBm. Channel R is for levelling in waveguide test sets and for error reduction using the fourth arm of the bridge. Bridge directivity is ~ -40 dB. Amplitude and frequency resolutions are 0·01 dB and 1 MHz with a maximum of 256 spots and a measurement time of 100—270 ms, depending on power level. Thus a typical 50 MHz sweep at full resolution takes 5 s to complete. Computer control is facilitated through a keyboard and printer/display unit. Two software packages are of interest for automated measurements:

● Transmission/reflection measurements
● Air-line use in enhanced accuracy measurements with the fourth arm of the bridge.

It is also possible to store up to 99 test set-ups for return runs in automatic production testing.[5]

6.4 Reflection bridges

When scalar analysers are applied to reflection measurements, a bridge or directional coupler is added to make a test set similar to that in Fig. 6.14, in which the analyser controls the source and a splitter gives a reference and test channel. Significant error occurs when the fraction of forward signal coupled to the reflection detector via the directivity of the bridge or directional coupler is comparable to DUT reflection. When they are equal the measured reflection

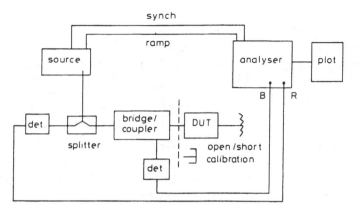

Fig. 6.14 *Scalar reflection test set*

amplitude can be zero or doubled, depending on the phase relationship of the two waves. These worst-case conditions are expressed as

True reflection $\qquad |\rho_D| = |D|$

Measured reflection $\qquad |\rho_{DM}| = 0$ or $2|D|$

where $|D|$ and $|\rho_D|$ are the directivity and reflection coefficient of the bridge and DUT, respectively. The effect is illustrated in Fig. 6.15, where, for a true $|\rho_D|$ of 30 dB, measured reflection coefficient is either ∞ dB or 24 dB.

In the bridge diagram of Fig. 6.16, a balanced detector is placed between junctions 1 and 2, a transmission line is connected at the test port between 2 and 4 and the forward wave enters at the port between 3 and 4. First, consider the line at the test port to be infinitely long, so that a forward wave sees the characteristic impedance R_0 of the line, and the bridge impedance between 3 and 4 matches the source impedance R_0 to give a forward wave voltage of $E/4$

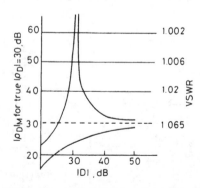

Fig. 6.15 *Showing the effect of coupler directivity on measured reflection coefficient*

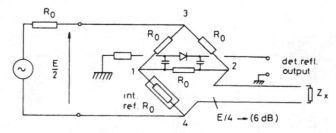

Fig. 6.16   *Bridge circuit for scalar measurements*

at the test port. When the line length is finite and terminated with an unmatched impedance Z_X, the voltage reflection coefficient is[7]

$$\rho_L = \frac{Z_X - R_0}{Z_X + R_0} \tag{6.9}$$

and the reflected wave at the test port is

$$V_R = \frac{E}{4}\frac{Z_X - R_0}{Z_X + R_0} \tag{6.10}$$

where, in eqn. 6.10, Z_X is strictly the transformed value of the terminating impedance at the test port.

A Thévenin equivalent circuit for the passage of the reflected wave through the bridge is given in Fig. 6.17, with an equivalent form beside it to show that detected voltage E_R across D equals half the reflected voltage. Therefore

$$E_R = \frac{V_R}{2} = \frac{E}{8}\frac{Z_X - R_0}{Z_X + R_0} \tag{6.11}$$

Since the input amplitude was $E/2$ there is an overall loss, due to voltage division in the bridge, of 12 dB between input forward wave and detected reflected wave. Examination of Figs. 6.16 and 6.17 shows that arm D is decoupled from the input wave and arm A from the reflected wave. Reflected output

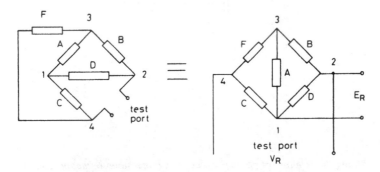

Fig. 6.17   *Equivalent bridge for reflected wave*

voltage across D is balanced off ground, but is converted in the detection circuit to a single-ended output by a balanced-to-unbalanced transforming network. Resistance C is an internal reference, defining the characteristic impedance of the bridge. Both reference and detector circuit are integral with the bridge, which thus conveniently becomes a small dual-purpose signal separator, with an in-built padding of -6 dB in forward and reverse directions, respectively. Fig. 6.13 shows how a DUT is connected at the test port for reflection, with a second detector at the DUT output for transmission.

An alternative to the reflection bridge, the directional coupler, does not suffer 12 dB two-way loss, but multiple reflection between source and DUT can introduce large uncertainty errors, that can be minimised by means of the matched levelling, reflection and transmission couplers, illustrated in Fig. 2.24. In a bridge the -6 dB loss per pass reduces multiple reflections and obviates the need for external levelling.

Signal paths in a bridge circuit are illustrated in, Fig. 6.18. There are two reflection paths: one forming a multiple loop, with the source; the other with the reflection-detector connector match. The former is negligible, because of the round path loss of 12 dB, and the latter is also extremely small because the reflection detector is usually integral with the bridge. A simplified flowgraph in Fig. 6.19 does not include multiple effects between test and detection ports, but does take account of source reflections in the input channel. L represents the 6 dB bridge loss and E_D the directivity coupling between forward and detection ports. The ρ_c are the port-connector reflection coefficients and E_T, E_L refer to transmission loss and load reflection at the transmission detector, respectively. The DUT is represented by its S-parameters.

Directivity and test-port-connector internal reflection add together at the detector port, and clearly limit effective directivity to the quality of the test-port connector, except that the hybrid nature of the bridge allows some cancellation of connector mismatch by careful choice of the impedance in arm A to produce a small reflection coupled to the detector port, but de-coupled from the test

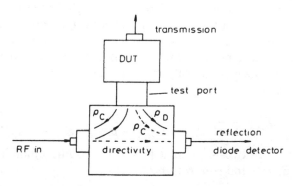

Fig. 6.18 *Showing flow paths in a bridge*

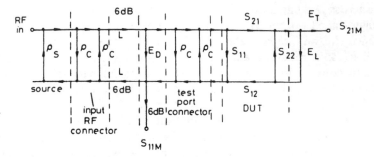

Fig. 6.19 *Flowgraph for bridge in scalar measurement*

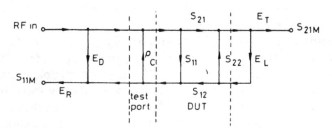

Fig. 6.20 *Reduced flowgraph for bridge*

port. By subsuming the parallel ρ_c and E_D paths and ignoring internal reflections due to sweeper mismatch, a simplified bridge-circuit flowgraph reduces to that shown in Fig. 6.20.

6.5 Reflection calibration

A method of calibration using air lines, to be described in later Sections, achieves accuracies comparable to, or better than, those available in vector analysers employing 12-term error models. Although air lines are equally successful in reflection and transmission, the underlying theory is essentially the same, and we can confine our discussion to reflection calibration only. But before introducing the air-line method we will first carry out an analysis of the conventional short/open-circuit calibration procedure. In Fig. 6.20 we have for reflection coefficient

$$S_{11M} = E_R E_D + \frac{E_R E_L (S_{21})^2 + E_R S_{11}}{1 - (\rho_c S_{11} + S_{22} E_L + \rho_c S_{21} E_L S_{12}) + S_{11} \rho_c S_{22} E_L} \quad (6.12)$$

where E_R is the 12 dB two-way bridge loss. If we assume a matched load at the DUT output, $E_L = 0$ and eqn. 6.12 becomes

$$S_{11M} = E_R E_D + \frac{E_R S_{11}}{1 - S_{11} \rho_c} \quad (6.13)$$

Phase in terms E_D, ρ_c and S_{11} causes variations of errors with frequency, which can be studied by writing them in complex form as

$$E_D = D \exp(j\phi_D)$$

$$\rho_c = R_c \exp(j\phi_c) \tag{6.14}$$

$$S_{11} = R_L \exp(j\phi_L)$$

where D, R_c, R_L are amplitudes and ϕ_D, ϕ_c, ϕ_L are corresponding phases. A normalisation procedure, whereby the common loss E_R is removed in calibration, gives, for the normalised measured reflection coefficient,

$$\frac{S_{11M}}{E_R} = D \exp(j\phi_D) + \frac{R_L \exp(j\phi_L)}{1 - R_L R_c \exp(j\phi_L + \phi_c)} \tag{6.15}$$

Diode detection is square law and gives as the detected output

$$\left|\frac{S_{11M}}{E_R}\right|^2 = \frac{\begin{aligned}D^2 + D^2 R_L^2 R_c^2 + R_L^2 - 2D^2 R_L R_c \cos(\phi_L + \phi_c) + 2R_L D \cos(\phi_L - \phi_D)\\ - 2DR_L^2 R_c \cos(\phi_c + \phi_D)\end{aligned}}{1 - 2R_L R_c \cos(\phi_L + \phi_c) + R_L^2 R_c^2} \tag{6.16}$$

True reflection coefficient R_L is obscured by fixed errors of the form D^2, $D^2 R_L^2 R_c^2$, and four oscillating error terms with phases $(\phi_L + \phi_c)$, $(\phi_L - \phi_D)$ and $(\phi_c + \phi_D)$. ϕ_L is the phase angle at the detector due to reflection at the DUT, and it includes the path between the two ports. ϕ_c and ϕ_D are similar quantities for the test-port-connector mismatch and the directivity path length. The rate at which these quantities change with frequency depends on their path lengths to the detection port. ϕ_c and ϕ_D move slowly with frequency, since their path lengths are typical of the bridge dimensions. If the DUT is attached close the test port, ϕ_L has a similar frequency dependence; but if a length of high-quality air line separates DUT and test port, it changes more rapidly. It is then sometimes possible to separate its effects from the others in order to reduce some of the error terms in eqn. 6.16.

Noting that $|E_R|^2$ is still to be measured in the short/open-circuit calibration we now separate it from $|S_{11M}|^2$, and set

$$R_L = 1, \quad \phi_L = \phi'_L \qquad \text{for a short circuit}$$

and

$$R_L = 1, \quad \phi_L = \phi'_L + 180° \qquad \text{for an open circuit}$$

at the test port, where ϕ'_L is the phase due to path length from test port to calibrating standard. The result on changing from a short- to an open-circuit is that cosinusoidal terms in eqn. 6.16 containing ϕ'_L change sign to give

$$\frac{|S_{11M}|^2_{sc/oc}}{|E_R|^2}$$

$$= \frac{\begin{array}{c} D^2 + D^2 R_L^2 R_c^2 + R_L^2 \pm 2D^2 R_L R_c \cos{(\phi'_L + \phi_c)} \mp 2R_L D \cos{(\phi'_L - \phi_D)} \\ - 2DR_L^2 R_c \cos{(\phi_c + \phi_D)} \end{array}}{1 \pm 2R_L R_c \cos{(\phi'_L + \phi_c)} + R_L^2 R_c^2}$$

$$(6.17)$$

The largest error is $2R_L D \cos{(\phi'_L - \phi_D)}$, but it varies with frequency due to the dependence of $(\phi'_L - \phi_D)$ on path length. It can be removed by taking the average of short- and open-circuit measured reflection coefficients, because of the 180° sign reversal. A worst-case estimate of the remaining error after averaging is found by using only the amplitudes of the oscillating terms in eqn. 6.17. To fourth order, worst-case average then becomes

$$\frac{|S_{11M}|^2_{sc} + |S_{11M}|^2_{oc}}{2}$$

$$= |E_R|^2 \frac{1 + D^2 + R_c^2 + 2D^2 R_c^2 \pm 2DR_c(1 + R_c + R_c^2) \pm 4DR_c(1 - R_c)}{(1 + R_c^2)^2}$$

$$(6.18)$$

Further reduction to second order gives three fixed terms and a single oscillating term $\pm 6DR$, which cannot be cancelled by changes in ϕ_L since it contains only ϕ_c and ϕ_D. Eqn. 6.18 is therefore the best reference level, $|E_R|_{meas}$, available through calibration; and to second order it is

$$|E_R|^2_{meas} = |E_R|^2(1 + D^2 + R_c^2 \pm 6DR_c) \qquad (6.19)$$

During calibration $|E_R|_{meas}$ is calculated from $|S_{11M}|_{sc}$ and $|S_{11M}|_{oc}$ for each frequency and stored as a reference level for subsequent elimination by ratioing with measured reflections from the DUT. This process is sometimes referred to as normalisation.

A graphical display of the measured results in Fig. 6.21 illustrates the ripple form of the oscillating errors, and also that crossovers due to the approximate anti-phase shift between short- and open-circuit ripples define an average line representing the mean level of $|E_R|_{meas}$. The slow-moving mean level is due to the variation with frequency of residual bridge losses. Its variation is found from eqn. 6.19, with typical directivity $|E_D| = 0.02$ (-34 dB) and connector reflection coefficient $|\rho_c| = 0.1$ (20 dB return loss at the test port), to be equivalent to a percentage error from -0.4 to $+2$.

Since a length of air line between the test port and a terminating short circuit increases the number of ripples for a given frequency range, it is possible to obtain an average calibration level from only the short circuit, and, in Fig. 6.22,

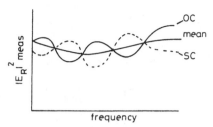

Fig. 6.21 *Open- and short-circuit calibration curve*

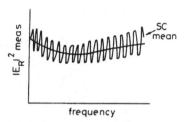

Fig. 6.22 *Short-circuit and air-line calibration curve*

the accuracy of an average line may exceed that from a set of crossover points, if the air line is long enough. The method is equivalent to setting to zero all the cosine terms containing ϕ'_L in eqn. 6.17 to give

$$|S_{11M}|^2_{sc} = |E_R|^2 \frac{1 + D^2 + D^2 R_c^2 \pm 2DR_c}{1 + R_c^2} \tag{6.20}$$

where worst-case form $\pm 2DR$ has been taken for the remaining oscillating term in $(\phi_c - \phi_D)$. To second order $|S_{11M}|^2_{sc}$ becomes

$$|S_{11M}|^2_{sc} = |E_R|^2 (1 + D^2 + R_c^2 \pm 2DR_c) \tag{6.21}$$

6.6 Reflection measurement

With the DUT at the test port, $|S_{11M}|^2$ in eqn. 6.16 is the measured reflection coefficient. In a worst-case version this is

$$|S_{11M}|^2 = |\rho_{DM}|^2$$
$$= \frac{D^2 + D^2 R_L^2 R_c^2 + R_L^2 \pm 2D^2 R_L R_c \pm 2R_L D \pm 2DR_L^2 R_c}{1 \pm 2R_L R_c + R_L^2 R_c^2} |E_R|^2 \tag{6.22}$$

$|E_R|$ can be determined by an open/short-circuit or short-circuit and air-line calibration to the accuracies of eqns. 6.19 or 6.21, respectively. Thus dividing eqn. 6.22 by $|E_R|^2_{meas}$ gives

$$|\rho_{DM}|^2 = R_L^2 \frac{1 + D^2 R_c^2 \pm 2D^2 R_c \pm 2D/R_L \pm 2DR_c}{(1 \pm 2R_L R_c)^2 (1 \pm DR_c^2)^2} \tag{6.23}$$

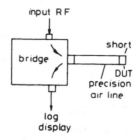

Fig. 6.23 *Showing precision air-line at the DUT*

where eqn. 6.21 has been used for $|E_R|^2_{meas}$. On applying a binomial expansion to the denominator and approximating to second order, the measured reflection coefficient of the DUT is

$$|\rho_{DM}|^2 = R_L^2 \left[1 \pm \frac{2D}{R_L} + \frac{D^2}{R_L^2} \right] \qquad (6.24)$$

First-order error $\pm 2D/R$ stems from variation of $(\phi_L - \phi_D)$ with frequency in eqn. 6.16. If an air line is placed between the test port and DUT to increase the number of ripples over a swept frequency range, an averaging process, similar to that employed with a short-circuit and air-line calibration, eliminates the rapid amplitude variations due to the linear term $\pm 2D/R$, and gives a reflection coefficient

$$|\rho_{DM}|^2 \approx R_L^2 \left[1 + \frac{D^2}{R_L^2} \right] \qquad (6.25)$$

Fig. 6.23 shows the experimental arrangement with a precision coaxial air line separating the test port from calibration short circuit or DUT. A graphical display for DUT ripples would be similar to the one in Fig. 6.21, but, in practice, computer software averages the data and prints out only the calculated reflection coefficient.

With ripple extraction as in eqn. 6.25, if $R_L \sim \sqrt{2D}$, the error ratio is 1·7 dB, which gives a reflection error <1 dB for a return loss within 3 dB of the bridge directivity. When there is no DUT ripple averaging, this degrades to <1 dB within 12 dB of the directivity. Directivity therefore sets an ultimate limit to the smallest measurable reflection coefficient, even when ripple averaging is used.

6.7 Directivity reference and error reduction

We now turn to an interesting method of error correction that allows reflection coefficients to be measured to levels below the directivity of the bridge. First introduced as a general method by Hollway and Somlo,[11] it is particularly effective in bridge circuits with deliberately high-directivity coupling.[9,10] The

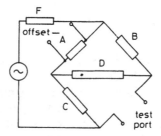

Fig. 6.24 *Showing offset through decoupled arm*

fourth arm A of the bridge in Fig. 6.24 can be used to increase the directivity coupling by a mechanism similiar to that for cancelling internal reflection at the test-port connector. If $D^2 \gg R^2$, it dominates fixed components in the numerator of eqn. 6.16. Any ripple of the measured resultant as a function of frequency becomes a disturbance of the directivity-coupled component rather than of the reflected component; and directivity becomes a fixed reference against which variations can be accurately measured. Previously we discarded ripple variations by averaging, but now we wish to measure their amplitude. No open/short-circuit calibration is necessary; only an air line between the test port and DUT, to aid determination of average levels and ripple amplitudes as a function of frequency.

The original expression for measured reflection coefficient of the DUT in eqn. 6.16 is separated into two parts. The running average consists of all the non-oscillating terms and the ripple is the sum of oscillating terms. Each can be expanded separately, using the binomial theorem on the denominator, to sixth and fifth order, respectively, to find the following equations:

Running average

$$|S_{11M}|^2_{av} = (D^2 + R_L^2 - R_L^4 R_c^2 \pm 2DR_L^2 R_c)|E_R|^2 \tag{6.26}$$

Ripple amplitude

$$|S_{11M}|^2_{ripple} = (2R_L D \pm 2D^2 R_L R_c)|E_R|^2 \tag{6.27}$$

The only oscillating term remaining depends on $(\phi_c + \phi_D)$ which is not affected by the air line and cannot be averaged out. It accounts for most of the slow-moving variation of the average and ripple amplitude with frequency, but can be negligible if particular attention is given to the quality of test port and air-line connectors.

An error-corrected reflection coefficient is found by dividing the ripple amplitude by half the square root of the average. Therefore, to third order,

$$|\rho_{DM}| = \frac{1}{2} \frac{|S_{11M}|^2_{ripple}}{|S_{11M}|^2_{av}} = R_L \left[1 - \frac{1}{2} \frac{R_L^2}{2D^2} (1 \pm 2DR_c) \pm DR_c \right] \tag{6.28}$$

For high-quality connectors at the test port $DR_c \rightarrow 0$ and

$$|\rho_{DM}| = R_L \left[1 - \frac{1}{2} \frac{R_L^2}{D^2} \right] \tag{6.29}$$

Provided $R_L \geqslant D/\sqrt{2}$, the error is less than -2 dB, or we may say the error is < -2 dB to within 3 dB of the directivity.

Ripple extraction from a directivity reference is most useful when the reflection coefficient is less than the normal directivity of the bridge, whereas error averaging with open/short-circuit calibration is well suited to the reverse case. There is an overlap, typically between -25 and -35 dB return loss, where either method might be appropriate. An effective limit to ripple extraction is set by the -60 dB reflection of the highest-precision air lines. Wiltron quote a computed accuracy to 0·02% in characteristic impedance for air lines with precision micro-finish and gold flashed for low loss. The lines have beadless connectors at the DUT end, in order to keep connector SWR below 1·002 in the best case, and a beaded N-type connector at the test port.[10] A maximum line length at this stringent specification restricts the lower frequency to about 2 GHz, and the method is not suitable when reflection coefficients move rapidly with frequency because of difficulty in separating the average from the ripples.

That completes our discussion of scalar analysis when only amplitude ratios are of interest. When phase information is required from amplitude measurements, it is necessary to increase the number of measurement ports. Whilst this may appear more complicated than using a vector analyser, it is balanced by the fact that the high accuracy of modern thermistor power heads, $\sim 0·001$ dB from $+10$ dBm to -20 dBm, allows phase measurement to within 2° up to 100 GHz, whereas the most advanced vector analysers do not extend beyond 60 GHz, and even at 20 GHz cannot improve on the 2° phase accuracy. Vector analysers are limited at high frequencies by reference-channel instability, the consequences of phase-locked harmonic down conversion and the complication of coherent detection. We turn, therefore, in the next Chapter to multiport reflectometers, and show that a minimum of six ports is necessary if we require both amplitude and phase.

6.8 References

1 WARNER, F. L.: 'Attenuation measurement' *in* BAILEY, A. E. (Ed.): 'Microwave measurement' (Peter Peregrinus, 1985) p. 133, and private communication

2 PIDGEON, R. E.: 'Signal detection, noise and dynamic reponse' *in* HOLLIS, J. S., LYON, T. J., and CLAYTON, L. (Eds.): 'Microwave antenna measurements' (Scientific Atlanta, 1970) pp. 4–17

3 GRIFFIN, E. J.: 'Detectors and detection for measurement' *in* BAILEY, A. E. (Ed.): 'Microwave measurement' (Peter Peregrinus, 1985) p. 91

4 'Microwave scalar network measurements.' Seminar, Hewlett–Packard, 5652-9236, May 1977; and 5954-1518, Feb. 1985

5 Wiltron Catalogue, 1984. Wiltron House, Pinehall Road, Crowthorne, Berks.

6 Hewlett–Packard Catalogue 1986. Hewlett–Packard, Elstree House, Elstree Way, Boreham-wood, Herts.

7 OLDFIELD, W. W.: 'Present day simplicity in broadband SWR measurements', *Wiltron Tech. Rev.* 1979, **1**, (1)

8 'Vector measurements of high frequency networks'. Seminar, Hewlett–Packard, 5952-9270

9 DUNWOODIE, D. E., and LACEY, P.: 'Why tolerate unnecessary measurement errors', *Wiltron Tech. Rev.*, June 1979, (5)

10 LACEY, P., and ANDRES, I.: 'New technique provides unparalleled accuracy and convenience for automated measurement of transmission and return loss', *Wiltron Tech. Rev.*, June 1984, (12)

11 SOMLO, P. I., and HUNTER, J. D.: 'Microwave impedance measurement' (Peter Peregrinus, 1985) p. 140

12 SPENLEY, P., and FOSTER, W: 'Automatic scalar analyser uses modern technology', *Microwave J.*, April 1982

13 Pink, J. J.: 'New scalar network analyser boasts powerful processing', *Microwave Sys. News & Commun. Technol.*, 1987, **17**, 5, pp. 8–16

6.9 Examples

1   A notch filter with a 20 dB attenuation at f_0 is used to analyse the spectrum of a swept carrier. The level at the output of the filter is 0 dB except at frequencies of f_0 and $f_0/2$, where it reduces by 6·8 dB and 1·0 dB, respectively. Assuming square-law detection of the filter output, what can you deduce about the spectrum of the swept carrier?

2   The directivity of a reflectometer coupler is -30 dB. What is the minimum reflection coefficient that can be measured to an accuracy of 10%

3   Discuss the relative merits of internal and external levelling in swept-frequency scalar analysis. How does ratioing a test and reference channel relate to the source reflection coefficient?

4   Complete the analysis to derive the sixth-order running average and fifth-order ripple-amplitude expressions in eqns. 6.26 and 6.27. Use the result to find the second-order expression for the reflection coefficient

$$|\rho_{DM}| = R_L \left[1 - \frac{1}{2} \frac{R_L^2}{D^2} \right]$$

5   The mutual coupling between two elements in a large phased array is to be measured by transmitting from one element and receiving at the other. The expected coupling level is ~ -40 dB, but site reflections may also be at the -50 dB level at the receive element. The nearest site reflection point is at about 20 m distant from the array. The element bandwidth is 300 MHz centred at 3·0 GHz. Estimate the maximum and minimum received signal levels. Suggest a method of separating the reflections from the required quantity, and state the conditions under which this would be possible.

Chapter 7

Six ports

7.1 Problem of phase measurement

Complete specification of a wave function includes amplitude and phase. Amplitude is usually derived from a power measurement with an accuracy dependent on impedance matching and power-meter error. Modern thermistor bridge devices have linearity of the order of 0·001 dB in a range[1] +10 dBm to −20 dBm. Phase, on the other hand, presents greater difficulties; among which are comparison with a stable reference and the avoidance of disturbing the measurand, e.g. when sampling probes are inserted into a channel. Vector analysers have overcome these problems by means of accurate synthesisers, phase-locked harmonic down convertors and coherent detectors, but as frequency increases, it becomes difficult to maintain accurate phase lock and great care must be exercised to avoid disturbance of cables and connectors which might significantly alter phase paths.

The familiar two-coupler reflectometer, illustrated in Fig. 7.1, requires that detectors A and B yield amplitude and phase if it is necessary to find the complex reflection coefficient of the device under test. To specify ρ_D two amplitudes and two phases are required, but, in principle, any four independent measurements in a test set should be capable of providing the same information. A simple reflectometer has four ports, if we include source and DUT connections. To allow four independent amplitude measurements, it can be extended to six ports by the addition of one more coupler with detectors P_3—P_6 at the positions indicated in Fig. 7.2. A short circuit on the middle coupler is

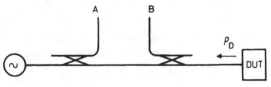

Fig. 7.1 *Four-port reflectometer*

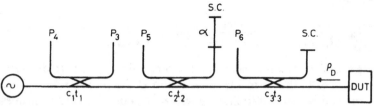

Fig. 7.2 *Six-port reflectometer*

connected through a variable line with phase path α and a second short circuit on the third coupler is fixed.[9] A simple flowgraph can be drawn if we ignore coupler directivities, subsume path lengths between components under coupling and transmission factors c_i, t_i and make each port arm of equal length except for the sliding short circuit on the middle coupler. From the flowgraph in Fig. 7.3

$$b_3 = c_1 a \tag{7.1}$$

$$\frac{b_4}{b_3} = t_1 t_2^2 t_3^2 \rho_D - t_1 t_2^2 c_3^2 - t_1^2 c_2^2 \exp(-j2\alpha) \tag{7.2}$$

$$\frac{b_5}{b_3} = t_1 t_2 t_3^2 \frac{c_2}{c_1} \rho_D - t_1 t_2 \frac{c_3^2 c_2}{c_1} - t_1 t_2 \frac{c_2}{c_1} \exp(-j2\alpha) \tag{7.3}$$

$$\frac{b_6}{b_3} = t_1 t_2 t_3 \frac{c_3}{c_1} \rho_D - t_1 t_2 t_3 \frac{c_3}{c_1} \tag{7.4}$$

It is convenient to isolate ρ_D in eqns. 7.2—7.4 to give

$$k_4 \left|\frac{b_4}{b_3}\right|^2 = \left|\rho_D - \frac{c_3^2}{t_3^2} - \frac{c_2^2}{t_2^2 t_3^2}(\cos 2\alpha - j \sin 2\alpha)\right|^2 = |W_4|^2 \tag{7.5}$$

$$k_5 \left|\frac{b_5}{b_3}\right|^2 = \left|\rho_D - \frac{c_3^2}{t_3^2} - \frac{1}{t_3^2}(\cos 2\alpha - j \sin 2\alpha)\right|^2 = |W_5|^2 \tag{7.6}$$

$$k_6 \left|\frac{b_5}{b_3}\right|^2 = |\rho_D - 1|^2 = |W_6|^2 \tag{7.7}$$

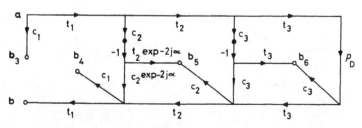

Fig. 7.3 *Flowgraph for six-port reflectometer*

where

$$k_4 = \frac{1}{t_1 t_2^2 t_3^2}, \qquad k_5 = \frac{c_1}{t_1 t_2 t_3^2 c_2}, \qquad k_6 = \frac{c_1}{t_1 t_2 t_3 c_3}$$

b_3 is the sampled forward wave, and the remaining b_4—b_6 are different samples of a mixture of both waves due to reflection at the short circuits. A calibration procedure can be devised to find values for the coupling and transmission-test-set coefficients in eqns. 7.1—7.7, but for the moment we will assume they are known. By taking ratios with respect to the forward wave, the number of equations has been reduced to three, with only two unknowns, the magnitude and phase of ρ_D. The modulus signs indicate that only power has been measured. It would therefore appear that a five-port network with three measurement ports should suffice to determine ρ_D from two simultaneous equations. For linear equations that would indeed be the case, but our non-linear quadratic equations can have ambiguous solutions that are resolvable only through a third equation. A graphical solution illustrates the nature of the ambiguity and how it can be resolved.

Consider first a reflection coefficient

$$\rho = u + jv \tag{7.8}$$

where u and v are its co-ordinates in the complex plane and $|u + jv| \leqslant 1$. The locus of ρ is a circle of radius $|u + jv|$ with its centre at zero. The centre of the circle is shifted to a point p if

$$\rho - p = u - u_p + j(v - v_p) \tag{7.9}$$

The modulus is found from

$$|\rho - p|^2 = (u - u_p)^2 + (v - v_p)^2 \tag{7.10}$$

which lies on a circle of radius $|\rho - p|$ with centre at $p(u_p, v_p)$. These circles are illustrated in Fig. 7.4.

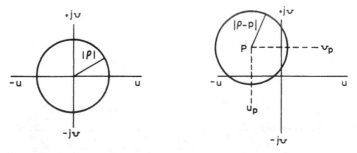

Fig. 7.4   *Reflection in the complex plane*

We can now show that eqns. 7.5—7.7 generate three circles in the complex plane, as shown in Fig. 7.5. The radii are $|W_4|$, $|W_5|$ and $|W_6|$ with centres at

$$\left| \frac{c_3^2}{t_3^2} + \frac{c_2^2}{t_2^2 t_3^2} \cos 2\alpha - j \frac{c_2^2}{t_2^2 t_3^2} \sin 2\alpha \right|$$

$$\left| \frac{c_3^2}{t_3^2} + \frac{1}{t_3^2} \cos 2\alpha - j \frac{1}{t_3^2} \sin 2\alpha \right| \qquad \text{and 1, respectively.}$$

The positions of the centres in the Figure are not obvious from the expressions for them, until an important design choice is made about the coupling factors of the directional couplers. A compromise is sought between maximising power coupled to the detectors and minimising return loss (**b** in Fig. 7.3) to the source. In this design,[9] the first and third couplers were chosen as 6 and 3 dB,

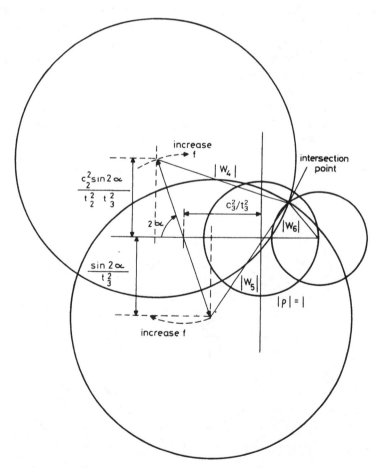

Fig. 7.5 *Graphical solution for six-port*

respectively, and the second was also 3 dB with a short-circuit offset of one eighth of a mean wavelength in the 26—40 GHz band to ensure a $\pi/2$ intersection of at least two of the circles. With this arrangement, detectors P_4, P_5 and P_6 received power over a range 20 dB upward from -20 dBm for all $|\rho_D| \leqslant 1$ when input power was just below 1 mW. For 3 dB couplers $c_2 = c_3 = -j/\sqrt{2}$, giving $c_2^2 = c_3^2 = -1/2$, thus accounting for the negative intercept at $-c_3^2/t_3^2$ and the centre positions in Fig. 7.5.

The loci of the radial vectors intersect at the solution for ρ_D, and, for least ambiguity, intersections should be as near perpendicular as possible. Adjustment of α by movement of the sliding short circuit gives a measure of control over the centres, but there is inevitably a bandwidth limitation because α is frequency dependent, as can be seen by following the dotted lines showing centre movements with frequency. Errors in the power-measuring heads or other calibration errors cause an area of confusion from triangulation of the intersection point, that is made larger if only two equations are available from three measurements. For instance, in Fig. 7.6, there are two intersecting circles, with radii $|W_1|$ and $|W_2|$, but only one of the intercept points P' and P'' is the correct solution. A third circle, with radius $|W_3|$ resolves the ambiguity to the smaller triangular area around P''.

Coupling and transmission factors, including the effects of mismatches between components, can be accounted for by calibration with known standards at the test port. We will not discuss calibration at this stage, except to note that standards used in the present case[3,9] were two known lengths of waveguide (spacers) and an electroformed short circuit. All had dowelled flanges to minimise misalignment on assembly to the test port. Seven further reflection

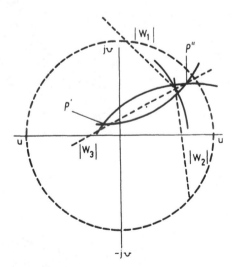

Fig. 7.6 *Showing ambiguity resolved by sixth port*

coefficients were provided by a probe inserted into a waveguide to different depths. Calibration is lengthy, and proceeds by reducing a six-port to a four-port to avoid solving non-linear equations and to take advantage of the proven four-port methods.[2] The extremely high accuracy of power measurements with thermistor heads has given results within 0·0008 of zero for a nominal 1·02 SWR load. Standard deviations of magnitude and phase, as determined by a least-squares fit of measured data to circles in the u, v plane, were 0·0005 and 4.6°, respectively. For a short-circuit load the uncertainty was estimated as $\pm 0·001$ in magnitude and $\pm 0·05(1 + 1/|\rho_L|)$ degrees in phase.[9] High accuracy is maintained to greater than 105 GHz without need of a synthesised frequency source. Another six-port network analyser, operating in the range 29—38 GHz, with two of the ports provided by inserting probes into a transmission line in the test set, gave a reflection measurement uncertainty of 0·006 in magnitude and 2° in phase; and two similar analysers, used in transmission measurement, gave an accuracy[4] of 0·1 dB.

Six-ports are therefore at least as accurate as vector analysers, but over a much wider frequency range, though they have bandwidth limitations. They are less easy to calibrate, mainly because ambiguities of solutions with non-linear equations, and a larger number of measurement ports, calls for at least ten standards compared to six for a similar reflection measurement on a vector analyser. Calibration methods, first described by Engen,[2,6,7] rely on solution of a set of quadratic equations to reduce the six-port to a four-port, and then to make a vector calibration of the four-port. The theory is complex and requires extensive software with relatively long computing times. In the next Section we will show how the quadratic equations can be linearised and solved explicitly both for calibration and measurement.

7.2 General six-port

In the previous Section we studied a particular example of a six-port to give practical substance to the general theory of this Section before extending it to dual six-port transmission measurements and other forms such as the multi-state reflectometer. In generalised form we begin with the incident and transmitted wave on a transmission line illustrated in Fig. 7.7, and place the six-port between the source and transmission line as in Fig. 7.8. Power meters at ports 0—3 in Fig. 7.9 measure the magnitudes of emerging waves $|b_0|$ to $|b_3|$, from which it is possible to derive both the magnitude and phase of reflection

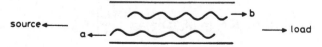

Fig. 7.7 *Forward and reverse waves*

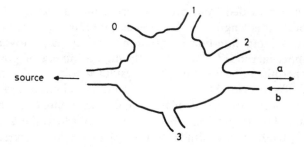

Fig. 7.8 *General six-port*

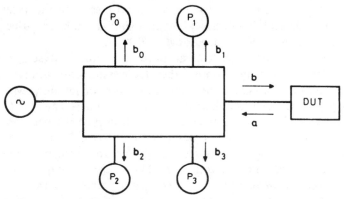

Fig. 7.9 *Showing source, DUT and measurement ports*

relative to the incident wave at the DUT. This can be seen by writing each emerging wave as a linear combination of a and b to give[1]

$$b_0 = A_0 a + B_0 b$$

$$b_1 = A_1 a + B_1 b$$

$$b_2 = A_2 a + B_2 b$$

$$b_3 = A_3 a + B_3 b \tag{7.11}$$

These can be verified by writing $a_0 = \rho_0 b_0$, $a_1 = \rho_1 b_1$ etc. for each port and substituting for the bs in the S-parameter equations to show that $A_0, A_1, \ldots, B_0, B_1, \ldots$ are linear combinations of the scattering and reflection coefficients at source and measuring ports. Power out of each port is found in terms of the DUT reflection coefficient as

$$P_0 = |b_0|^2 = |A_0 a + B_0 b|^2 = |A_0|^2 |b|^2 \left| \rho_D + \frac{B_0}{A_0} \right|^2$$

$$= |A_0|^2 |b|^2 |\rho_D - q_0|^2$$

$$P_1 = |b_1|^2 = |A_1 a + B_1 b|^2 = |A_1|^2 |b|^2 |\rho_D - q_1|^2$$

$$P_2 = |b_2|^2 = |A_2a + B_2b|^2 = |A_2|^2|b|^2|\rho_D - q_2|^2$$
$$P_3 = |b_3|^2 = |A_3a + B_3b|^2 = |A_3|^2|b|^2|\rho_D - q_3|^2 \qquad (7.12)$$

These equations can be re-arranged to form the following ratios:

$$\frac{P_i}{P_0} = \frac{|A_i|^2}{|A_0|^2}\frac{|\rho_D - q_i|^2}{|\rho_D - q_0|} \qquad (7.13)$$

where

$$i = 1, 2, 3 \qquad (7.14)$$

alternatively

$$\left|\frac{\rho_D - q_1}{\rho_D - q_0}\right|^2 = \frac{|A_0|^2}{|A_1|}\frac{P_1}{P_0} = p_1$$

$$\left|\frac{\rho_D - q_2}{\rho_D - q_0}\right|^2 = \frac{|A_0|^2}{|A_2|}\frac{P_2}{P_0} = p_2$$

$$\left|\frac{\rho_D - q_3}{\rho_D - q_0}\right|^2 = \frac{|A_0|^2}{|A_3|}\frac{P_3}{P_0} = p_3 \qquad (7.15)$$

A linearised set of equations is obtained by taking complex conjugates of ρ_D and q_i to give

$$\left|\frac{\rho_D - q_i}{\rho_D - q_0}\right|^2 = \frac{\rho_D - q_i}{\rho_D - q_0}\left[\frac{\rho_D - q_i}{\rho_D - q_0}\right]^* = \frac{(\rho_D - q_i)(\rho_D^* - q_i^*)}{(\rho_D - q_0)(\rho_D^* - q_0^*)} = p_i \qquad (7.16)$$

which on expansion becomes a linear combination of $\rho_D\rho_D^*$, ρ_D and ρ_D^* for $i = 1, 2, 3$ in

$$(1 - p_1)\rho_D\rho_D^* + (p_1q_0^* - q_1^*)\rho_D + (p_1q_0 - q_1)\rho_D^* + q_1q_1^* - p_1q_0q_0^* = 0$$
$$(1 - p_2)\rho_D\rho_D^* + (p_2q_0^* - q_2^*)\rho_D + (p_2q_0 - q_2)\rho_D^* + q_2q_2^* - p_2q_0q_0^* = 0$$
$$(1 - p_3)\rho_D\rho_D^* + (p_3q_0^* - q_3^*)\rho_D + (p_3q_0 - q_3)\rho_D^* + q_3q_3^* - p_3q_0q_0^* = 0$$
$$(7.17)$$

The p_i are a matter of measurement, but q_i are determined by calibration as constants of the six-port. Since calibration is subject to the normal errors of measurement, the q_i, at best, are mean values with RMS errors. But they do not have equal weighting in the linear equations, so that errors in $\rho_D\rho_D^*$, ρ_D and ρ_D^* vary according to the sequence of a solution; i.e. according to which two equations are first solved. A cyclic procedure is therefore adopted, beginning with three different pairs of the equations to first eliminate $\rho_D\rho_D^*$ and then ρ_D^* to give three solutions in the form

$$^1\rho_D = \frac{C_2D_1 - C_1D_2}{C_2C_1^* - C_1C_2^*}$$

$$^2\rho_D = \frac{C_3D_2 - C_2D_3}{C_3C_2^* - C_2C_3^*}$$

$$^3\rho_D = \frac{C_1D_3 - C_3D_1}{C_1C_3^* - C_3C_1^*} \qquad (7.18)$$

where

$$C_1 = \frac{p_1 q_0 - q_1}{1 - p_1} - \frac{p_2 q_0 - q_2}{1 - p_2}, \quad D_1 = \frac{p_1 |q_0|^2 - |q_1|^2}{1 - p_1} - \frac{p_2 |q_0|^2 - |q_2|^2}{1 - p_2}$$

$$C_2 = \frac{p_2 q_0 - q_2}{1 - p_2} - \frac{p_3 q_0 - q_3}{1 - p_3}, \quad D_2 = \frac{p_2 |q_0|^2 - |q_2|^2}{1 - p_2} - \frac{p_3 |q_0|^2 - |q_3|^2}{1 - p_3}$$

$$C_3 = \frac{p_3 q_0 - q_3}{1 - p_3} - \frac{p_1 q_0 - q_1}{1 - p_1}, \quad D_3 = \frac{p_3 |q_0|^2 - |q_3|^2}{1 - p_3} - \frac{p_1 |q_0|^2 - |q_1|^2}{1 - p_1}$$

This cyclic solution serves to reduce the ambiguity previously associated with the intersection error of the three circles of a graphical solution, with the q_i as complex numbers defining centres of circles in the p-plane. In this computer equivalent of the three-circle intersection, a calculation time of 0·24 s matches well with the power-head settling time[1] of 0·175 s.

7.3 Calibration of general six-port

Calibration determines the q-points and coupling ratios A_0, A_i in eqns. 7.15, which we rewrite as

$$\left| \frac{\rho_{Sj} - q_i}{\rho_{Sj} - q_0} \right|^2 = r_i p_{ij}$$

where

$$r_i = \left| \frac{A_0}{A_i} \right|, \quad p_{ij} = \frac{P_{ij}}{P_{0j}}, \quad i = 1, 2, 3$$

and ρ_{Sj} is the known reflection coefficient for the jth standard. On re-arranging, we have

$$r_i p_{ij} |\rho_{Sj} - q_0|^2 = |\rho_{Sj} - q_i|^2 \tag{7.19}$$

If r_i is known, four standards are sufficient to find the q_i from known and measured ρ_{Sj}. To find r_i a fifth standard is required. This generates the following 15 equations from eqn. 7.19:

$$r_i p_{i1} \{ q_0 q_0^* - \rho_{S1}^* q_0 - \rho_{S1} q_0^* + \rho_{S1} \rho_{S1}^* \} = q_i q_i^* - \rho_{S1}^* q_i - \rho_{S1} q_i^* + \rho_{S1} \rho_{S1}^*$$

$$r_i p_{i2} \{ q_0 q_0^* - \rho_{S2}^* q_0 - \rho_{S2} q_0^* + \rho_{S2} \rho_{S2}^* \} = q_i q_i^* - \rho_{S2}^* q_i - \rho_{S2} q_i^* + \rho_{S2} \rho_{S2}^*$$

$$\cdot \quad \cdot \quad \cdot \quad \cdot \quad \cdot \quad \cdot \quad \cdot \quad \cdot \quad \cdot \quad \cdot \quad \cdot \quad \cdot \quad \cdot \quad \cdot \quad \cdot \quad \cdot \quad \cdot \quad \cdot \quad \cdot \quad \cdot$$

$$r_i p_{i5} \{ q_0 q_0^* - \rho_{S5}^* q_0 - \rho_{S5} q_0^* + \rho_{S5} \rho_{S5}^* \} = q_i q_i^* - \rho_{S5}^* q_i - \rho_{S5} q_i^* + \rho_{S5} \rho_{S5}^*$$

with

$$i = 1, 2, 3 \tag{7.20}$$

These are solved in a cyclic manner similar to the previous method with a DUT in place, and the result is a best estimate of the q-points and coupling ratios. Calibration error occurs because the source output may vary with the reflection from each standard load, and in consequence the $r_i p_{ij}$ may not be truly independent of one another. Coupling factors A_0, A_i can be determined separately by connecting a standard power head to the test port, measuring its complex reflection coefficient and applying its calibration factor (see Chapter 8) to find the incident power $|b|^2$. The A_0 and A_i follow from eqn. 7.12 and corrected ratios r_i are calculated from $|A_0/A_i|$ for $i = 1, 2, 3$. This completes a full characterisation of the six-port.

7.4 Transmission dual six-port

Two six-ports are necessary for transmission measurements with the DUT connected between them and a single split source, as shown in Fig. 7.10. Input and output waves are related through DUT S-parameters as

$$a_1 = S_{11}b_1 + S_{12}b_2$$
$$a_2 = S_{21}b_1 + S_{22}b_2 \qquad (7.21)$$

But measured reflection coefficients are

$$\rho_1 = \frac{a_1}{b_1}, \qquad \rho_2 = \frac{a_2}{b_2} \qquad (7.22)$$

where b_1 and b_2 are waves leaving the six-ports. From eqns. 7.21 and 7.22,

$$\rho_1 = S_{11} + S_{12}\frac{b_2}{b_1}$$
$$\rho_2 = S_{21}\frac{b_1}{b_2} + S_{22} \qquad (7.23)$$

ρ_1, ρ_2 may be greater than unity because of the wave transmitted through the DUT. b_1/b_2 can be eliminated to derive an equation in terms of the measured

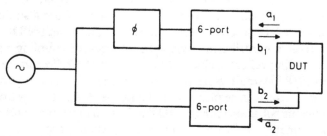

Fig. 7.10 *Transmission six-ports*

ρ_1, ρ_2 and the *S*-parameters. By adjusting the phase shifter in the upper channel to obtain different ratios b_1/b_2, three more equations are generated and the total of four solved for the *S*-parameters of the DUT.

7.5 Six-ports and vector network analysers

Six-ports have yet to achieve the popularity of vector analysers, in spite of their high accuracies and excellent high-frequency performance. This is probably due to the complexities of earlier theories hindering an easy understanding of their operation, and to the formidable amount of computing to extract a result.[6] The use of dedicated microprocessor chips may ensure a rapid development in the future.

Magnitude and phase accuracies of vector analysers at millimetre wavelengths are similar to those at low frequencies when a good synthesised source is available;[5] but six-ports derive their phase accuracy from power accuracy and therefore do not need synthesised sources, thus giving them an advantage at frequencies above 60 GHz. Real-time operation of a six-port is limited by excessive computing time and the response time of power heads. If each output port is linked and switched between a power head and diode detector, diode voltages can be stored at, say 3 dB intervals, against measured levels in the power head, to build a look-up table for real-time testing using diode response times. Real-time operation becomes software dependent against calibration time as a trade-off.

7.6 Multi-state reflectometers

In a six-port, power ratios are derived with respect to one of the ports as P_i/P_0 with $i = 1, 2, 3$, whereas in a four-port only one power ratio is available from two ports, P_0 and P_1. Since this ratio depends on path lengths and coupling factors, it is possible to find a different ratio by altering the state of the reflectometer. This is conveniently done by adjusting line lengths, coupling factors, reflection levels etc. in different parts of the reflectometer test set. One example is illustrated[8,9] in Fig. 7.11, in which the detector at P_0 indicates incident power, and the other at P_1 is associated with the sum of power from a sliding reflector in the fourth arm of the second directional coupler and from the DUT. Both P_0 and P_1 also receive contributions via other directivity coupling paths. Each setting of the sliding reflector changes the state of the reflectometer to give a new value to P_1/P_0. A multi-state reflectometer may simulate a six-port if just three states are selected in sequence and the ratios P_i/P_0 are substituted in the three simultaneous equations for the DUT reflection amplitude and phase. The parallel measurement in a six-port becomes a sequential procedure in the multi-state, with trade-off between test-set size and measurement time. Similar considerations apply to calibration, since each reflectometer state requires separate

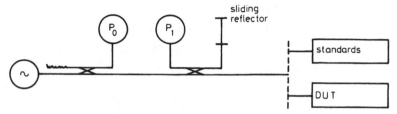

Fig. 7.11 *Multi-state reflectometer*

calibration. Advantages of a multi-state are that source power divides between only three ports, thus conserving power at millimetre wavelengths; and path lengths are shorter so that effects of junction reflections are slower-moving functions of frequency.

In the particular example chosen for Fig. 7.11, the sliding reflector is a short circuit with a microwave absorber attached to give a reflection coefficient of about 0·3. This increases the phase sensitivity by making the reflected level into P_1 comparable with the coupled level from the DUT. Calibration and measurement depend on a full S-parameter matrix description of all coupling and transmission gains, but, for our purposes, a simplified flowgraph of the most significant nodes and paths will suffice for an understanding of circuit behaviour and its relation to the six-port. In Fig. 7.12, b_0 and b_1 are emergent waves at P_0 and P_1; ρ_D is the DUT and ρ_R the sliding reflection coefficient. Directional couplers have coupling C_0, C_1 and directivities D_0 and D_1. The level of approximation is similar to that for Fig. 7.3. We begin by applying the non-touching loop rule to find the ratios b_1/a and b_0/a, where a is the input wave at the source, as

$$\frac{b_1}{a} = \frac{C_1(1 + D_1\rho_R + D_1C_1^2\rho_R)\rho_D + C_1D_1 + C_1\rho_R}{\Delta} \tag{7.24}$$

$$\frac{b_0}{a} = \frac{C_0D_0(1 + C_1^2D_1\rho_R)\rho_D + C_0(1 + C_1^2D_0\rho_R)}{\Delta} \tag{7.25}$$

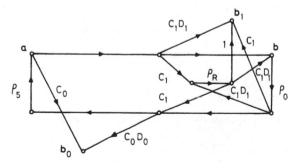

Fig. 7.12 *Flowgraph for multi-state reflectometer*

The required ratio is b_1/b_0, or

$$\frac{b_1}{b_0} = \frac{C_1[1 + D_1\rho_R(1 + C_1^2)]\rho_D + C_1(D_1 + \rho_R)}{C_0D_0(1 + C_1^2D_1\rho_R)\rho_D + C_0(1 + C_1^2D_0\rho_R)} \tag{7.26}$$

Eqns. 7.15 give, for the six-port,

$$p_i = \left|\frac{A_0}{A_i}\right|^2 \frac{P_i}{P_0} = \left|\frac{A_0}{A_i}\right|^2 \left|\frac{b_1}{b_0}\right|^2 = \left|\frac{\rho_D - q_i}{\rho_D - q_{0i}}\right|^2 \tag{7.27}$$

and for the ith setting of the sliding reflector in the multi-port, eqn. 7.26 can be re-arranged as

$$\left|\frac{b_1}{b_0}\right|_i^2 = \left\{\frac{C_1[1 + D_1\rho_{Ri}(1 + C_1^2)]}{C_0D_0(1 + C_1^2D_1\rho_{Ri})}\right\}^2 \left|\frac{\rho_D - q_i}{\rho_D - q_{0i}}\right|^2 \tag{7.28}$$

The ith state of the multi-port is equivalent to the ith port of the six-port if

$$\frac{A_i}{A_0} = \frac{C_1[1 + D_1\rho_{Ri}(1 + C_1^2)]}{C_0D_0(1 + C_1^2D_1\rho_{Ri})} \tag{7.29}$$

$$q_i = \frac{C_1(D_1 + \rho_{Ri})}{C_1[1 + D_1\rho_{Ri}(1 + C_1^2)]} \tag{7.30}$$

$$q_{0i} = \frac{C_0(1 + C_1^2D_0\rho_{Ri})}{C_0D_0(1 + C_1^2D_1\rho_{Ri})} \tag{7.31}$$

For a six-port $q_{0i} = q_0$, since it does not vary with the particular ratio taken, but in the multi-state it changes with each ratio because of an inherent change of state.

Equivalences, such as those in eqns. 7.29—7.31, indicate that the six-port calibration procedures previously described can be employed for the states of multi-ports. A more direct approach, taking advantage of some special features of the multi-state, is illustrated graphically as follows. We begin with a measurement sequence to find ρ_{Ri} by re-arranging eqn. 7.28 for the ith state as

$$\left(\frac{P_1}{P_0}\right)_i = \left|\frac{b_1}{b_0}\right|_i^2 \left|\frac{C_1(D_1 + \rho_{Ri})}{C_0(1 + C_1^2D_0\rho_{Ri})}\right|^2 \left|\frac{\dfrac{1 + D_1\rho_{Ri}(1 + C_1^2)}{D_1 + \rho_{Ri}}\rho_D + 1}{\dfrac{D_0(1 + C_1^2D_1\rho_{Ri})}{1 + C_1^2D_0\rho_{Ri}}\rho_D + 1}\right|^2 \tag{7.32}$$

and writing

$$L_i = \left|\frac{C_1(D_1 + \rho_{Ri})}{C_0(1 + C_1^2D_0\rho_{Ri})}\right|$$

a real quantity, for the transmission loss depending mainly on the coupling C_0,

$$R_i \exp(j\alpha_i) = \frac{1 + D_1\rho_{Ri}(1 + C_1^2)}{D_1 + \rho_{Ri}}$$

a complex quantity depending mainly on the phase and amplitude of the sliding reflector, and

$$S_i \exp(j\beta_i) = \frac{D_0(1 + C_1^2 D_1 \rho_{Ri})}{1 + C_1^2 D_0 \rho_{Ri}}$$

a complex quantity depending mainly on the directivity D_0.

If ρ_D is written in amplitude and phase form as $|\rho_D| \exp(j\theta)$, eqn. 7.32 becomes

$$\left(\frac{P_1}{P_0}\right)_i = L_i \left| \frac{R_i \exp(j\alpha_i)|\rho_D| \exp(j\theta) + 1}{S_i \exp(j\beta_i)|\rho_D| \exp(j\theta) + 1} \right|^2$$

which on re-arranging and taking the modulus is

$$\left(\frac{P_1}{P_0 L}\right)_i = \frac{1 + |\rho_D|^2 R_i^2 + 2|\rho_D| R_i \cos(\alpha_i + \theta)}{1 + |\rho_D|^2 S_i^2 + 2|\rho_D| S_i \cos(\beta_i + \theta)} \tag{7.33}$$

The quantities L_i, R_i, S_i, α_i and β_i are to be found by calibration, whereas $|\rho_D|$ and θ are to be found from measured power ratios of a number of states. A graphical solution is based on the fact that both numerator and denominator of eqn. 7.33 are cosine-rule expressions for the third side of a triangle whose other two sides are 1 and $|\rho_D| R_i$ and 1 and $|\rho_D| S_i$, respectively. These are drawn in Fig. 7.13 for the ith state of the reflectometer, where the two sides are labelled $\sqrt{P_1}$ and $\sqrt{P_0 L_i}$ since their ratio is given by the right-hand side of eqn. 7.33. The loci of $|\rho_D| R_i$ and $|\rho_D| S_i$ will depart from circles to some extent, but are defined during calibration. In a measurement the states are chosen to distribute α_i and

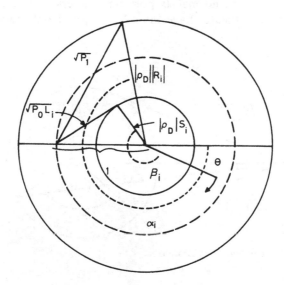

Fig. 7.13 *Graphical solution for multi-state*

β_i evenly through 360°, so that $|\rho_D|$ and θ can be adjusted for least-squares coincidence with calibration loci involving R_i and S_i. Unlike a solution for a six-port, the number of states need not be limited to three, thus allowing enhanced error correction. The sliding reflector covers a 360° range if it moves through half a wavelength.

To find L_i, the loss constant for each state, the test port is terminated with a 'perfect' load through three separate spacers, and the power ratio is measured for each one. In practice, a good high-quality load has reflection coefficient $|\rho_L| < 0.01$. If port P_0 is from a high-quality directional coupler, $D_0 \sim 35$ dB, $S_i \sim 0.05$ and the denominator in eqn. 7.33 tends to 1. For a given state, or setting of the sliding reflector, the three spacers give three equations

$$(P_1/P_0)_i' = L_i\{1 + |\rho_L|^2 R_i^2 + 2|\rho_L|R_i \cos (\alpha_i + \theta')\}$$

$$(P_1/P_0)_i'' = L_i\{1 + |\rho_L|^2 R_i^2 + 2|\rho_L|R_i \cos (\alpha_i + \theta'')\}$$

$$(P_1/P_0)_i''' = L_i\{1 + |\rho_L|^2 R_i^2 + 2|\rho_L|R_i \cos (\alpha_i + \theta''')\} \qquad (7.34)$$

These are three triangles, related as shown in Fig. 7.14, with a common side $\sqrt{L_i}$, equal sides $|\rho_S|R_i \sqrt{L_i}$ and unequal sides proportional to the square root of the measured power ratios. A least-squares fit determines L_i. The remaining constants R_i, S_i, α_i and β_i can be found by a direct least-squares solution of eqn. 7.33 using the experimental L_i and known loads at the test port. The minimum number of loads is a short circuit with three spacers for each i-state, though redundant use of more spacers gives greater accuracy.

A complete six-port equivalence is possible by re-arranging a multi-state reflectometer as indicated in the second example at the end of this Chapter. Calibration is then as for a normal six-port, but in a repeated sequential fashion for all the states. Mutli-states have the advantages of only two nearly equal

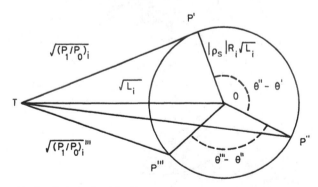

Fig. 7.14 *Graphical solution of calibration*

$$TP'' = \sqrt{\left(\frac{P_1}{P_0}\right)_i''}$$

main paths at the reflection port P_1, fewer components and, because of the reduced path lengths, increased stability at higher frequencies. Typical accuracies are better than 0·0001 in amplitude and ~2° in phase for an unlevelled source with frequency stability of ±200 kHz at 10 GHz. The chief disadvantage is a longer measurement time.

Finally we show a table comparing six-ports (including multi-states) with automatic vector network analysers.

	Six-ports	*ANAs*
Resolution	0·001 dB 2°	0·02—10 dB from 0 to 80 dB 0·2°—20° from reference
Source	Non-synthesised	Synthesised
High frequency	Good to > 100 GHz	60—100 GHz
Theory	Very difficult	Difficult
Calibration time	Long	Less long
Operation	Not real time (unless diodes calibrated)	Real time

Accuracy, as distinct from resolution, depends on the accuracy of the calibration standards, and can only be inferred from the results of many measurements. Further development of the Hewlett–Packard 8510B has extended measurements[5,10] from 60 to 100 GHz.

7.7 References

1 LABAAR, R.: 'The exact solution of the six-port equations', *Microwave J.*, Sept. 1984, pp. 219–228

2 ENGEN, G.F.: 'Calibrating the six-port reflectometer by means of sliding terminations', *IEEE Trans.*, 1978, **MTT-26**, pp. 951–957

3 HILL, L. D.: 'Six-port reflectometer for the 75—105 GHz band,' *IEE Proc.*, 1985, **132H**, pp. 141–143

4 FRAMPTON, A.: 'Microwave network analysers for millimetric bands', *Microwave J.*, April 1982, pp. 89–96

5 'Procedures explained for mm-wave vector measurements with the Hewlett–Packard HP8510A analyser', *Microwave Syst. News*, Dec. 1984, pp. 65–90

6 ENGEN, G. F.: 'The six-port reflectometer: An alternative network analyser', *IEEE Trans.*, 1977, **MTT-25**, pp. 1075–1080

7 ENGEN, G. F.: 'An improved circuit for implementing the six-port technique of microwave measurements', *ibid.*, pp. 1080–1083

8 OLDFIELD, L. C., IDE, J. P., and GRIFFIN, E. J.: 'A multistate reflectometer', *IEEE Trans.* 1985, **IM-34**, pp. 198–201
9 GRIFFIN, E. J.: 'Six-port reflectometer circuit comprising three directional couplers', *Electron. Lett.*, 1982, **18**, pp. 491–493
10 *Test and Measurement News, Hewlett–Packard*, May/June 1987

7.8 Examples

1 List the chief sources of errors in a six-port network analyser.

2 Show that the multi-state arrangement in Fig. 7.15 is equivalent to a six-port reflectometer, by proving that q_0 is invariant with state i. What is the advantage of this equivalence compared to the multi-state described in Section 7.6?

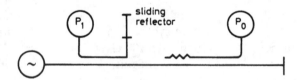

Fig. 7.15 *Multi-state arrangement for example 2*

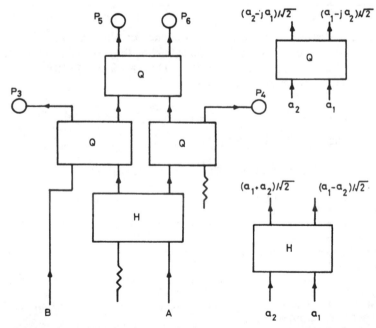

Fig. 7.16 *Vector voltmeter comparing amplitude and phase of inputs at ports A and B (example 4)*

3 Show in the general six-port that the four emergent waves at the measurement ports can be expressed as a linear combination of the forward and reflected waves at the test port.

4 The circuit in the Fig. 7.16 is a vector voltmeter, that compares the amplitude and phase of the inputs at ports A and B. Show how, by the use of a 3 dB directional coupler, that it is possible to measure the reflection coefficient of a DUT. Select port A as incident power and port B as reflected power. Show that the real and imaginary parts of the reflection coefficient ρ_L are given by

$$\frac{P_5 - P_6}{P_4} = 2\sqrt{2}\,\mathrm{Re}\,(\rho_L)$$

$$\frac{P_5 + P_6 - P_3 - P_4}{2P_4} = 2\sqrt{2}\,\mathrm{Im}\,(\rho_L)$$

where P_1 to P_6 are the power levels at each port.

Chapter 8

Power measurement

8.1 Characteristics of microwave sources

The power available from a microwave signal source depends on a number of factors, among which are frequency, loading impedance, ambient temperature, modulation applied, etc. Because all these factors are dependent on one another, they interact with non-linear effects with greater or lesser significance according to the type of oscillator and its particular application. For example, power available is a function of the source impedance, but this, in turn, depends on power level. Similarly, the noise generated along with the signal is determined by operating conditions, which include power level. Noise also depends on frequency stability and bandwidth, having amplitude-, frequency-, and phase-modulation characteristics. Power is usually required at a fixed frequency, or in a finite band of frequencies, but unwanted harmonics, again principally dependent on power level, are always potentially present in the output.

Power measurement is not straightforward, and has therefore received the close attention of national standards laboratories, both in the devising of measurement procedures and of standard sensors. In transferring these standards to industrial laboratories and production lines, uncertainties associated with the active source device have to be separated from those of the passive fixture components. The fixture, consisting of the output connector and the passive circuit components of the oscillator, can usually be characterised by replacing the active device with three known loads and treating it as a two-terminal-pair network. The procedure is known as 'unterminating' and leads to an equivalent circuit similar to the error correction models of vector network analysis. Use of this calibrated model to determine the active device characteristics is known as de-embedding. With a knowledge of the device S-parameters and its fixture behaviour at all frequencies of interest, it is possible to some extent to analyse the behaviour of a source when connected to an external circuit. For instance, the operating frequency can be well defined from the small-signal S-parameters, but power output cannot.[1] The determination of large-signal

S-parameters at a particular frequency and bias can successfully predict oscillator performance, but involves extensive measurement with power level as a running parameter. It has generally been found for large signals that impedance or admittance models and Smith-chart concepts are more effective than S-parameters.

8.2 The Rieke diagram

The output power of an oscillator changes with the load impedance at its terminals. When constant power and frequency loci are plotted on a Smith chart of the loading impedance, the result is called a Rieke diagram. Its derivation can be outlined using the circuit example[2] in Fig. 8.1 of a source consisting of an active device with negative resistance $-R(A)$, device and fixture-dependent reactive elements L, C, circuit resistance R_S and a higher-order mode loss represented by R_M. The device resistance $-R(A)$ depends on the signal amplitude A, as do reactances L and C so far as they are device dependent. If we assume a low harmonic loss $R_M \gg Z_L$, where Z_L is the impedance of a load connected at the source output, the circuit will oscillate at the fundamental frequency when the negative resistance just cancels the circuit losses R_S and R_L. Thus

$$R_L - R(A) + R_S = 0 \tag{8.1}$$

The remaining resonant circuit has a total EMF of zero, as in Fig. 8.2. Thus

$$\left(\omega L - \frac{1}{\omega C}\right) + X_L = 0 \tag{8.2}$$

gives the signal frequency ω.

The device resistance $R(A)$ depends on the bias conditions and R_S cannot easily be separated from it by measurement. If A represents a current amplitude the output power appears in R_L as

$$P_L = \frac{1}{2} R_L A^2 \tag{8.3}$$

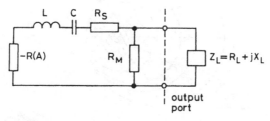

Fig. 8.1 *Model of oscillator*

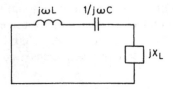

Fig. 8.2 *Resonant circuit*

If $R_L > R_S$, eqn. 8.1 gives $R_L \approx R(A)$ and the output power is

$$P_L \approx \frac{1}{2} R(A) A^2 \qquad (8.4)$$

In the limit when R_L becomes very large, eqn. 8.1 cannot be satisfied and oscillation does not occur. Thus, within the approximation of eqn. 8.4, constant-power contours on a Smith chart of Z_L coincide with constant-resistance loci, except in regions where R_L is too large for oscillations to occur. Eqn. 8.2 indicates that constant-frequency contours coincide with constant-load-reactance loci. The combined result is shown in Fig. 8.3 where the shaded circle includes the values of R_L which are too large for oscillation.

These plots of constant frequency and power loci on a Smith chart of load impedance are called a Rieke diagram. The ideal equivalent circuit of Fig. 8.1 gives a very untypical example of practical diagrams drawn from measured results using real oscillators, whose output loads are varied over the oscillating frequency range. Even the ideal circuit shows considerable distortion if harmonic loss is included by making R_M comparable to $|Z_L|$. Power is then absorbed into higher-order modes capable of existing in only limited frequency ranges. Constant power and frequency contours become discontinuous when mode changes occur and the operating point may jump between the different stable states of overlapping contours multi-valued in power and frequency. The most convenient method of varying the load impedance is to place a sliding screw tuner between the oscillator output port and a well-matched power meter.

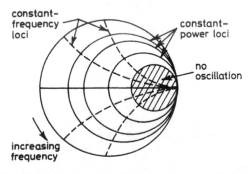

Fig. 8.3 *Rieke diagram for elementary circuit*

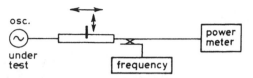

Fig. 8.4 *Sliding-screw tuning*

Any point on the Smith chart can be found by adjusting the position and depth of the screw in the output transmission line. The general arrangement is shown in Fig. 8.4 with a power meter in the main line and a sample coupled to a frequency meter. A choked sliding carriage holds the probe as positions and depths are found for constant power and frequency. Fig. 8.5 shows a coaxial arrangement with a total displacement along the line of $\lambda/2$ and having a flat external surface on which the probe carriage, also shown in the Figure, may slide. Provided the probe insertion can be varied from zero to near contact with the inner conductor, a setting can be found for any point on the Smith chart. A similar probe carriage can be used for waveguide, but probe insertion range extends across the total narrow dimension.

Measurements are susceptible to errors due to mismatch reflections between the sliding tuner and the power meter. Near unstable regions even slight probe movements cause large jumps in power and frequency, making the method difficult and error prone. Changes in power level due to loading are sometimes referred to as load pulling. We met an example of this in the previous Chapter on six-ports, where a change of standard at the test port could cause a small change in source power level. Load pulling may also cause a frequency change. This can most easily be demonstrated by first considering a perfect match between source and load. The constant power contour, passing through the origin on the Smith chart, is labelled P_0 in Fig. 8.6. Slight adjustment of the load will cause a small reflection ρ, and the load impedance relates to the Z_0 load as

$$Z_L = Z_0 \frac{1+\rho}{1-\rho} \tag{8.5}$$

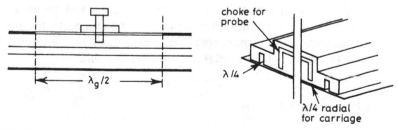

Fig. 8.5 *Sliding-screw tuner*

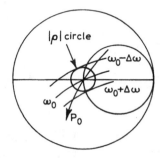

Fig. 8.6 *Showing load pulling*

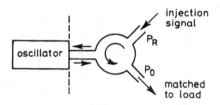

Fig. 8.7 *Injection locking*

If P_R is the reflected power

$$|\rho| = \sqrt{\frac{P_R}{P_0}}$$

(8.6)

If the small reflection is phased, its locus is a circle of radius $|\rho|$. The constant-frequency contours drawn in Fig. 8.6 are tangential to the circle at $\omega_0 \pm \Delta\omega$, where ω_0 is the resonant frequency with zero reflection. Thus both power and frequency are changed by introducing a small reflection at the load.

Frequency stabilisation by injection locking takes advantage of load pulling by locking a high-power oscillator to a low-power frequency reference. By varying the phase of the injected signal the controlled oscillator can be locked to the frequency accuracy of the reference in the range $\omega_0 \pm \Delta\omega$. An isolator separates the three components as shown in Fig. 8.7.

This simplified description of the use of Rieke diagrams has highlighted the importance of empirically derived data in the investigation and design of oscillators, and shows how errors in performance may be related to the physical properties of internal active components. But a measurement of output power is subject also to errors in the external measurement circuit.

8.3 Errors in the external circuit

The external circuit begins at the transmission line connecting the source output to a power detector. It includes the circuits placed close to the detecting element

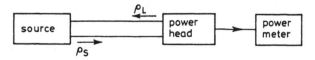

Fig. 8.8 *Measuring power from a source*

in the power head and also in the power meter. A typical power-measurement connection is illustrated in Fig. 8.8. The most readily available power detector is a square-law diode, and in common with many other transducers its output is usually a direct current, though there is often signal conditioning in the head circuits, for instance, to produce a low-frequency chopped DC input to the power meter.

The chief sources of microwave error are:

● Match to the power head
● Uncertainty due to multiple reflections between the power head and source
● Power-head calibration factor and efficiency

In addition there are errors arising in the signal-processing circuits in the power head and meter. In the following Sections we will examine each type of error in detail.

8.3.1 Power-head match

A flowgraph of the connection between source and power head is given[3] in Fig. 8.9. b_S is the wave emerging from the source, or alternatively it is the forward wave at the source into a Z_0 load. When the load reflects, the forward wave becomes b'_S due to the addition of the source-reflected wave $\rho_S a_S$. Thus

$$b'_S = b_S + \rho_S a_S \tag{8.7}$$

An alternative way of modelling the process is through the Thévenin equivalent circuit in Fig. 8.10. e_S is the source EMF and Z_S accounts for the mismatch of the source to a Z_0 line. If V_S is the voltage across Z_0, then we have seen in Chapter 3 that one definition of b_S, the wave into a Z_0 match, is

$$b_S = \frac{V_S}{\sqrt{Z_0}} \tag{8.8}$$

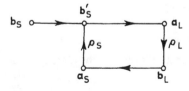

Fig. 8.9 *Source and load reflection flowgraph*

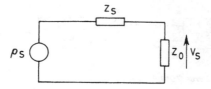

Fig. 8.10 *Source and load impedance circuit*

But the source is not matched to the transmission line because $Z_S \neq Z_0$, or $\rho_S \neq 0$. Therefore

$$V_S = e_S \frac{Z_0}{Z_S + Z_0}$$

or

$$b_S = e_S \frac{\sqrt{Z_0}}{Z_S + Z_0} \tag{8.9}$$

with

$$\rho_S = \frac{Z_S - Z_0}{Z_S + Z_0} \tag{8.10}$$

In Chapter 3 we saw that, for a maximum transfer of power, the source and load should be conjugately matched. The power in the load can be found by subtracting the reflected from the forward power. In Fig. 8.9, the incident and reflected waves at the load are

$$a_L = \frac{b_S}{1 - \rho_S \rho_L} \tag{8.11}$$

$$b_L = \frac{b_S \rho_L}{1 - \rho_S \rho_L} \tag{8.12}$$

The net power dissipated by the load is

$$P_L = \frac{1}{2} |a_L|^2 - \frac{1}{2} |b_L|^2$$

$$= \frac{1}{2} |b_S|^2 \frac{1 - |\rho_L|^2}{|1 - \rho_S \rho_L|^2} \tag{8.13}$$

Conjugate match occurs when

$$\rho_S = \rho_L^* \tag{8.14}$$

and the maximum power available is found from eqn. 8.13 as

$$P_{av} = \frac{\frac{1}{2} |b_S|^2}{1 - |\rho_S|^2} \tag{8.15}$$

The conjugate mismatch loss is defined as

$$\frac{P_L}{P_{av}} = \frac{(1 - |\rho_S|^2)(1 - |\rho_L|^2)}{|1 - \rho_S\rho_L|^2}$$ (8.16)

and is always $\leqslant 1$. On the other hand Z_0 match occurs when $\rho_L = 0$ or

$$P_{Z_0} = \frac{1}{2}|b_s|^2$$ (8.17)

and gives a Z_0 mismatch loss

$$\frac{P_L}{P_Z} = \frac{1 - |\rho_L|^2}{|1 - \rho_S\rho_L|^2}$$ (8.18)

which can be > 1 if P_L is conjugate.

8.3.2 Uncertainty due to multiple reflections

The denominator in eqns. 8.16 and 8.18 is a source of uncertainty, since usually only the moduli of ρ_L and ρ_S are known, whereas the product $\rho_L\rho_S$ also depends on the phasing due to transmission path length between the source and power head. The range of uncertainty is given by the worst-case maximum and minimum $(1 \pm |\rho_L||\rho_S|)^2$. A similar kind of uncertainty, due to reflections between the probe and source or power head, sets accuracy limits for a sliding-screw tuner. Screws are rarely used with modern heads, because screw residual reflection is not insignificant compared to head match. Thus Z_0 matching is normally used.

8.3.3 Power-head calibration factor and efficiency

A power head ideally collects all the power dissipated in the effective load presented by the element and its circuitry. But we have already seen that imperfect matching means that only a fraction $(1 - |\rho_L|^2)$ of the incident power is absorbed in the load. However, not even all this fraction finally reaches the detection element, because some is lost in the transmission-medium walls and some is radiated into space or leaks past the element into the instrumentation. Power-head efficiency η_e, defined as the ratio of the power collected in the detection element to the total power dissipated, is therefore reduced. Of the incident power P_i, $(1 - |\rho_L|^2)P_i$ is not reflected, but only $\eta_e(1 - |\rho_L|^2)P_i$ of this is absorbed in the detection element. The overall ratio of incident power to that absorbed in the element, called the calibration factor K_e, is therefore

$$K_e = \eta_e(1 - |\rho_L|^2)$$ (8.19)

Power-head manufacturers usually supply calibration factors at specified frequencies that are traceable to national standards, and modern power meters can be set to correct both efficiency and mismatch loss through the single factor K_e.

If P_M is the measured power, correction gives the incident power

$$P_i = \frac{P_M}{K_e} \tag{8.20}$$

But from eqn. 8.11 the incident power is

$$P_i = \frac{1}{2}|a_L|^2 = \frac{\frac{1}{2}|b_S|^2}{|1 - \rho_S\rho_L|^2} \tag{8.21}$$

But $\frac{1}{2}|b_S|^2$ is the Z_0 matched power P_{Z_0} or

$$P_i = \frac{P_{Z_0}}{|1 - \rho_S\rho_L|^2} \tag{8.22}$$

The wanted quantity is P_{Z_0}, but even after application of the calibration factor K_e to the measured power P_M, the best estimate is P_i with an uncertainty of $(1 \pm |\rho_L||\rho_S|)^2$.

8.4 Power-head elements

There are many ways of measuring power, but this Section is confined to just three of the most important general-purpose laboratory methods based on the following detecting elements:

● Thermistor
● Thermocouple
● Diode detector

Between them, they cover a power range from -70 dBm to a few watts, and are similar in use though based on different detecting principles. The following sections give descriptions of the power heads and meters based on each element.

8.4.1 Thermistor power meters
Elements that respond to power absorption by changing their resistance are known as bolometers. There are two well-known types—barretters and thermistors. The former consist of a thin piece of wire with a positive temperature coefficient of resistance, whereas the latter are semiconductors with negative coefficient. Because of their low coefficients, small power changes are detectable with barretters only if they have very thin wires, with the result that at high power levels they operate very close to burn-out and are subject to accidental destruction. Even so, they are easily made and have the advantage of greater immunity to ambient temperature changes. However, they are not mechanically robust and, with their fast thermal response, are not well suited to measuring average power.[13] For these reasons they are not now in common use, though they are still to be found in standards laboratories for measuring low-milliwatt

powers in waveguide. In thermistors the disadvantage of high drift with temperature change, and consequent increase in compensating circuitry, is balanced by a wider power range and better resistance to damage.

The thermistor element is a small bead of metallic oxides, typically 0·4 mm in diameter with 0·03 mm supporting wires. In a rectangular-waveguide construction the head is supported between a vertical post and horizontal bar with the electrical field vertical, as illustrated in Fig. 8.11. A compensating bead is thermally coupled to the RF detection bead by being attached to the bar, but not exposed to the incident RF power. Coaxial versions are also available.[3] This kind of mounting ensures maximum absorption in the element, low reflection, low resistive and dielectric losses and good RF shielding. The power range is from 1 μW to 10 mW, with typical SWR from 1·5 to 2·0.

Thermistors have negative temperature coefficients of resistance with highly non-linear characteristics similar to those outlined in Fig. 8.12. They are mechanically rugged and difficult to damage with RF overload, but are not suitable for direct-reading instruments because of the large variations in characteristics from element to element. A substitution method, in which DC power is exchanged with RF power, maintains a fixed operating point to avoid non-linear effects, whilst also providing temperature compensation against the unexposed bead attached to the bar in Fig. 8.11. Two DC bridges are required, one for each thermistor bead.

For convenience in connecting an RF detector into a DC bridge, two thermistors, each of resistance R_d are in parallel to the RF and in series for the DC bridge, thus obviating the need for an isolating RF choke and maximising the frequency range of the mount. The temperature-compensating circuit consists of a second pair of thermistors (attached to the bar) in series but not exposed to RF. In the circuit arrangement shown in Fig. 8.13, the bypass capacitor C_b ensures that the two thermistors in the detection circuit are in parallel across the RF input to give a 50 Ω termination in the case of a coaxial mount, and in series in one arm of the upper DC bridge in Fig. 8.14. The lower bridge contains a series pair of compensating thermistors.[7]

When the RF power increases in the exposed elements, the feedback amplifier reduces the DC power in the thermistor to just compensate for the increase in

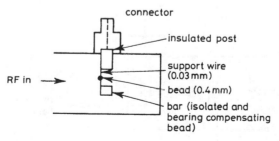

Fig. 8.11 *Thermistor mount*

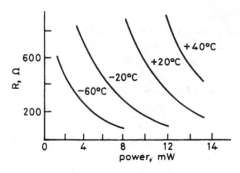

Fig. 8.12 *Thermistor characteristics*

RF power. Temperature changes affect both bridges equally, causing no difference output. A sum-and-difference calculation is indicated in Fig. 8.14 because we are measuring power changes. Thus, if V_c is the output from each bridge when there is no RF power, the measured output power is zero, or

$$P_M = \frac{V_c^2}{4R} - \frac{V_c^2}{4R} = 0$$

where R is the load resistance at each amplifier output. When RF power is present, the DC output in one circuit changes to give

$$P_M = \frac{V_c^2}{4R} - \frac{V_{cRF}^2}{4R} = \frac{1}{4R}(V_c^2 - V_{cRF}^2) \tag{8.23}$$

The measured power P_M is proportional to the input RF power, and can be calculated as the product of the sum and difference of the amplifier output voltages by writing eqn. 8.23 as

$$P_M = \frac{1}{4R}(V_c - V_{cRF})(V_c + V_{cRF}) \tag{8.24}$$

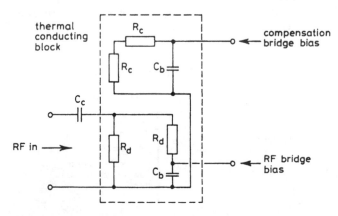

Fig. 8.13 *Thermistor compensation*

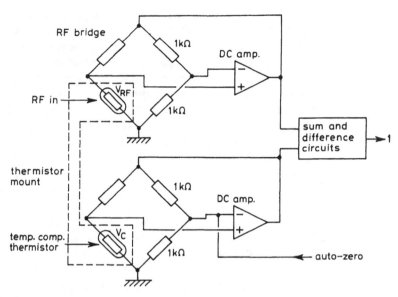

Fig. 8.14 *Thermistor bridge circuits*

Typical operating ranges using coaxial mounts are from 10 MHz to 18 GHz. The reflection coefficient is ∼0·2 over most of the range, and calibration factors from 0·7 to 0·99 are available, depending on the manufacturer. Waveguide thermistor mounts cover frequencies to greater than 140 GHz.

8.4.2 Thermocouple power meters

In a thermocouple, consisting of two dissimilar metal wires joined at one end and connected to a meter at the other, thermoelectric voltages are generated if the junction temperature is higher than at the connections to the meter. Simple wire thermocouples, e.g. constructed of antimony and bismuth, have little use at microwave frequencies because of the difficulty of absorbing power into the hot junction. A resistance must be associated with the absorbing junction to match the Z_0 impedance of the power-carrying transmission line. By combining thin-film technology with semiconductor material it is possible to fabricate thermocouples with calibration factors comparable to thermistor mounts, but with much better reflection coefficients,[3] typically 0·1. Power range is from 100 nW to 1 W, though this can be extended upwards by the use of calibrated attenuators; but minimum detectable power is limited by Johnson noise in the element. The chief characteristics of the thermocouple power meter can be summed up as

- Very wide dynamic range
- High thermal stability
- Frequency range DC to 40 GHz
- High accuracy ∼1%
- Fast response time
- 300% short-term overload

Construction of an element, as used in the Hewlett–Packard HP8480 series power sensor, is shown in Fig. 8.15. Two thermocouples are deposited on a single silicon chip to provide a series/parallel circuit arrangement for separation of RF and DC circuits. A thin-film resistor of tantalum nitride next to the hot junction matches the stripline Z_0 at the beam-lead connection. It is separated from the silicon by a silicon dioxide layer, except near the centre of the chip where there is a small hole in the dioxide layer. The centre of the chip is very thin and the outside edges thick, so that heat absorption in the resistor raises the centre temperature to form a hot junction. Careful shaping of the resistor deposit concentrates current density, and hence heat, at the chip centre.

Connection of the beam leads to co-planar microstrip lines is shown in Fig. 8.16. In this instance construction follows the Marconi 6900 series of power heads.[4,5,6] The total size is approximately 7×4 mm, with the thermocouples connected to the RF via a blocking capacitor (not shown). Coupling capacitors for a series/parallel, DC/RF configuration are shown in Fig. 8.17, but note that in Fig. 8.16 a DC connection is made direct to one thermocouple, whereas the other is grounded to RF by the 10 pF and 10 000 pF capacitors. Broad-band matching is achieved by distributed bonding pins on the 10 pF capacitor and matching chokes at the RF input.

Power-head design depends on the frequency and power range. For instance, lower frequencies require a larger blocking capacitor, which can compromise high-frequency performance. The mass and shape of the resistive load has an important bearing on both dynamic range and matched bandwidth. Drifting due to ambient-temperature variations is reduced to very low levels because hot and cold junctions generate EMFs of opposite polarity, thus giving no net output when the ambient temperature changes. Temperature dependence of the convective cooling of the assembly complex can have a significant effect on RF to DC sensitivity, but careful design can reduce this to less than 0·1% per deg C.

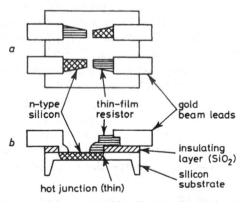

Fig. 8.15 *Semiconductor thermocouple (After Hewlett–Packard, ref. 3)*
  a Plan view
  b Side view

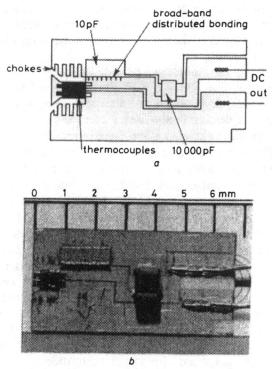

Fig. 8.16   *Stripline thermocouple mount (Photo courtesy, Marconi Instruments Ltd.)*

Power-sensor DC output is fed to a power meter which consists of a very low noise, high-gain DC amplifier and a meter. In use, the very low minimum input, close to the noise level in the DC output of less than $1.5\,\mu V$ on the most sensitive range, necessitates the use of a chopper amplifier system to avoid drift. It has also been found necessary to pay particular attention to temperature uniformity of the circuits connected immediately to the head. For this reason,

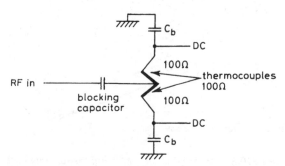

Fig. 8.17   *Showing decoupling of DC from RF*

heads now contain the sensor, the chopper and a first-stage amplifier, as in the example shown in Fig. 8.18. Even then, thermoelectric effects in the connecting paths have to be alleviated by the use of a common metal (gold) in all DC paths within the chopper circuit.

Thermocouples require calibration, provided by a 50 MHz precision source in the power meter outlined[4] in Fig. 8.19. Power-sensor DC output is square-waveform modulated, pre-amplified and filtered to remove noise and spurious signals. Synchronous demodulation uses a common driver to modulate and demodulate the sensor input, with a low-noise amplifier and filter between the two processes. Response time and bandwidth of the integrating amplifier are traded according to the noise level, effectively averaging for longer when the noise is higher. Gain control and stability are maintained in the feedback network. The power-head compensator corrects the head for square-law behaviour and allows for a range of different power heads to be connected. Microprocessor control of component adjustments makes the changes virtually invisible to the operator.

In the calibrator, which provides precision power levels in the range 1—100 mW at 50 MHz, the comparator and diode feedback ensure a constant input to a 50 Ω power head against the accurate DC reference. Typical power accuracy at 10 mW is $\pm 0.7\%$. A local power reference is essential because the power meter is an open-loop system, susceptible to the usual drift problems that can only be corrected by frequent re-calibration. Again, microprocessor storage and control reduces the tedium of repeated calibration.

Thin-film semiconductor thermocouple power sensors have very low reflection coefficients, thus minimising measurement uncertainty and allowing accurate broad-band wide-dynamic-range measurements. When proper attention is

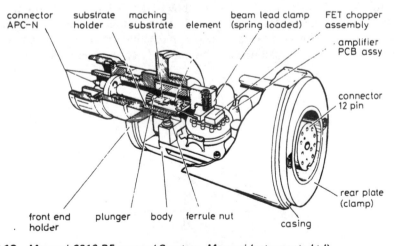

Fig. 8.18 *Marconi 6910 RF sensor (Courtesy, Marconi Instruments Ltd)*

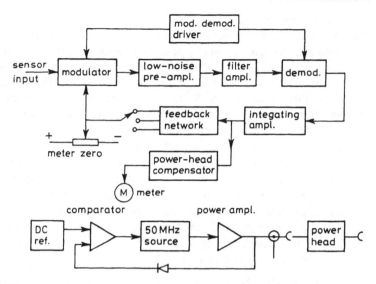

Fig. 8.19 *Power-head meter and calibrator (After Luskow, ref. 4)*

given to the component shapes and layout, their thermal drift is considerably less than either a thermistor element or diode detector. The effects of a sudden change of temperature, such as occurs when the heads are grasped by hand, are markedly different, as shown in Fig. 8.20. These curves are merely representative; however, they do show the great advantage of the thermocouple over other types. But the major disadvantage of the thermocouple is its very low output, of the order of 160 nV for 1 μW of applied power.

The final general-purpose sensor we shall consider, the low-barrier Schottky diode, is some 3000 times more efficient than the thermocouple, or about 50 nV for 100 pW.

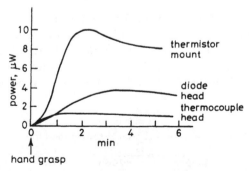

Fig. 8.20 *Temperature sensitivity of different heads (After Hewlett–Packard, ref. 3)*

8.4.3 Diode-detector power meters

Diode detectors, used in scalar-network analysis and as level detectors in power-monitoring applications, generally operate in a range from -50 dBm to $+20$ dBm. The square-law range is from -70 dBm to -20 dBm, above which they become linear. Combined with a thermocouple head the total dynamic range is 90 dB, as illustrated in Fig. 8.21. Their accuracy is no better than ± 0.5 dB, though they can be characterised by measurement and microprocessor corrected. Diode and thermocouple heads are similar in use and both require open-loop calibration with an accurate source, usually at 50 MHz.

Diodes, used as detectors with zero bias, have a resistance given by the origin slope of the I/V characteristic, a spreading or bulk resistance of the material and a junction capacitance. The equivalent circuit shown in Fig. 8.22 is a simplified model of a diode, in which we shall assume the spreading resistance R_S is negligible compared to R_0, and that the frequency range allows C_0 to be ignored. A strong temperature dependence is indicated by the exponential form of the diode equation, already met in eqn. 6.3:

$$i = I_S [\exp (ev/nkT) - 1] \tag{8.25}$$

where i is the diode current for voltage v, e is the charge on an electron, k is Boltzmann's constant and T the absolute temperature. I_S is the reverse saturation current, as $\exp (eV/nkT) \to 0$ for negative voltage. n is typically $1 \cdot 1$ in Schottky barrier diodes[9] and e/nkT is typically 25 mV at 290°K.

Differentiating eqn. 8.25 with respect to i and setting $v = 0$ gives the origin resistance as

$$\left.\frac{dv}{dt}\right|_{v=0} = \frac{nkT}{I_S e} = R_0 \tag{8.26}$$

In a coaxial system, matching R_0 to 50 Ω would give the maximum power conversion and detector sensitivity. Now on substituting for e/nkt,

$$R_0 \approx \frac{25}{I_S} \quad \text{at } 290°K$$

which gives $I_S = 0.5$ mA for $R_0 = 50$ Ω. The highest reverse saturation current I_S found in point-contact diodes is ~ 10 μA, giving a minimum practical R_0 of 2.5 kΩ. But these diodes are fragile and unstable, and cannot normally be used for accurate power measurements. PN and metal–semiconductor junctions are highly stable but have an I_S of 1 nA and 1 μA, respectively. In Fig. 8.23

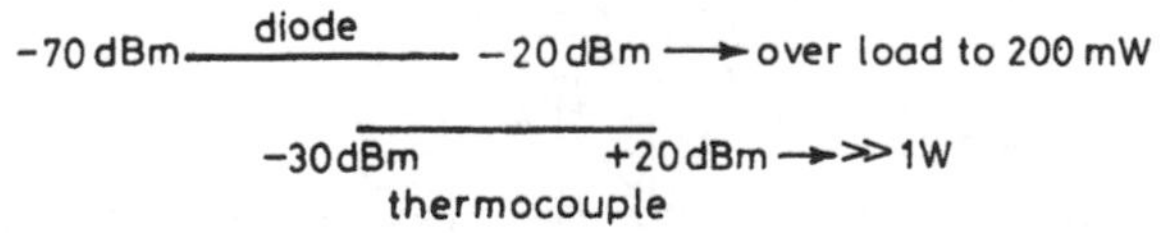

-70 dBm————diode————-20 dBm $\longrightarrow$ over load to 200 mW

-30 dBm————————$+20$ dBm $\longrightarrow \gg$ 1W
thermocouple

Fig. 8.21 *Power ranges*

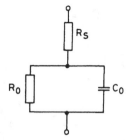

Fig. 8.22 *Equivalent circuit of detector diode*
R_s = spreading resistance
R_0 = origin resistance
C_0 = junction capacitance

comparison of a PN junction and Schottky diode shows the greatly increased reverse current and reduced forward threshold of the normal Schottky metal–semiconductor diode. In the latter case an energy-level diagram in Fig. 8.24 shows the metal work function ϕ_M as greater than the n-type semiconductor work function ϕ_S. In reverse bias, electrons have to overcome the sharply discontinuous Schottky barrier. It is possible to reduce the height of the barrier, by altering the junction surface states, to achieve a 10 μA reverse current, typical of the point-contact diode, but with the rugged repeatability of a Schottky metal–semiconductor diode.[9,10] The resulting characteristic is shown in Fig. 8.25.

Since R_0 cannot be reduced below 2·5 kΩ even in a Schottky barrier diode, a separate matching resistor shunts the detector in Fig. 8.26 to match the generator R_g to the detector circuit. This implies that R_0 should be much greater than R_g, but by a second differentiation of eqn. 8.25 we can see that the rate of

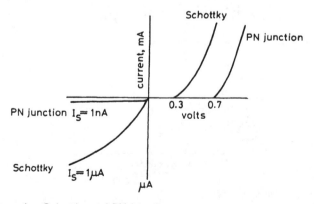

Fig. 8.23 *Comparing Schottky and PN junction*

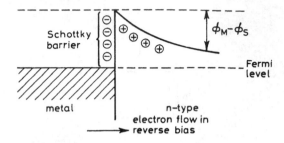

Fig. 8.24 *Schottky barrier at metal semiconductor junction*

change of the origin resistance is proportional to R_0, or

$$\frac{dR_0}{dT} = \frac{nk}{I_s e} = \frac{R_0}{T} \tag{8.27}$$

This requires that R_0 should be as small as possible. In practice, a reasonable compromise is given by $R_0 = 2.5 \text{ k}\Omega$.

For good high-frequency performance the junction capacitor C_0 should be as low as possible. A planar epitaxial construction (Fig. 8.27), with a mesa shape and small junction area, reduces C_0 to about 0.1 pF for operation to 18 GHz. The epitaxial layer is lightly doped compared to the heavily doped silicon

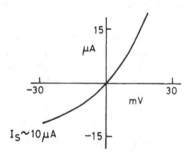

Fig. 8.25 *Low-barrier-junction characteristics*

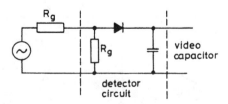

Fig. 8.26 *Source and detector circuit*

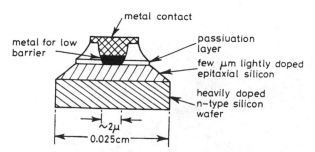

Fig. 8.27   *Mesa construction of low-barrier Schottky diode (After Hewlett–Packard, ref. 3)*

substrate. An oxidised passivation layer insulates the epitaxial layer from the upper layers, except in the centre where a small hole is etched for the metal–semiconductor junction. The choice of junction metal is critical for low barrier production.[3] Broad-band performance is obtained by embedding the diode in a compensating matching circuit on a sapphire substrate using thin-film techniques. This is shown in circuit form in Fig. 8.28, where R_M ($= 50\,\Omega$), L_C, R_C are compensating components; L_S, C_S are stray effects and the diode chip is inside the dashed lines.[9] Finally, the sensor housing has to be carefully designed to minimise temperature gradients across the diode by spreading heat uniformly to both ends of the package.

Apart from some changes at the input to the chopper to provide a higher input impedance to the diode, power meters for diodes and thermocouples are the same, including the 50 MHz calibration reference output. Thermocouple and diode heads are interchangeable, as can be seen by the similarity of the Marconi 6910 and 6920 heads in Figs. 8.18 and 8.29, respectively. Calibration may be more important with diodes because of the larger variation of sensitivity with temperature. For instance, levels may change by 1 dB over a temperature range of 0—50°C. As we have seen, the diode is particularly suitable for low power levels to 100 pW, yet it can withstand overloads at 200 mW, or 23 dB above its maximum accurate level.

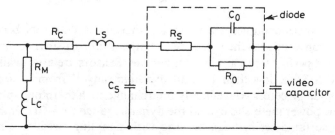

Fig. 8.28   *Equivalent circuit of Schottky diode*

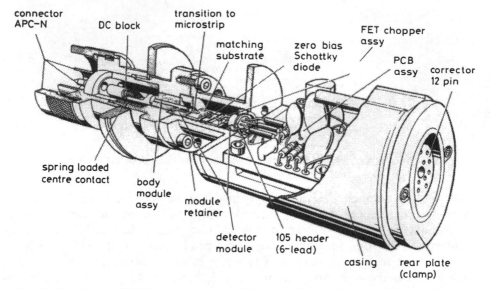

connector
APC–N   DC block

transition to
microstrip

matching
substrate

zero bias
Schottky
diode

FET chopper
assy

PCB
assy   corrector
              12 pin

spring loaded
centre contact   body
                    module
                    assy   module
                              retainer

detector   105 header
module   (6–lead)

casing   rear plate
                (clamp)

Fig. 8.29 *Marconi 6920 low-power sensor (Courtesy, Marconi Instruments Ltd.)*

Errors in the power head are usually quoted in worst-case sum of uncertainties as $\sim \pm 2\%$ to 10 GHz. Resolution is $\sim 0{\cdot}01\%$, accuracy of the head $\sim \pm 0{\cdot}5\%$ and VSWR $1{\cdot}1$ to $1{\cdot}4$ from 10 MHz to 20 GHz. Noise performance for the Marconi head in Fig. 8.29 is quoted as $< 50\ \mu\text{V}$ peak to peak in a 400 kHz bandwidth with a CW input to produce a 100 mV output. This compares with $< 200\ \mu\text{V}$ for a point-contact diode in similar circumstances.

It is interesting to note that at millimetre wavelengths up to 200 GHz the Schottky barrier diode takes the form of the original cat's whisker crystal detector. These are the earliest detectors for RF and are notoriously inaccurate. However, they have exceedingly low capacitance, and therefore greater sensitivity, at high frequencies.

Total measurement error is a combination of mismatch uncertainty ($\pm 0{\cdot}1$ dB), calibration-factor error ($\pm 0{\cdot}15$ dB) and instrumentation error ($\pm 0{\cdot}1$ dB). A worst-case addition of these as powers gives $\pm 0{\cdot}35$ dB, and a root sum of the squares $0{\cdot}12$ dB.

Increasingly, thermocouples are replacing thermistor detectors because of their greater convenience, their reduced temperature sensitivity, improved match and good performance at high frequencies. Sensors are now available up to 50 GHz with SWR less than $1{\cdot}5$ and dynamic range[11] from -30 dBm to $+20$ dBm. Diode detectors conveniently interchange with thermocouple heads, using the same power meters to extend the dynamic range to -70 dBm. Any of the detectors so far described can be used with sampling circuits to measure higher powers, normally beyond their operating range.

8.5 High-power measurements

The three high-power methods to be discussed in this Section are:

- Direct calorimetric methods
- Substitution methods using water flow
- Sampling via a directional coupler to a thermistor (substitution) or thermo-couple (direct)

These methods are usually associated with average power over a period of time, but sometimes the source output is a train of pulses such as from a high-power radar transmitter. The power at the peak of each pulse is then required, and must be calculated from a measured average by also finding the pulse-repetition frequency (PRF) and pulse width τ. The peak/average power ratio in a regular train of rectangular pulses is

$$\frac{P_{pk}}{P_{av}} = \frac{1}{\text{PRF} \times \tau} \tag{8.28}$$

Pulse-repetition frequency and width can be measured by examining a small sample in an oscilloscope after detection.

The distinction between direct and substitution methods is that RF power in the former case is found directly, e.g. by the rise in temperature caused by its absorption in water; and in the latter, indirectly by substitution of DC power to maintain a temperature constant at the place of absorption. The broad division made in the following account of some of the more important methods is between those where the full power interacts with the detector and those where only samples are taken, usually via directional couplers in the main power line.

8.5.1 Direct water calorimeter

A direct conversion of RF energy to measurable heat is possible by supplying power to a coil immersed in flowing water, as in Fig. 8.30. The rise in temperature ΔT is measured in a thermocouple whose cold and hot junctions are in the appropriate water channels to and from the RF coil. If S is the specific gravity, C the specific heat of water and F the flow rate, the power absorbed is

$$P = KFSC \, \Delta T \tag{8.29}$$

where K is a constant ideally equal to 1. Errors occur due to heat loss and flow friction loss. These can be kept to a minimum by adopting a substitution method. In this case, a DC heater is placed in the flow line to keep the temperature difference constant. Thus, when the RF is turned on, the DC input is reduced by an amount equal to the RF power.[12] The method is suitable for power levels from 5 to 20 kW at X-band and up to 100 kW at S-band.

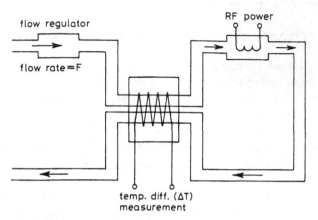

Fig. 8.30 *Flow calorimeter*

8.5.2 Power-ratio water calorimeter

An alternative version of the substitution water calorimeter adds to, rather than subtracts, DC from the RF power, in a ratio of 1 : 10. It employs a bridge circuit with all its elements immersed in water flowing in a tube. There are two DC heaters and an RF power input. The relative positions of the components in Fig. 8.31 is an essential feature of the operation. The bridge is first balanced with no RF or DC power at the input flow temperature T_1. When RF power enters, only R_1 is unaffected by the temperature rise, since it is upstream from the heaters. The bridge could be re-balanced by raising the temperature of either R_1 or R_4, because the voltage at A is higher than at B due to the relatively lower value of R_1. Increasing R_4 raises the potential at B to re-balance the bridge, but the amount by which R_4 should be raised depends on the ratio of R_5 to R_4. R_4 is placed in the flow where its temperature alone can be affected by a DC power heater balance, P_B. A separate DC heater close the the RF coil is for calibrating the instrument. We will show that by making R_4 ten times greater

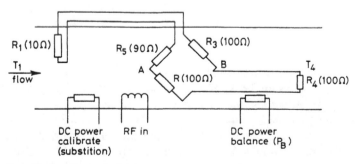

Fig. 8.31 *Substitution flow calorimeter*

than R_1, only one tenth of the RF heat is required at the DC input to re-balance the bridge.

Suppose the bridge has been re-balanced by raising the temperature of R_4 by ΔT_4. If the remaining components of the bridge are at T, resistor R_1 is at $T - \Delta T_1$ and R_4 is at $T + \Delta T_4$. Corresponding changes in resistance are

$$\Delta R_1 = R_1 \alpha \, \Delta T_1$$

$$\Delta R_4 = R_4 \alpha \, \Delta T_4 \tag{8.30}$$

where α is the temperature coefficient of resistance. Since the bridge is in balance and $R = R_5 + R_1 = R_4 = R_3$

$$\frac{R_5 + R_1 - \Delta R_1}{R} = \frac{R}{R_4 + \Delta R_4}$$

Therefore

$$(R - \Delta R_1)(R + \Delta R_1) = R^2$$

which gives

$$\Delta R_1 = \Delta R_4 \qquad \text{if } \Delta R_1 \, \Delta R_4 \to 0 \tag{8.31}$$

Substituting eqn. 8.31 in 8.30, we have

$$\frac{\Delta T_4}{\Delta T_1} = \frac{R_1}{R_4} \tag{8.32}$$

But $\Delta T_4/\Delta T_1$ is the ratio of the DC balance to the RF power, P_B/P_{RF}. Thus if $R_1 = 0{\cdot}1 R_4$ the DC balance power is only $0{\cdot}1 P_{RF}$.

This method has the advantage of a sensitive null to ensure precise adjustment and a simple effective calibration over a range of power levels. Among the

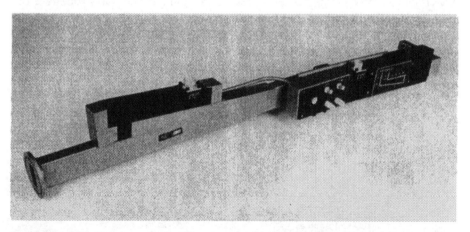

Fig. 8.32 *Mid-Century substitution calorimeter (Courtesy, Mid-Century Microwavegear Limited)*

disadvantages are distributive heat losses that may cause temperature variations among the bridge resistors, and the limiting effects of adding more DC power at higher RF power levels.

A Mid-Century Microwavegear *S*-band high-power head, shown in Fig. 8.32, and based on this power-ratio method, measures mean powers of several 100 W, and has a 5° bend in the *H*-plane of the waveguide, allowing entry of a borosilicate glass tube that runs along the waveguide axis. The tube and flowing water column form a cut-off circular waveguide to give low reflection and radiation loss.

8.5.3 Sampling

When a low-power sample is coupled from a high-power line, standard laboratory power heads may be employed as in Fig. 8.33, where a thermistor element implies that this is also a substitution method. Most of the power is absorbed in a load. Mean power is given by the thermistor, and the oscilloscope is required only for the peak power of a pulsed source. The spectrum analyser checks out-of-band spectral components including amplitude and phase noise, because in some applications, such as MTI radar, these quantities are as important to overall performance as the in-band signal characteristics. The coupling factor to the main line must be accurately known, and great care taken with the matching of test-set components. Error calculation is more complicated because the uncertainty relates to multiple reflections between the source, directional-coupler ports and absorbing load. We will look more closely at these kinds of errors in a later Section on automatic power sensor calibration.

8.5.4 Direct pulse power

Peak pulse power can be measured directly by comparison with a known CW power source. Where high power is present, sampling is necessary to reduce the level from both the pulsed and CW source, whereas Fig. 8.34 shows only one source sampled. Pulses are detected for display on the oscilloscope, and then compared with the exchanged CW signal at the same detector. After adjusting the CW level to correspond with the estimated peak of the pulse, the CW power

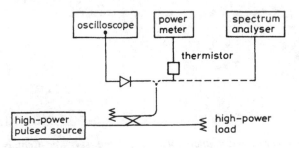

Fig. 8.33 *Sampling power meter*

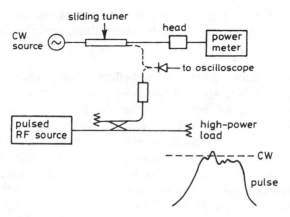

Fig. 8.34 *Peak-pulse-power comparison with CW*

is connected directly to the power-meter head. If great accuracy is required a slide tuner may be necessary because of the different match of the diode and head.

8.5.5 Notch wattmeter

In a radar of very low duty cycle, the average power is too low compared to the peak for satisfactory measurement. The notch wattmeter adds a CW signal to an attenuated sample of the pulsed power, but can be triggered to switch it off for the period of each pulse. An almost continuous level is then displayed on an oscilloscope connected to a diode detector when the CW power has been adjusted to equal the pulse power. Non-rectangular pulses can be measured this way, but judgment must be exercised about the 'average peak' since the 'notched' portion takes the pulse shape. Operation, illustrated in Fig. 8.35, depends on a pulse generator synchronising both the CW and pulsed sources to

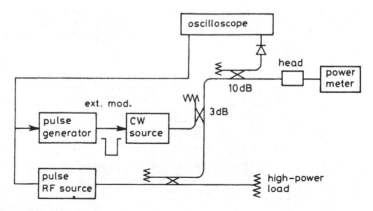

Fig. 8.35 *Notch wattmeter*

the oscilloscope display. The increased number of components, particularly the three directional couplers, increases the uncertainty, typically to 28%.[13]

8.5.6 Diode peak detector

This simple but highly inaccurate method effectively replaces the CW comparison source with a DC voltage. The diode detection circuit of Fig. 8.36 responds to the peak level in a RF pulse, provided the time constant CR during the forward conduction cycle is much shorter than the pulse length. If the diode is slightly forward biased, the capacitor is able to discharge rapidly between pulses but holds close to the peak value during a pulse. The choice of CR and bias voltage determines the circuit's ability to follow a pulse envelope. Typically such circuits can be made to measure peak levels in pulses $>0.25\,\mu s$ in duration, at frequencies from 50 MHz to 2 GHz and PRFs up to 1.5 MHz. A block

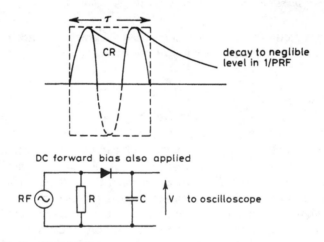

Fig. 8.36 *Diode peak detection*

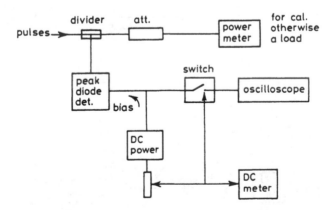

Fig. 8.37 *Diode peak detector*

diagram of the complete peak detector is given in Fig. 8.37. The instrument must first be calibrated with a CW source, but in operation a comparison DC voltage is adjusted to equal the peak detected power by alternately switching them at the oscilloscope input, whilst power is read from a DC meter. Accuracy is no better than $\pm 1 \cdot 5$ dB and the chief advantage is ease of use.[3]

8.6 Automatic measurements with power meters

We have already seen how accurate power meters with six-port analysers give both phase and amplitude information. Thermistor mounts have generally been used in that case because of their extreme linearity.[14] Power meters are also used with scalar analysers, and have the advantage of wide dynamic range, ~ 95 dB in some circumstances, and the high accuracy available from modern heads with low reflection coefficients. Worst-case uncertainties, almost entirely due to mismatch reflections, are $\sim \pm 0 \cdot 3$ dB, with instrument errors less than $\pm 0 \cdot 1$ dB. Measurement time is longer than with diodes, but is usually less than 1 s per frequency, including power-meter settling time and averaging to minimise noise. The very large dynamic range makes it essential to reduce source harmonics and spurious output to a minimum by the use of tracking YIG filters or fundamental oscillators.

In examples of automatic measurements using power heads, given in Sections 8.6.1 and 8.6.3, we first look at a conventional reflection/transmission test set and then see how a multi-state reflectometer can be adapted for the transfer of calibration standards between power sensors.

8.6.1 Attenuation measurements
The reflection/transmission test set in Fig. 8.38 follows conventional arrangements we have already analysed in earlier Chapters, with the exception that power sensors replace the diode detectors. This reduces reflection uncertainties

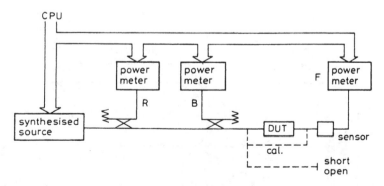

Fig. 8.38 *Reflection/transmission test set with power heads*

to a minimum because of the high accuracy and low reflection coefficient of the heads. A conventional measurement of loss in an attenuator is made by first removing the DUT for a through measurement, and then repeating the process with the DUT in place. As before, the effective source reflection is reduced to the coupler directivity by ratioing with the sampled incident power. An estimation of reflection errors depends on the scalar-error formula

$$\Delta\rho_{max} = A + B|\rho_L| + C|\rho_L|^2 \qquad (8.33)$$

already discussed in Section 2.9 on scalar calibration. From measurements on short and open circuits over the operating frequency range, it is possible to find the constants A, B and C. They are then stored in the computer for future use.

Accurate determination of attenuation using diode detectors depends on substitution measurement against a standard attenuator to keep the diode output constant. With power sensors this is not necessary since direct-power ratios can be accurately observed. Removing an extra component from the test set reduces uncertainty by eliminating a source of multiple reflections. A simplified flowgraph for the measurement is given in Fig. 8.39. The S-parameters refer to the attenuator under test, ρ_L is the reflection coefficient of the power head, ρ_S the source reflection coefficient; and the remaining paths refer to the incident wave R the reflected wave B and the usual coupling factors of the test set, including reflection loops at power detectors 1 and 2. The expression relating A_1 and B_1 at the input to the attenuator, already given in eqn. 2.30 as

$$A_1 = \frac{RT}{C_1}(1 - \rho_{D1}\rho_1) - (\rho_0 - TD_1)B_1 \qquad (8.34)$$

can be represented by the reduced flowgraph of Fig. 40a or after further topological reduction by Fig. 40b. There is then a simple relationship between the forward wave into the sensor F and the sampled incident wave R of the form

$$F = R\frac{T_E}{1 - \rho_E\rho_L} \qquad (8.35)$$

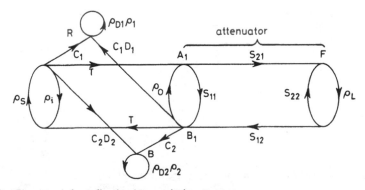

Fig. 8.39 *Flowgraph for reflection/transmission test set*

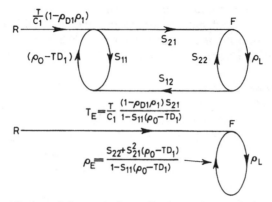

Fig. 8.40 *Reduced flowgraph for equivalent reflection and transmission coefficients*

where

$$T_E = \frac{T}{C_1} \frac{(1 - \rho_{D1}\rho_{D2})S_{21}}{1 - S_{11}(\rho_0 - TD_1)}$$

the effective transmission coefficient and

$$\rho_E = S_{22} + \frac{S_{21}^2(\rho_0 - TD_1)}{1 - S_{11}(\rho_0 - TD_1)}$$

the effective source reflection coefficient. S_{21} is the required attenuation of the attenuator.

A measurement sequence begins with a through calibration so that, if F_c is the calibration forward wave,

$$\frac{F_c}{R} = \frac{T}{C_1} \frac{(1 - \rho_{D1}\rho_{D2})}{1 - (\rho_0 - TD_1)\rho_L} \tag{8.36}$$

since $S_{11} = S_{22} = 0$ and $S_{21} = 1$.

Next the ratio F/R is measured with the attenuator inserted, and finally F/F_c is calculated to give

$$\frac{F}{F_c} = S_{21} \frac{1 - (\rho_0 - TD_1)\rho_L}{1 - S_{22}\rho_L - (\rho_0 - TD_1)[S_{11} - (S_{22} + S_{21}^2)\rho_L]} \tag{8.37}$$

The ratio F/F_c depends on the reading of two power meters, each with typical accuracy of ± 0.02 dB, and could therefore be in error by ± 0.04 dB. If the power-head reflection coefficient $|\rho_L|$ is 0.1, the directivity D and connector reflection coefficient $|\rho_0|$ are -35 dB and S_{11}, S_{22} are 0.1, the remaining terms in eqn. 8.37 give a worst-case uncertainty error of ± 0.15 dB. The total error is therefore ± 0.19 dB. If ρ_L is reduced to negligible proportions by using a high-quality thermocouple sensor, eqn. 8.37 approximates to

$$\frac{F}{F_c} = S_{21} \frac{1}{1 - (\rho_0 - TD_1)S_{11}} \tag{8.38}$$

With ρ_0 and D_1 as before, the uncertainty reduces to 0·03 dB. This very approximate analysis has ignored many other sources of error in practical test sets, and has therefore given optimistic results. However, it has served to highlight the advantages of using power sensors with low SWR and high accuracy.

8.6.2 Power-sensor calibration

Power-head standards supplied by manufacturers can be used in direct comparison methods to calibrate power sensors in laboratories or on production lines. It is a matter of transferring the calibration factor of the standard to the test power head. National standards laboratories use accurate calorimetric methods to produce standards with uncertainty limits[15,16] of $\pm 1·5\%$. An accurate automatic-network-analyser calibration has $\pm 3\%$ uncertainty, whereas a transfer calibration from a standard, using scalar techniques similar to those in Section 8.6.2 might have $\pm 4\%$ uncertainty.[17] In the modified transmission/reflection test set of Fig. 8.41 the DUT is a test power head. First the standard head is inserted and its power reading and reflection coefficient noted. The procedure is next repeated with the test head in place. From these results the test-head calibration factor and efficiency can be found. Measurement uncertainties and errors are minimised using stored data obtained during the usual calibration procedures for a transmission/reflection test set. Such data, in the form of T_E and ρ_E, as defined in eqn. 8.35, are often measured by the manufacturer using vector analysis, and supplied as part of the system software.

If P_{MS}, P_{MT} are the powers measured by the standard and test heads, respectively, and P_{iS}, P_{iT} are the incident powers on the sensors in each case, the calibration factors are

$$K_S = \frac{P_{MS}}{P_{iS}} \quad \text{for the standard head}$$

$$K_T = \frac{P_{MT}}{P_{iT}} \quad \text{for the test head} \tag{8.39}$$

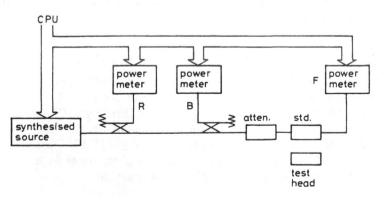

Fig. 8.41 *Reflectometer for power-head calibration*

But from eqn. 8.35 the incident wave F at the sensor is F_S or F_T, depending on which head is inserted at the test port. If ρ_S and ρ_T are the standard and test-head reflection coefficients, again applying eqn. 8.35, the ratio of the incident waves is

$$\frac{F_S}{F_T} = \frac{R_S}{R_T} \frac{1 - \rho_E \rho_T}{1 - \rho_E \rho_S} \tag{8.40}$$

Taking the ratio of eqn. 8.39, noting that

$$\frac{P_{iS}}{P_{iT}} = \left(\frac{F_S}{F_T}\right)^2$$

and substituting from eqn. 8.40, we have in worst-case form

$$\frac{K_T}{K_S} = \left(\frac{P_{MT}}{P_{MS}}\right)\left(\frac{R_S}{R_T}\right)^2 \left(\frac{1 \pm |\rho_E||\rho_T|}{1 \mp |\rho_E||\rho_S|}\right)^2 \tag{8.41}$$

K_T, the calibration factor of the test power head, can therefore be calculated from the power ratio of the two heads inserted in turn and the corresponding power ratio $(R_S/R_T)^2$ as measured at the incident wave sampling port. The uncertainty factor

$$\left(\frac{1 \pm |\rho_E||\rho_T|}{1 \mp |\rho_E||\rho_S|}\right)^2$$

cannot be reduced by this kind of scalar measurement because it is incapable of determining the phase of the reflection coefficients, although knowledge of the amplitude $|\rho_T|$ is sufficient to calculate the efficiency η_T from the relation

$$K_T = \eta_T(1 - |\rho_T|^2) \tag{8.42}$$

8.6.3 Power-sensor calibration with multi-state reflectometers

A multi-state reflectometer entails little more complexity in power-standard transfer than a simple reflectometer, but, in also providing phase information during its calibration programme, it eliminates the multiple-reflection uncertainty of a conventional scalar reflectometer.[18] As before, first the standard and then the unknown or test head is placed at the test port; see Fig. 8.42 and also

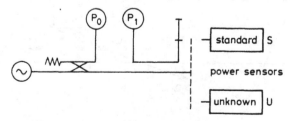

Fig. 8.42 *Multi-state reflectometer for power-head calibration*

Section 7.6 for an explanation of the multi-state connections and output ports. For each head the ratio P_0/P is measured, where P is power output at the test port. The reflection coefficients ρ_S and ρ_T are also measured for the standard and test heads, respectively.

From the flowgraph of Fig. 8.43 we can find the ratio of the incident wave b at the test-port power sensor and the output wave b_0 at port P_0. Application of the non-touching loop rule gives

$$\frac{b_0}{b} = \frac{C_0 D_0 (1 + C_1^2 D_1 \rho_R) \rho_D + C_0 (1 + C_1^2 D_0 \rho_R)}{1 + C_1^2 D_1 \rho_R}$$

$$= \frac{C_0 (1 + C_1^2 D_0 \rho_R)}{1 + C_1^2 D_1 \rho_R} \left[\frac{D_0 (1 + C_1^2 D_1 \rho_R)}{1 + C_1^2 D_0 \rho_R} \rho_D + 1 \right] \tag{8.43}$$

ρ_D now refers to the test- or standard-head reflection coefficient, and the other terms are as in Section 7.6. Similarly, the factors in the expression can be written as before in terms of a modulus and exponent to give

$$\frac{P_0}{P} = \left|\frac{b_0}{b}\right|^2 = M |S_i \exp(j\beta_i)|\rho_D| \exp(j\theta) + 1|^2 \tag{8.44}$$

with

$$M = \left|\frac{C_0 (1 + C_1^2 D_0 \rho_R)}{1 + C_1^2 D_1 \rho_R}\right|^2$$

and

$$S = \left|\frac{D_0 (1 + C_1^2 D_1 \rho_R)}{1 + C_1^2 D_0 \rho_R}\right|$$

In the measurement P_0/P becomes $(P_0/P)_M$, and is related to the true incident power through the calibration factors K_0 and K of the heads at port P_0 and the

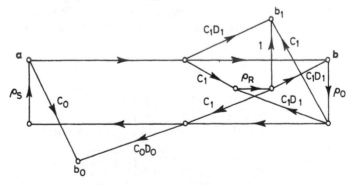

Fig. 8.43 *Flowgraph for multi-state power-head calibration*

test port, respectively. Therefore

$$\left(\frac{P_0}{P}\right)_M = \frac{K_0 P_0}{KP} = \frac{K_0}{K}\left|\frac{b_0}{b}\right|^2$$

or

$$\left(\frac{P_0}{P}\right)_M = \frac{P_{0M}}{P_M} = \frac{K_0}{K}M[1 + |\rho_D|S^2 + 2|\rho_D|S\cos(\beta + \theta)] \qquad (8.45)$$

For the standard at the test port $K = K_S$, and for the test head $K = K_T$. The corresponding power ratios are $(P_0/P)_{MS}$ and $(P_0/P)_{MT}$, and when these are ratioed we have for the measured test calibration factor

$$K_T = K_S \left(\frac{P_0}{P}\right)_{MS}\left(\frac{P}{P_0}\right)_{MT}\left[\frac{1 + |\rho_T|^2 S^2 + 2|\rho_T|S\cos(\beta + \theta_T)}{1 + |\rho_S|^2 S^2 + 2|\rho_S|S\cos(\beta + \theta_S)}\right] \qquad (8.46)$$

S, β are found from calibration of the multi-state. $|\rho_T|$, $|\rho_S|$, θ_T, θ_S are measured separately for each head, though $|\rho_S|$ can be checked from the known calibration factor K_S and efficiency η_S of the standard head. From the measured quantities η_T can be found by substitution in eqn. 8.42.

Vector determination of ρ_T and $S\exp(j\beta)$ removes the uncertainty error, leaving the head and calibration standards inaccuracies as the only significant errors.

8.7 Summary

Three sensors have been described covering a wide dynamic range from -70 dBm to $+20$ dBm or even greater with the assistance of attenuators. Speed and simplicity, particularly in peak power measurements, are found best with diode detectors, as are very low power levels. Thermistors operate in closed-loop circuits, and are therefore preferred where ambient temperature changes become large, unless special care is taken to ensure temperature equalisation in the thermocouple and diode heads. Thermocouples have the lowest SWR. Calibration of power sensors depends on standard heads with calibration factors and efficiencies traceable to national standards laboratories. These secondary standards are compared directly with laboratory or production heads in order to maintain accurate test equipment.

8.8 References

1 PENGELLY, R. S.: 'Microwave field-effect transistors—Theory, design and applications (Research Studies Press, John Wiley, 1984) p. 245
2 KUROKAWA, K.: 'Microwave solid state oscillator circuits' *in* HOWES, M. J., and MORGAN, D. V. (Eds.): 'Microwave devices' (John Wiley, 1978) pp. 209–265

3 'Fundamentals of RF and microwave power measurements'. Hewlett–Packard Application Note 64-1, Aug. 1977

4 LUSKOW, A. A.: 'Microwave power meter for the military environment', *Marconi Instrumentation*, 1977, **15**, (6)

5 MCALLISTER, P.: 'Accuracy improvements in power measurement, instruments'. 13th European Microwave Conference, Nuremberg, W. Germany, Sept. 1983

6 PARTRIDGE, R. W.: 'Automated RF power measurement'. Marconi Instruments, Instrument Application Note no. 52

7 ENGEN, G. F.: 'A self-balancing direct-current bridge for accurate bolometric power measurements', *J. Res. NBS*, 1957, **59**, pp. 101–105

8 'Power meter—New designs add accuracy and convenience', *Microwaves*, 1974, **13**, (11)

9 SZENTE, P. A., ADAM, S., and RILEY, R. B.: 'Low barrier Schottky diode detectors', *Microwave J.*, 1976, **19**, (2)

10 ADIR, BAR-LEV: 'Semiconductor and electronic devices' (Prentice-Hall, 1984) p. 114

11 *Hewlett–Packard Measurement and Computation News*, July/Aug. 1985, p. 3

12 RUMFELT, A., and ELWELL, L.: 'Radio frequency power measurement', *Proc. IEEE*, June 1967, **55**

13 ADAM, S. F.: 'Microwave theory and applications (Prentice-Hall, 1969) p. 224

14 ORFORD, G. R., and ABBOTT, N. P.: 'Some recent measurements of linearity of thermistor power meters'. *IEE Colloquium Digest* 49, 1981, pp. 5/1–5/9

15 'Bolometric power meters'. *IEEE Standard Application Guide* 470, 1972

16 'Electrothermic power meters'. IEEE Standard 544, 1976

17 'Extended applications of automatic power meters'. Hewlett–Packard Application Note 64-2, Sept. 1978

18 OLDFIELD, L. C., IDE, J. P., and GRIFFIN, E. J.: 'A multistate reflectometer', *IEEE Trans.*, 1985, **IM034**, pp. 198–201

8.9 Examples

1   A power sensor has a VSWR of 1·2 and an efficiency of 0·9. It is connected to a source with reflection coefficient of 0·1. The measured power is 8·7 mW. What is the uncertainty range of this measurement in milliwatts?

2   A temperature-compensated thermistor head is connected through a long transmission line to a constant power source which has a VSWR of 1·2 and a nominal output at 5·0 mW. The reading changes from 5·1 to 4·9 mW over a long period of continuous measurement during which the temperature changes from 15°C to 25°C. The source is temperature controlled to ±0·5°C. Suggest a possible cause of the change in measured power, and estimate the minimum magnitude of the power-head reflection coefficient.

3   What factors contribute to the total error in power measurement using a thermocouple head?

4   Calculate the peak pulse power if the average power in a sampling power measurement is 1·2 mW and the PRF is 3 kHz with a 1 μs pulse. Sampling is via a 20 dB coupler.

5   In measurement of RF power using the substitution water calorimeter de-

scribed in Section 8.5.2, the input and output water temperatures are 10°C and 50°C, respectively, when the flow rate is 0·25 litre/min. What is the RF power input, assuming the bridge is balanced?

6   Complete the step, following eqn. 8.35, in the flowgraph analysis to find the effective transmission and reflection coefficient when an attenuator is included in the power-head reflectometer.

Chapter 9

Noise

9.1 Kinds of noise

The definition of noise as 'any sound unwanted by the hearer' is not entirely satisfactory, but this is not a perfect world and noise is the consequence. Noise in electrical circuits can be classified as follows:

- Interference, e.g. lightning, man-made sparks, an unwanted radio station
- Mechanical disturbance of the circuit, such as vibrations causing electrical changes
- Hum from power supplies
- Fluctuation noise caused by thermal motions of charge carriers in resistors etc., or random particle flow in conducting media

The first three can, in principle, be reduced to zero by screening and careful mechanical and electrical design. In this Chapter we concentrate on the last, fluctuation noise, which reduces to zero only at absolute-zero temperature where no communication takes place. Since thermal and random noise therefore accompanies all signals, its measurement is an important contribution to device and system design. Absolute measurement of noise power is possible, but its ratio to a given signal is of greater significance in specifying the performance of most amplifying or signal-conversion devices.

9.2 Fluctuation noise

Historically this noise was recognised as a sound like soft breathing emerging from radio receivers. On a meter it is observed as small random fluctuations of the pointer. It was an early observation on high-input-impedance valve amplifiers that noise measured at the output increased with the ohmic value of an input shunt resistor. This observation is reviewed in Fig. 9.1, where, as R is increased, the noise, measured on a power meter at the output of an amplifier,

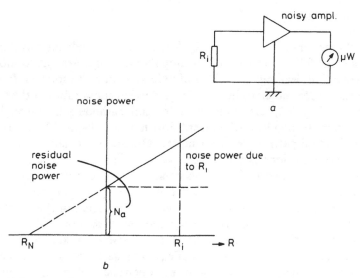

a

b

Fig. 9.1 *Noise dependence on amplifier input resistance*

rises from a residual level when $R = 0$. The residual noise power apparently
originates in the amplifier, whereas the resistor appears to add extra noise
proportionally according to its ohmic value. This simple presentation ignores
the effect of input match and output saturation in the amplifier, and is approx-
imatley correct only for high-input-impedance amplifiers and for a small range
of R and output power. However, it remains true that amplifiers themselves
generate noise, and, since noise is also apparently present in resistors, it may
sometimes be convenient to represent amplifier noise by an equivalent resistance R_N at the input of a noiseless amplifier. Thus in Fig. 9.1b the intercept
R_N is the equivalent noise resistor, which, if placed at the input of a noise-free
amplifier, as in Fig. 9.2, produces the residual noise power N_a at the output. It
is in series with R when calculating the noise contribution of the total circuit,
but is not present during signal-flow calculations. The idea of equivalent resis-
tances for specifying noise power is based on the work of Johnson and
Nyquist.[1,3]

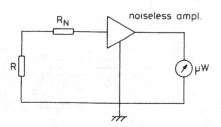

Fig. 9.2 *Equivalent noise resistance of an amplifier*

9.3 Johnson noise

Measurements, similar to those in Fig. 9.1, were first made on valve amplifiers by Johnson in 1928. The effects of the amplifier and input-resistance noise were clearly separated, leading to the idea of independent noise sources. In fact, the noise types are also different in kind. In gridded vacuum tubes of the type used by Johnson (Fig. 9.3), electrons from the hot cathode pass through a grid before striking the anode. Such current flow has been likened by Schottky to a rain of individual particles, or shot noise.[2] On the other hand, resistor noise, due to random motions and collisions between electron carriers in the bulk material, is commonly called Johnson noise. In practice, other forms of noise, including shot noise, are expressed in terms of an equivalent Johnson resistor. R_N in Fig. 9.2 therefore represents shot noise originating in the amplifier, but expressed as equivalent Johnson noise at the amplifier input.

Johnson's experimental results have been given an analytical form by Nyquist[3] by considering two resistors of equal value connected one at each end of a coaxial transmission line of length L. When the switches S in Fig. 9.4 are open, the line is matched at each end if each resistance R equals the characteristic impedance of the transmission line. The line is lossless, and both it and the connections at the ends perfectly screened. All the resistor noise energy emitted at one end is absorbed in the resistor at the other end. Noise energy is assumed to be emitted by charge fluctuations due to thermal agitation of conduction

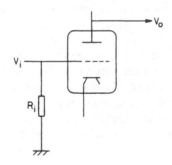

Fig. 9.3 *Johnson noise in a valve*

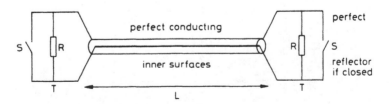

Fig. 9.4 *Coaxial resonator for calculating Johnson noise*

electrons in the resistors, and transferred end to end by electromagnetic waves set up in the coaxial line. Finally, the whole system is kept in equilibrium by maintaining both resistors at the same constant temperature. Each therefore delivers the same power to the other.

If both switches are simultaneously short-circuited, the energy in the line is trapped between two perfectly reflecting terminations. The line begins oscillating in its normal modes, all assumed to be TEM waves with a fundamental frequency

$$f_1 = \frac{c}{2L} \tag{9.1}$$

and harmonics

$$f_1 = \frac{2c}{2L}, f_3 = \frac{3c}{2L}, \ldots, f_n = \frac{nc}{2L} \tag{9.2}$$

where c is the velocity of light.

The trapped energy is distributed among the modes or energy levels according to quantum theory, but, at frequencies up to 100 GHz, the distribution is approximated by a continuum governed by the principle of equipartition. This states that the thermal energy of a system is equally divided among its degrees of freedom, with $kT/2$ in each degree. k is Boltzmann's constant and T the absolute temperature. There are two degrees of freedom per mode since each has an electric- and magnetic-field storage exchange.

The number of modes between two frequencies f_m and f_n is found from

$$f_n - f_m = \frac{nc}{2L} - \frac{mc}{2L} \tag{9.3}$$

or

$$n - m = \Delta f \frac{2L}{c} \tag{9.4}$$

Since there is kT energy per mode, the electromagnetic energy stored in a frequency range $\Delta f = (f_n - f_m)$ is

$$W = (n - m)kT = kT \, \Delta f \frac{2L}{c} \tag{9.5}$$

Half this energy is supplied by each resistor; so the energy in the line due to one resistor is

$$\frac{W}{2} = kT \, \Delta f \frac{L}{c} \tag{9.6}$$

But L/c is the time for a wave to traverse the line length; therefore the rate of noise-energy generation is

$$\frac{W}{2L/c} = N = kT \, \Delta f \tag{9.7}$$

Since any power delivered to the line is totally absorbed by the other resistor, this is the power available into a matched load. When reactive components are involved, N is the noise power available to a conjugate match. Boltzmann's constant $k = 1.38 \times 10^{-23}$ J deg K^{-1}; so for a 1 Hz bandwidth $N = 1.38 \times 10^{-23}$ W deg K^{-1} Hz^{-1}, or in dB referenced to 1 mW, $N = -198.6$ dBm deg K^{-1} Hz^{-1}; whence at a standard room temperature of 290°K, the Johnson noise power per hertz is -174 dBm.

An alternative form of eqn. 9.7 based on an equivalent noise EMF is often more convenient in circuit analysis. We consider a matched pair of resistors connected in parallel as in Fig. 9.5a, each exchanging equal noise power with the other. For instance, since resistor 1 delivers power $\overline{i^2}R$ to resistor 2, if the mean-square current flow due to resistor 1 is $\overline{i^2}$, current from resistor 1 can be modelled by inserting a series-voltage generator, as in Fig. 9.5b. The noise power available from resistor 1 is

$$N = \overline{i_S^2}R = \frac{\overline{e^2}}{4R^2} R = kT \, \Delta f$$

or

$$\overline{e^2} = 4kTR \, \Delta f \tag{9.8}$$

where $\sqrt{\overline{e^2}}$ is the Thévenin equivalent RMS noise voltage of a Johnson resistor.

Another useful equivalence follows from the current model in Fig. 9.5c, in which the instantaneous noise-current generator is in parallel with resistor 1 such that

$$\overline{i_S^2} = \frac{\overline{i^2}}{2^2} \tag{9.9}$$

and the power absorbed in resistor 2 is

$$N = \frac{\overline{i^2}}{2^2} R = kT \, \Delta f$$

or

$$\overline{i^2} = 4kTG \, \Delta f \tag{9.10}$$

where $\sqrt{\overline{i^2}}$ is the Norton equivalent RMS noise current in conductance $G = 1/R$.

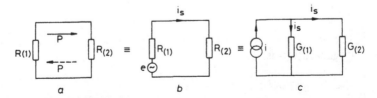

Fig. 9.5 *Available noise power from a resistor*

9.4 Shot, flicker and diode noise

The mean-square current due to a random arrival of electrons at the anode of a temperature-limited diode with an average current I is

$$\overline{i^2} = 2eI\,\Delta f \tag{9.11}$$

where e is the charge on an electron and Δf the bandwidth. A full derivation for thermionic shot noise can be found in the literature.[4,5,6] Shot noise can also occur in semiconductor junctions, but it is difficult to separate from other forms of noise such as flicker. Flicker noise usually occurs at low frequencies below 1 kHz, though in GaAs FETs it extends to about 100 MHz. It increases with decreasing frequency and is referred to as $1/f$ noise. Though its origins are not well understood, there is evidence that charge-carrier trapping and recombination at surface imperfections in semiconductor junctions is a major cause.[6]

A zero-bias diode generates noise equivalent to its resistance, but when current flows across the junction, energies are higher than thermal levels. For instance, electrons crossing a Schottky barrier are referred to as 'hot electrons'.[6] Spreading resistance, shot emission across the barrier and flicker effects all contribute to diode noise in a complex way, producing very variable levels in a cat's whisker, but with greater predictability in Schottky diodes. Typical figures for diode noise output are -50 to -55 dBm in a 100 MHz bandwidth.[7]

9.5 Noise figure and noise temperature

Noise accompanies any signal at the input to a two-port device, such as an amplifier or attenuator. It may arise entirely from Johnson effects in the source output resistance or be increased above that by other forms of noise. A measure of signal quality is its signal/noise ratio S_S/N_S, where S_S, N_S refer to the source of signal and noise powers, respectively, at the input to a two-port network. If signal and noise are equally affected by the network, this ratio is unchanged at the output. However, there will be extra added noise due to the network components—Johnson for passive resistive components, and other kinds, such as shot or flicker, for active devices—and this reduces the signal/noise ratio S_0/N_0 at the output.

A measure of this signal/noise degradation in a two-port network is given by the noise figure

$$F = \frac{S_S/N_S}{S_0/N_0} \tag{9.12}$$

Noting that the power gain (or loss) of the network is

$$G = \frac{S_0}{S_S} \tag{9.13}$$

we have the alternative form

$$F = \frac{N_0}{GN_S} \qquad (9.14)$$

or

$$F = \frac{\text{total noise power in output}}{\text{noise power in output arising from signal source alone}}$$

If N_a is the noise at the network output due to the network, the total output noise is

$$N_0 = N_a + GN_S$$

and

$$F = \frac{N_a + GN_S}{GN_S} \qquad (9.15)$$

But

$$N_S = kT_S B_N \qquad (9.16)$$

where T_S is the temperature of the input noise source and B_N is the bandwidth of the noise. Some care is necessary regarding B_N, because the noise source itself may have a different bandwidth from the network. For instance, if the network has a smaller bandwidth than the source, noise bandwidth is really associated with the gain G rather than N_S.

It is sometimes convenient to define an effective noise bandwidth when broad-band noise passes through a filter or band-limited amplifier. To take account of frequency shaping of the pass-band gain an effective noise band-width is defined by integrating the gain G_f at frequency f over the frequency range, and normalising to the peak gain G_0 to give

$$B_N = \frac{1}{G_0} \int_{f_1}^{f_2} G_f \, df \qquad (9.17)$$

This B_N can then be used in eqn. 9.16 if G becomes G_0 in eqn. 9.15.

Noise figure depends on T_S, the source temperature; so it would be difficult to compare different amplifiers unless they have a common input noise temperature. For this reason F is always quoted for a standard input temperature of 290°K. Eqn. 9.15 can be written in terms of noise temperature as

$$F = \frac{GkT_e B_N + GkT_0 B_N}{GkT_0 B_N} \qquad (9.18)$$

where $T_0 = T_S = 290°K$ for the standard temperature and T_e is the effective input noise temperature to account for internally generated network noise. On cancelling terms in eqn. 9.18 the noise figure is

$$F = \frac{T_e + T_0}{T_0} \qquad (9.19)$$

An alternative method of calculating noise figure is through an equivalent Johnson resistance R_e in series with the source resistance R_S. In Fig. 9.6a, since R_e is not a real resistance, the mean-square noise current is

$$\frac{4kT_0B_N(R_S + R_e)}{(R_S + R_i)^2}$$

and the total noise power absorbed in R_i is

$$\frac{4kT_0B_N(R_S + R_e)}{(R_S + R_i)^2} R_i$$

and the noise power from R_S alone is

$$\frac{4kT_0B_NR_S}{(R_S + R_i)^2} R_i$$

Taking the ratio and cancelling common terms we have the noise figure as

$$F = \frac{R_e + R_S}{R_S} \qquad (9.20)$$

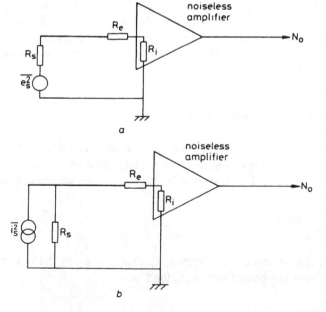

Fig. 9.6 *Thévenin and Norton equivalent sources with equivalent Johnson resistance*

R_i, being cancelled, can take any value, and a quicker calculation simply assumes R_i is an open circuit and adds the equivalent mean-square noise EMFs to give the same answer from

$$F = \frac{4kT_0B_N(R_e + R_S)}{4kT_0B_NR_S} \tag{9.21}$$

Similarly, if the source noise is written in equivalent Norton form, F is given directly by assuming R_i is a short circuit. Thus from Fig. 9.6b the ratio of short-circuited mean-square currents is

$$F = \frac{\overline{i_S^2} + \dfrac{4kT_0B_NR_e}{R_S^2}}{\overline{i_S^2}}$$

But

$$\overline{i_S^2} = 4kT_0B_NG_S$$

where G_S is the source conductance. Therefore

$$F = \frac{R_S + R_e}{R_S} \tag{9.22}$$

Noise calculations in networks can therefore be approached either from the notion of an increase in input effective noise temperature or of an equivalent extra input resistance to account for internally generated noise. Generally speaking, the concept of effective noise temperature is most useful at high frequencies involving wave propagation and distributed components, whereas equivalent Johnson resistance is more useful when dealing with low frequencies and lumped elements. Often both concepts are useful simultaneously, and the greatest care must be taken about whether temperature or resistance is held constant.

9.6 Source effective noise temperature

Measurement of noise temperature depends on the availability of known noise-temperature sources. Input match at source and network can cause errors in the observed temperature. For a conjugate match between source and network, the power available is

$$P_{av} = \frac{\frac{1}{2}|b_S|^2}{1 - |\rho_S|^2} . \tag{9.23}$$

where b_S is the wave into a matched Z_0 line and ρ_S is the source reflection coefficient. Now the power into a Z_0 load is

$$P_{Z_0} = \frac{1}{2}|b_S|^2 \tag{9.24}$$

which, in eqn. 9.23, gives

$$P_{Z_0} = P_{av}(1 - |\rho_S|^2)$$

If T_{N_e} is the effective noise temperature into a Z_0 load and T_a is the available noise temperature into a conjugate match, these powers can be expressed as equivalent Johnson noise, and we have

$$kT_{N_e}B_N = kT_aB_N(1 - |\rho_S|^2)$$

or

$$T_{N_e} = T_a(1 - |\rho_S|^2) \tag{9.25}$$

Available noise temperature T_a is generally quoted in specifications, particularly when source-match conditions can change with demanded temperature.[13]

Source-noise temperatures are generally greater than 290°K, and are expressed as an excess noise ratio in decibels as

$$\text{ENR} = 10 \log\left[\frac{T_{N_e} - T_e}{T_0}\right] \tag{9.26}$$

9.7 Noise in amplifiers and attenuators

Noise generated near the input to an amplifier chain appears with full gain at the output, whereas noise originating further down the chain is amplified only partially. It is therefore essential to minimise noise in the front-end circuits. Attenuators reduce signals and noise in the same ratio, but an excess noise is radiated into the output due to the absorption loss in the circuit. It is therefore essential to reduce attenuation in transmission media to a minimum if low noise conditions are required.

9.7.1 Amplifiers in cascade

Amplifiers with gains G_1 and G_2 are shown cascaded in Fig. 9.7. We can calculate the total noise at the output for input noise temperature T_S if we let T_{e12} be the overall effective input noise temperature of the two amplifiers in cascade. Therefore

$$N_0 = G_1G_2kB_N(T_S + T_{e12}) \tag{9.27}$$

The same noise can be calculated from the effective noise temperatures for each

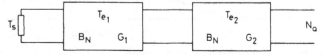

Fig. 9.7 *Noise temperature in cascaded two-ports*

stage, T_{e1} and T_{e2}, respectively, as

$$N_0 = G_1 G_2 k B_N (T_S + T_{e1}) + G_2 k B_N T_{e2} \tag{9.28}$$

Equating eqns. 9.27 and 9.28 gives

$$T_{e12} = T_{e1} + \frac{T_{e2}}{G_1} \tag{9.29}$$

For n cascaded amplifiers this becomes

$$T_{e1n} = T_{e1} + \frac{T_{e2}}{G_1} + \frac{T_{e3}}{G_1 G_2} + \cdots + \frac{T_{en}}{G_1 G_2 \cdots G_{n-1}} \tag{9.30}$$

Successive contributions are decreased by the gains of preceding amplifiers in the chain. From eqn. 9.19,

$$T_e = T_0(F - 1)$$

and substituting in eqn. 9.30, we have the noise figure for n cascaded amplifiers:

$$F_n = F_1 + \frac{F_2 - 1}{G_1} + \frac{F_3 - 1}{G_1 G_2} + \cdots + \frac{F_n - 1}{G_1 G_2 \cdots G_{n-1}} \tag{9.31}$$

If the amplifiers are all identical

$$F_n = F + \frac{F - 1}{G} + \frac{F - 1}{G^2} + \cdots + \frac{F - 1}{G^{n-1}}$$

$$= 1 + \frac{F - 1}{1 - 1/G} \tag{9.32}$$

$(F - 1)/(1 - 1/G)$ is called the 'noise measure' of the network of cascaded amplifiers. It can also refer to single amplifiers, and is used to minimise the overall noise figure by placing amplifiers with lower noise measures in ascending order in a chain.

9.7.2 Effective noise temperature of an attenuator

Suppose a noise source at temperature T_S is matched to an attenuator also at temperature T_S, which in turn feeds a matched noise power meter as in Fig. 9.8. The attenuator loss is

$$L = \frac{P_S}{P_0} \tag{9.33}$$

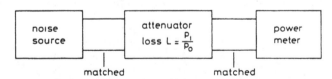

Fig. 9.8 *Effective noise temperature of an attenuator*

where P_S and P_0 are the input and output powers. Noise power input is kT_SB and a portion kT_SB/L reaches the power meter. The remaining portion $(1 - 1/L)kT_SB$ is absorbed in the attenuator. But both source and attenuator remain in thermal equilibrium at temperature T_S and the absorbed power is therefore re-radiated to the output with effective output temperature T_{eo}, given by

$$\left(1 - \frac{1}{L}\right)kT_SB = kT_{eo}B \tag{9.34}$$

or

$$T_{eo} = \left(1 - \frac{1}{L}\right)T_S \tag{9.35}$$

If this is referred to the input of the attenuator, the effective input temperature at ambient temperature T_S is

$$T_e = LT_{eo} = (L - 1)T_S \tag{9.36}$$

Thus the physical and output temperatures of an attenuator with infinite loss are equal.

9.8 Gain and noise-temperature optimisation

In principle an amplifier should introduce a minimum of extra noise whilst having the highest possible gain. Minimisation of noise measure

$$M = \frac{F - 1}{1 - 1/G}$$

combines these requirements into a single quality factor.[12] Noise measure could be minimised by decreasing F and increasing G if they were independent physical quantities, but it is not generally possible to maximise gain as well as minimise noise figure. Each depends on input matching conditions in different ways, so that a compromise is sought according to which factor is currently of greatest importance. For instance, a front-end low-noise amplifier has optimised noise figure at the expense of gain, whereas a power amplifier, at the end of a cascaded chain, adds little overall noise and is therefore optimised for maximum gain.

Measurable quantities in the noise measure are noise figure and gain. If a noise source at temperature T_S is connected at the input of an amplifier, the total noise temperature is $T_S + T_e$, where T_e is the amplifier effective temperature. Noise figure, referred to T_S, is

$$F_S = \frac{T_S + T_e}{T_S} \tag{9.37}$$

T_S may have arisen from a number of possible sources, such as a resistor, from sky temperature in an antenna etc., or from a noise source specifically designed for a noise-figure measurement. We have already seen that the standard reference temperature is $T_0 = 290°K$, and we will see later how T_S is reduced to T_0 in a method using hot and cold noise sources.

Matching between noise source and amplifier depends on the output impedance of the source, assuming the amplifier input impedance is fixed by design. Eqn. 9.25 shows how source effective noise temperature depends on source reflection, or match, to a Z_0 line. In a similar way the observed value of source temperature T_S, at an amplifier input, is a function of input match or source impedance. In particular, there is one source impedance, corresponding to a single point on the Smith chart, at which noise figure will be a minimum. Otherwise, Smith-chart loci of source impedances for constant noise figures appear as circles with centres on a line from the centre of the chart to the point for minimum noise figure F_M. In practice the source of noise is the same circuit from which an input signal emanates. Amplifier gain also depends on the same source impedance, but the Smith-chart point of maximum gain, occurring at conjugate match between source and amplifier input impedance, rarely coincides with the point for minimum noise figure, though there is a similar set of constant-gain circles with centres on a radial line to the point for maximum gain. Thus, from two sets of overlapping circles on a Smith chart, compromise choices of matching impedances are quickly obtainable, but, in fact, these circles have first to be found from a long and tedious series of measurements. In the following Sections we give the theoretical background to modern noise measurements from which these constant gain and noise-figure circles can be constructed.

9.9 Noise figure of a linear two-port

In Johnson's early experiments, valve circuits had high input impedances and current could be ignored, but, in general, a two-port network has an input current and voltage. A convenient way to include both is to write down input voltage and current, V_1, I_1, as linear combinations of the output voltage and current, V_2, I_2. In this A, B, C, D parameter form noise voltage and current, E and I, due to the network, are simply added as equivalent generators in Fig. 9.9 to give

$$V_1 = AV_2 + BI_2 + E$$

$$I_1 = CV_2 + DV_2 + I \qquad (9.38)$$

There is a degree of correlation between E and I so far as their originating components in the network are common, but they are also uncorrelated to the extent that components contribute exclusively to either voltage or current.

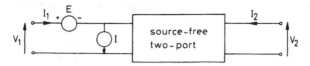

Fig. 9.9 *Noise model for two-port*

Noise parameters in lumped-component circuit analysis are conveniently expressed in terms of equivalent noise resistance or conductance, so that E and I in mean-square form are

$$\overline{e^2} = 4kT_0R_N\,\Delta f \tag{9.39}$$

$$\overline{i^2} = 4kT_0G_N\,\Delta f \tag{9.40}$$

where

R_N = equivalent Johnson resistance
G_N = equivalent Johnson conductance
Δf = circuit bandwidth

An external noise source connected to the input port is shown as I_S with admittance Y_S in Fig. 9.10. The total current is found by considering the components to be either instantaneous values or as a summation of Fourier amplitudes measured over a long period of time.[9] Following the derivation in eqn. 9.22, a noise figure may be calculated by short-circuiting the output terminals of the network in Fig. 9.10,

$$I_{total} = I_S + \dot{I} + Y_SE \tag{9.41}$$

The total mean-square fluctuating current is the average of the square of I_{total} for all Fourier components of the fluctuating quantities. Thus, if I_S is uncorrelated with I and E, in mean-square form we have

$$\overline{i^2_{total}} = \overline{i^2_S} + \overline{|i + Y_Se|^2}$$
$$= \overline{i^2_S} + \overline{i^2} + |Y_S|^2\overline{e^2} + Y^*_S\overline{ie^*} + Y_S\overline{i^*e} \tag{9.42}$$

where the star indicates a complex conjugate.

Since source noise is represented as a Norton equivalent circuit, noise figure can be calculated by the ratio of short-circuited currents as in eqn. 9.22. Thus,

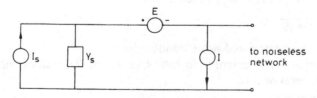

Fig. 9.10 *Input equivalent circuit to a noiseless two-port*

if the source is short-circuited, mean-square noise current due to the source alone is $\overline{i_S^2}$ and the noise figure is

$$F = \frac{\overline{i_{total}^2}}{\overline{i_S^2}}$$

$$= 1 + \frac{\overline{|i + Y_S e|^2}}{\overline{i_S^2}} \tag{9.43}$$

If the source admittance is

$$Y_S = G_S + jB_S \tag{9.44}$$

its noise current is related to conductance by the Nyquist formula

$$\overline{i_S^2} = 4kT_S G_S \, \Delta f \tag{9.45}$$

where G_S is a real conductance and T_S should be reduced to T_0 for the final noise figure.

Partial correlation of the internal equivalent noise generators can be accounted for by separating the current into two components: one correlated and the other uncorrelated with the noise voltage. Therefore, if i_u is the uncorrelated current, the average products are

$$\overline{ei^*} = 0 \tag{9.46}$$

$$\overline{(i - i_u)i_u^*} = 0 \tag{9.47}$$

If we define a correlation admittance

$$Y_c = G_c + jB_c \tag{9.48}$$

with $i - i_u = Y_c e$, as the correlated noise current, the cross-product fluctuations are

$$\overline{ei^*} = \overline{e(i - i_u)^*} = Y_c^* \overline{e^2} \tag{9.49}$$

$$\overline{e^*i} = \overline{e^*(i - i_u)} = Y_c \overline{e^2} \tag{9.50}$$

also

$$\overline{|i - i_u|^2} = |Y_c|^2 \overline{e^2} = 4kT_S |Y_c|^2 R_N \, \Delta f \tag{9.51}$$

where $\overline{e^2}$ is obtained from eqn. 9.39, and instead of using the total current in eqn. 9.40, we have the uncorrelated portion

$$\overline{i_u^2} = 4kT_S G_u \, \Delta f \tag{9.52}$$

where G_u is an uncorrelated noise conductance.

Total internal noise current $\overline{i^2}$ in eqn. 9.42 is now the sum of correlated and uncorrelated components.

$$\overline{i^2} = \overline{|i - i_u|^2} + \overline{i_u^2}$$

or, on substituting from eqns. 9.51 and 9.52,

$$\overline{i^2} = 4kT_S[|Y_c|^2 R_N + G_u] \, \Delta f \tag{9.53}$$

Eqns. 9.49, 9.50 and 9.53, substituted in eqn. 9.42, give

$$\overline{i^2_{total}} = \overline{i^2_S} + 4kT_S[|Y_c|^2 R_N + G_u] \, \Delta f + |Y_S|^2 \overline{e^2} + Y_S^* Y_c \overline{e^2} + Y_S Y_c^* \overline{e^2}$$

On dividing by $\overline{i^2_S}$ from eqn. 9.45 and substituting from eqn. 9.39, the noise figure is

$$F = 1 + \frac{G_u}{G_S} + \frac{R_N}{G_S} |Y_S + Y_c|^2 \tag{9.54}$$

or

$$F = 1 + \frac{G_u}{G_S} + \frac{R_N}{G_S} [(G_S + G_c)^2 + (B_S + B_c)^2] \tag{9.55}$$

The four unknown quantities G_u, R_N, G_c, B_c have to be found by experiments based on measuring F for different source impedances $G_S + jB_S$. But since the purpose is to find the minimum value of F as well as the loci of constant noise figure, it is useful to re-arrange eqn. 9.55 with minimum noise figure F_M as a factor. The orthogonal nature of real and imaginary parts gives one condition for minimum noise figure as

$$B_S = -B_c = B_M \tag{9.56}$$

On differentiating the remaining real terms with respect to G_S and setting to zero, we have

$$G_S = \left[\frac{G_u + R_N G_c}{R_N} \right]^{1/2} = G_M \tag{9.57}$$

When B_M and G_M are substituted back in eqn. 9.55, the minimum noise figure is

$$F_M = 1 + 2R_N(G_c + G_M) \tag{9.58}$$

Other noise figures can now be expressed in terms of source admittance for minimum noise figure to give

$$F = F_M + \frac{R_N}{G_S} [(G_S - G_M)^2 + (B_S - B_M)^2] \tag{9.59}$$

There are four real parameters, F_M, G_M, B_M and R_N, to be determined by experiment. When these are known, the noise figure for any source impedance follows from eqn. 9.59. The first three parameters are found together on determining F_M on a trial-and-error basis that depends on varying the source impedance Y_S by means of a sliding tuner until a minimum F_M is established. A separate measurement of the corresponding Y_S then gives G_M and B_M. One additional measurement[9] of F for any source admittance other than Y_M gives R_N.

It is a simple matter to show that eqn. 9.59 describes circles of constant noise figure on a Smith chart, but before doing that we will first calculate an expression for the power gain of a noiseless two-port.

9.10 Gain circles for a linear noiseless two-port

A flowgraph for the source-free two-port, repeated in Fig. 9.11 and already analysed in Section 3.5.2, will be a useful starting point for the present discussion.[10,14] Again, if b'_S is the source wave into a Z_0 match, ρ_S is the source reflection coefficient and P_{av} is the power into a conjugate match,

$$P_{av} = \frac{\frac{1}{2}|b'_S|}{1 - |\rho_S|^2} \tag{9.60}$$

The power into the output load with reflection coefficient ρ_L is

$$P_L = \frac{1}{2}|b_2|^2(1 - |\rho_L|^2) \tag{9.61}$$

Power gain is the ratio of the power delivered to the load to the power available from the source, or

$$G_p = \frac{|b_2|^2}{|b'_S|^2}(1 - |\rho_S|^2)(1 - |\rho_L|^2) \tag{9.62}$$

G_p is used for power gain in this Chapter to avoid confusion with conductance in other equations. Using the non-touching loop rule for $|b_2|^2/|b'_S|^2$ and setting $|S_{12}|^2$ to zero, we have for the unilateral gain

$$G_{pu} = \frac{|S_{21}|^2(1 - |\rho_S|^2)(1 - |\rho_L|^2)}{|1 - S_{11}\rho_S|^2|1 - S_{22}\rho_L|^2} \tag{9.63}$$

This unilateral gain can be separated into three parts with

$$G_{ps} = \frac{1 - |\rho_S|^2}{|1 - S_{11}\rho_S|^2} \quad \text{as the input gain} \tag{9.64}$$

$|S_{21}|^2$ as the forward gain of the matched network

and

$$G_{pL} = \frac{1 - |\rho_L|^2}{|1 - S_{22}\rho_L|^2} \quad \text{as the output gain}$$

Conjugate matching at input and output, with $\rho_S = S_{11}^*$ and $\rho_L = S_{22}^*$ gives the maximum available gain. Within present assumptions, maximising $|S_{11}|^2$ or G_{pL} is of no consequence to noise figure because both signal and noise are equally affected by them. Maximising G_{ps}, on the other hand, may well degrade noise figure since each has different dependence on source reflection coefficient. In

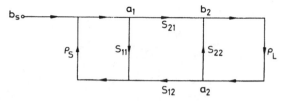

Fig. 9.11 *Flowgraph of two-port for gain calculation*

order to determine the relation between noise figure and gain, their constant loci are plotted on the same Smith chart. We will first find constant gain loci by writing eqn. 9.64 as

$$\frac{1}{G_{ps}}(1 - |\rho_S|^2) = 1 - S_{11}\rho_S - S_{11}^*\rho_S^* + |S_{11}|^2|\rho_S|^2$$

and adding and subtracting $|S_{11}|^2$, $|\rho_S|^2$ on the right-hand side to give

$$\frac{1}{G_{ps}}(1 - |\rho_S|^2) = 1 - |\rho_S|^2 - |S_{11}|^2 + |S_{11}|^2|\rho_S|^2 + |\rho_S|^2 - S_{11}\rho_S - S_{11}^*\rho_S^* + |S_{11}|^2$$

$$= (1 - |\rho_S|^2)(1 - |S_{11}|^2) + |\rho_S - S_{11}^*|^2$$

But from eqn. 9.64 the maximum input gain occurs when $\rho_S = S_{11}^*$, and

$$(1 - |S_{11}|^2) = \frac{1}{G_{ps\,max}}$$

which gives

$$\left(\frac{1}{G_{ps}} - \frac{1}{G_{ps\,max}}\right)(1 - |\rho_S|^2) = |\rho_S - S_{11}^*|^2 \qquad (9.65)$$

Let

$$g = \frac{1}{G_{ps}} - \frac{1}{G_{ps\,max}}$$

$$\rho_S = \rho_R + j\rho_I \quad \text{and} \quad S_{11}^* = S_R + jS_I$$

to give in eqn. 9.65

$$g - g\rho_R^2 - g\rho_I^2 = (\rho_R - S_R)^2 + (\rho_I - S_I)^2$$

or

$$\left(\rho_R - \frac{S_R}{g+1}\right)^2 + \left(\rho_I - \frac{S_I}{g+1}\right)^2 = \frac{g}{g+1} - \frac{g|S_{11}|^2}{(g+1)^2} \qquad (9.66)$$

This is the equation of a circle with centre at

$$u = \frac{S_R}{g+1}, \qquad v = \frac{S_I}{g+1}$$

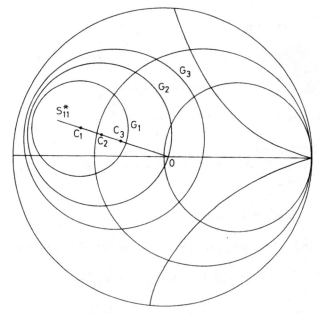

Fig. 9.12 *Constant-gain circles on Smith chart*

and radius

$$r = \sqrt{\frac{g}{g+1}\left(1 - \frac{|S_{11}|^2}{g+1}\right)} \tag{9.67}$$

These constant-gain circles lie on a line from the centre of the Smith chart to the S_{11}^* point, and their distance from the centre is

$$d = \frac{|S_{11}|}{g+1} = \frac{G_{ps}G_{ps\,max}|S_{11}|}{G_{ps\,max} - G_{ps} + G_{ps}G_{ps\,max}} \tag{9.68}$$

At maximum gain $g = 0$, $\rho_S = S_{11}^*$ and $r = 0$, giving a single point at the maximum available gain.

Some representative constant-gain circles shown in Fig. 9.12 have centres C_1 to C_3 on the radial line OS^*, and decreasing circle diameters with increasing gain.

9.11 Noise circles for a linear two-port

We can show, by re-arranging eqn. 9.59, how constant noise figures also form circular loci on a Smith chart. Thus if

$$(G_S - G_M)^2 + (B_S - B_M)^2 = |Y_S - Y_M|^2 \tag{9.69}$$

eqn. 9.59 becomes

$$\frac{F - F_M}{R_N} = \frac{Y_S - Y_M}{G_S} \tag{9.70}$$

where G_S now refers to conductance.

At microwave frequencies, since reflection coefficients are the measurable quantities rather than lumped constants, it is more appropriate to convert eqn. 9.70 to a reflection-coefficient form using the following relationships:

$$y_S = \frac{Y_S}{Y_0}, \qquad y_M = \frac{Y_M}{Y_0} \tag{9.71}$$

where Y_0 is the characteristic admittance of the line between source and two-port input

$$\rho_M = \frac{1 - y_M}{1 + y_M} \qquad \text{source reflection coefficient for} \atop \text{minimum noise figure} \tag{9.72}$$

$$\frac{4g_S}{|1 + y_S|^2} = 1 - |\rho_S|^2 \tag{9.73}$$

where ρ_S is the source reflection coefficient when its normalised conductance is $g_S = G_S / Y_0$.

Eqn. 9.73 can be verified from

$$1 - |\rho_S|^2 = 1 - \left|\frac{1 - y_S}{1 + y_S}\right|^2 = \frac{2(y_S + y_S^*)}{|1 - y_S|^2}$$

and substituting

$$y_S + y_S^* = 2g_S$$

In normalised admittance form eqn. 9.70 is

$$\frac{F - F_M}{R_N} = \frac{|y_S - y_M|^2}{Z_0 g_S} \tag{9.74}$$

where Z_0 is the line characteristic impedance. By addition and subtraction of terms, this can be arranged in the following way:

$$\frac{(F - F_M)Z_0}{R_N} = \frac{|1 + y_S - y_M - y_S y_M - 1 + y_S - y_M + y_S y_M|^2}{4g_S}$$

$$= \frac{|(1 + y_S)(1 - y_M) - (1 - y_S)(1 + y_M)|^2}{(1 - |\rho_S|^2)|1 + y_S|^2}$$

where eqn. 9.73 has been used to substitute for g_S. With $|1 + y_M|^2 = 4/|1 + \rho_S|^2$, we have

$$\frac{(F - F_M)}{4r_N} = \frac{|\rho_M - \rho_S|^2}{(1 - |\rho_S|^2)|1 + \rho_M|^2} \tag{9.75}$$

which is the reflection coefficient form of eqn. 9.59.

Direct comparison with the gain equation is possible if we write

$$N = \frac{(F - F_M)}{4r_N} |1 + \rho_M|^2 \tag{9.76}$$

to give

$$N(1 - |\rho_S|^2) = |\rho_M - \rho_S|^2 \tag{9.77}$$

Noise circles of constant N have centres on a radial line between the centre of the Smith chart and ρ_M at distances

$$d_F = \frac{|\rho_M|}{1 + N} \tag{9.78}$$

and with radii

$$r_F = \sqrt{\left(\frac{N}{N+1}\right)\left(1 - \frac{|\rho_M|}{N+1}\right)} \tag{9.79}$$

These follow by direct analogy with eqn. 9.65 if g replaces N and S_{11}^* replaces ρ_M.

When both gain and noise circles are plotted on the same Smith chart (Fig. 9.13), their intersections give values of reflection coefficient and source impedance for each gain/noise-figure pair; and the selection of an optimum source impedance is a compromise between low noise figure and high gain.

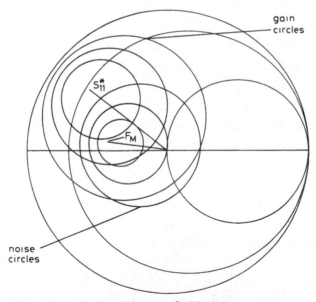

Fig. 9.13 *Constant-gain and noise circles on a Smith chart*

Circles have to be found by measurement for each frequency in the operating bandwidth, but, once known for a particular two-port, they can be rapidly applied to a wide range of designs. Manual methods of measuring so many operating conditions are tedious and costly in the use of skilled operators, but automatic methods, partly based on the algorithms of eqns. 9.65 and 9.75, lower the costs by reducing the time and de-skilling the labour. A full discussion of automatic procedures follows in a later Section, after we have completed some further analyses and described a new method, not yet commercially available, of measuring noise figure at microwave frequencies.

9.12 Noise figure from noise wave

Wave functions are particularly appropriate to microwaves because they relate directly to reflection and transmission coefficients in a system where lumped constants can rarely be determined. Signal/noise ratios are therefore more directly determined if both are expressed as wave functions. Power in a noise wave is associated with a noise temperature T_e, so that eqn. 9.75 can be written as[11]

$$T_e = T_{eM} + \frac{4T_0 r_N |\rho_M - \rho_S|^2}{(1 - |\rho_S|^2)|1 + \rho_M|^2} \tag{9.80}$$

where

$$F = \frac{T_e + T_0}{T_0} \quad \text{and} \quad F_M = \frac{T_{eM} + T_0}{T_0}$$

The total noise in a lumped-constant A, B, C, D parameter circuit was found by short-circuiting the input port. With a noise-wave equivalent circuit a matched load is substituted for the device and the total noise is the sum of the noise waves incident on it. Fig. 9.14 shows a noise source with reflection coefficient ρ_S and noise-wave output b_{Ns} connected to the input of a noiseless network whose internal noise is represented by forward and reverse waves a_N and b_N, respectively. The total noise-wave input to a matched load is

$$a_{Ns} = a_N + \rho_S b_N + b_{Ns} \tag{9.81}$$

Since a_N and b_N are partly correlated in the network and b_{Ns} is uncorrelated with either, the input power is

$$\begin{aligned}
\overline{|a_{Ns}|^2} &= \overline{|a_N + \rho_S b_N + b_{Ns}|^2} \\
&= \overline{|a_N|^2} + |\rho_S|^2 \overline{|b_N|^2} + \overline{|\rho_S a_N b_N^* + \rho_S^* a_N^* b_N|} + \overline{|b_{Ns}|^2} \\
&= \overline{|a_N|^2} + |\rho_S|^2 \overline{|b_N|^2} + 2 \operatorname{Re}(\rho_S \overline{a_N^* b_N}) + \overline{|b_{Ns}|^2}
\end{aligned} \tag{9.82}$$

Each power term is associated with a noise wave propagating in a Z_0 matched

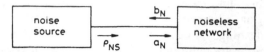

Fig. 9.14 *Equivalent noise waves at input to a noiseless network*

transmission line with noise temperatures defined by the system bandwidth through

$$\overline{|a_{Ns}|^2} = kT_{Ns}\,\Delta f$$

$$\overline{|a_N|^2} = kT_a\,\Delta f$$

$$\overline{|b_N|^2} = kT_b\,\Delta f \tag{9.83}$$

The complex quantity $\overline{a_N^* b_N}$ has a noise temperature T_c, with phase ϕ_c, if

$$\overline{a_N^* b_N} = kT_c\,\Delta f \exp j\phi_c \tag{9.84}$$

Therefore, if $\rho_S = |\rho_S|\exp j\phi_S$

$$\mathrm{Re}\,(\rho_S \overline{a_N^* b_N}) = kT_c\,\Delta f |\rho_S| \cos(\phi_S + \phi_c) \tag{9.85}$$

Finally, since $|b_{Ns}|$ is the power from the source into a Z_0 matched line and $kT_S\,\Delta f$ is the power into a conjugate match,

$$\overline{|b_{Ns}|^2} = (1 - |\rho_S|^2)kT_S\,\Delta f \tag{9.86}$$

where T_S is the noise temperature of the source.

Substitution of eqns. 9.83—9.86 in eqn. 9.82 gives, for the noise temperature at the input to the noiseless network,

$$T_{Ns} = T_a + |\rho_S|^2 T_b + 2T_c|\rho_S|\cos(\phi_S + \phi_c) + T_S(1 - |\rho_S|^2) \tag{9.87}$$

Noise properties of the two-port are now expressed as T_a, T_b, T_c and ϕ_c instead of G_u, R_N, G_c and B_c for the lumped constant model. These unknown parameters can be found from calibration measurements with known sources. Thus if we set $|\rho_S|^2 = 1$, $T_S(1 - |\rho_S|^2)$ is zero and

$$T_{Ns} = T_a + T_b + 2T_c\cos(\phi_S + \phi_c) \tag{9.88}$$

Now if ϕ_S is made variable and T_{Ns} is measured on a power meter, the mean value of T_{Ns} is $T_a + T_b$ and the amplitude of the variable term is $2T_c$. These are illustrated in Fig. 9.15. A variable ϕ_S noise source with $|\rho_S| = 1$ can be made by weakly coupling (-40 dB) a noise source to a variable length of shorted line connected at the input to a DUT. This is illustrated in Fig. 9.16. T_a and T_b can be found by next placing a matched load in the input transmission line as shown in Fig. 9.17, where the 60 dB padding effectively matches the noise source to the network. If $|\rho_S| = 0$, only T_a and T_S are left in eqn. 9.87 to give

$$T_{Ns} = T_a + T_S \tag{9.89}$$

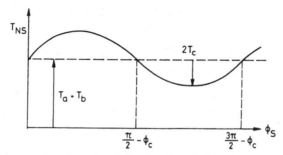

Fig. 9.15 *Noise temperature from input-line length variation*

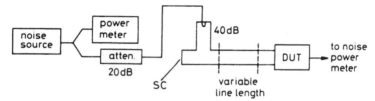

Fig. 9.16 *Noise test set with short-circuit source*

Fig. 9.17 *Noise test set with matched source*

Since T_S can be found from the padded noise temperature of the source, T_a, and hence T_b, are also known. The fourth constant ϕ_c follows from the crossover points of the oscillating and mean levels in Fig. 9.15.

Even though calibration constants define a noisy network, they do not give directly the effective input noise temperature. To find T_e we first measure noise at the network and divide by the network gain to give, according to eqn. 9.87,

$$T_{Ns} = T_a + |\rho_S|^2 T_b + 2T_c |\rho_S| \cos (\phi_S + \phi_c) + T_S(1 - |\rho_S|^2)$$

Next we remove the network and measure the noise temperature from the source. This is

$$T'_{Ns} = T_S(1 - |\rho_S|^2) \tag{9.90}$$

Finally we increase source noise temperature to make T'_{Ns} equal to the measured temperature with the network in place. If T_N is the increase in source temperature,

$$T_{Ns} = T_S(1 - |\rho_S|^2) + T_N(1 - |\rho_S|^2) \tag{9.91}$$

or

$$T_N(1 - |\rho_S|^2) = T_a + |\rho_S|^2 T_b + 2T_c |\rho_S| \cos (\phi_S + \phi_c) \tag{9.92}$$

T_N is therefore the effective available noise temperature at the input to the equivalent noiseless network. Therefore

$$T_e = T_N = \frac{T_a + |\rho_s|^2 T_b + 2T_c |\rho_s| \cos(\phi_s + \phi_c)}{(1 - |\rho_s|^2)} \tag{9.93}$$

This is the equivalent of eqn. 9.80, but is not in a form suitable for deriving minimum noise temperature and optimum source-reflection coefficient. However, it is possible to re-arrange the original expression as

$$T_e = T_{eM} + T_d \frac{|\rho_M - \rho_s|^2}{1 - |\rho_s|^2} \tag{9.94}$$

with $T_d = 4T_0 r_N / |1 + \rho_M|^2$ and T_{eM} as the minimum noise temperature, and to identify terms with eqn. 9.93 as follows:[11]

$$T_d = \frac{1}{2}(T_{ab} + \sqrt{T_{ab}^2 - 4T_c^2}), \qquad T_{ab} = T_a + T_b$$

$$T_{eM} = T_d - T_b, \qquad F_M = \frac{T_{eM} + T_0}{T_0}$$

$$|\rho_M| = \frac{T_c}{T_d}, \qquad \phi_M = \pi - \phi_c \tag{9.95}$$

Equivalence of eqns. 9.93 and 9.94 shows that a calibration of noise waves leads through eqns. 9.95 to a direct determination of T_{eM}, without the necessity of finding the minimum noise figure by a trial-and-error variation of source match. However, current practice, even at microwave frequencies, continues to follow the latter method.

9.13 Noise-figure measurement

Noise figure is often found from noise power at the output of a device under test when noise sources of known power are successively connected at the input. Measurement success is influenced by a number of factors, such as input match and reading-to-reading instrument stability, but above all it depends on the calibration accuracy of the noise source. As we shall see, modern noise sources depend on avalanche breakdown in a carefully designed reverse-biased PN junction.[16] The two input noise temperatures correspond to the ON and OFF states of the diodes. While OFF, the diode is at room temperature and often referred to as a cold source with temperature T_c. Hot electrons in the ON state increase the effective temperature by one or two decades. This hot source, at temperature T_h, has excess noise ratios (ENRs) from 5 to 40 dB, depending on the application. Generally speaking the lower the noise figure to be measured, the lower the ENR for greatest accuracy.

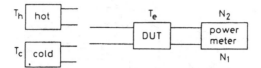

Fig. 9.18 *Noise temperature measurement with hot and cold noise sources*

Noise at the output is the amplified sum of two parts: one due to the source, the other from the DUT itself. The practice of referring both to the input of the DUT allows them to be added as noise temperatures. Thus if, in Fig. 9.18, N_1, the noise power at the cold temperature, is divided by N_2 for the hot temperature, a Y factor is defined as[21]

$$Y = \frac{N_2}{N_1} = \frac{kT_eB_NG + kT_hB_NG}{kT_eB_NG + kT_cB_NG}$$

$$= \frac{T_e + T_h}{T_e + T_c} \tag{9.96}$$

with symbols as defined in Section 9.5. On re-arranging we have for T_e

$$T_e = \frac{T_h - YT_c}{Y - 1} \tag{9.97}$$

and on substituting in eqn. 9.19, we have for the noise figure,

$$F = \frac{T_h - YT_c + YT_0 - T_0}{T_0(Y - 1)} \tag{9.98}$$

Neither measurement is made at the standard temperature, but noise figure is referred to it in order to compare devices against a common standard.

9.14 Solid-state noise source

A PN diode in reverse avalanche breakdown exhibits an excess noise power with available power inversely proportional to diode current. Unless special care is taken, the onset of breakdown and the resulting noise spectrum varies from diode to diode, and even from successive switchings of the same diode. If the breakdown area is made small and is surrounded by a guard ring, noise temperature can be controlled by bias current to an accuracy[15-17] of ± 0.1 dB. In the arrangement shown in Fig. 9.19 the guard ring has a higher breakdown voltage than the anode, thus confining it to a small central area.

The broad-band, typically 10 MHz to 18 GHz, range of avalanche sources imposes a severe matching problem, because, as we have already seen, noise figure is a sensitive function of source impedance. For convenience and efficiency, it is therefore desirable to have a match as near Z_0 as possible, usually

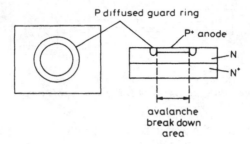

Fig. 9.19 *PN-junction noise diode construction*

50 or 70 Ω. This is achieved by an embedding circuit illustrated in Fig. 9.20, in which the diode is mounted in a microstrip circuit matched to a 3·5 mm coaxial output. A 6 dB pad is placed between the external connector and the noise cartridge to give an extra 12 dB return loss. Typical source match for a Hewlett–Packard 346B noise source is shown in Fig. 9.21.

Each noise source is calibrated against a standard traceable to fundamental standards at national standards laboratories. Hot and cold standards, usually in the form of heated resistors, are often used as transfer standards by manufac-

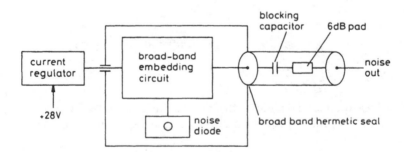

Fig. 9.20 *Embedding and output circuit*

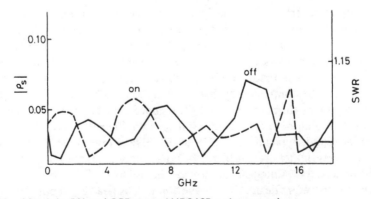

Fig. 9.21 *Match in ON and OFF states (HP346B noise source)*

turers. These, in turn, are periodically checked at the national standards laboratories. Only selected solid-state sources are directly comparable with traceable standards, and they in turn are used for production-line calibration. Hot and cold precision standards will be discussed in a later Section.

9.15 Automatic measurement

In the automatic instrumentation system illustrated in Fig. 9.22, a microprocessor controls a synthesiser, local-oscillator tuning, amplifier auto-ranging and noise-source switching.[15] It also prompts the operator to connect in the DUT or make a through connection for direct measurement of source noise. Final detection is at 20 MHz, where signal conditioning can be most effective in maintaining accuracy over a wide band. The first mixer, with a microwave synthesiser as local oscillator, converts the noise signal to the instrument baseband of 10—1500 MHz. Subsequent down-conversion takes place under microprocessor control to ensure frequency accuracy and linearity. Thus amplitude compression in the amplifiers is avoided by auto-ranging them through distributed attenuators, controlled in a feedback-level detection circuit. The final noise-power detector is a Schottky diode with a dynamic range from 0 to −22 dBm and deviation from linearity of less than 0·04 dB. At baseband the input SWR of the noise meter is < 1·5, its noise figure is 3·5 dB at 10 MHz, rising to 5·0 dB at 1500 MHz, and its gain is 20 dB ± 1 dB.

If the noise-meter noise figure is F_2, its contribution to overall measured noise is reduced by the gain G of the DUT, to give an overall noise figure of

$$F_{12} = F_1 + \frac{F_2 - 1}{G} \tag{9.99}$$

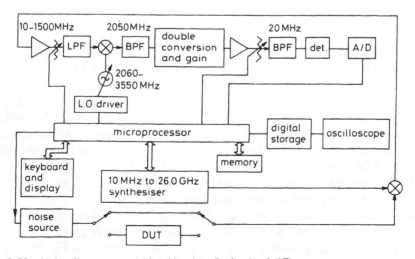

Fig. 9.22 *Noise-figure meter (After Hewlett–Packard, ref. 15)*

where F_1 is the required noise figure of the device under test. This causes a worst-case second-stage contribution of 0·02 at 1500 MHz for a DUT with a gain of 20 dB, or +0·05 dB error in F_1 when measuring a 3 dB noise figure.

In calibration the test ports are connected straight through to the noise source, whilst at each frequency, outputs at two noise levels are determined and stored automatically. F_2 is then calculated from the Y factors and stored for calculating second-stage noise during a similar set of measurements with the DUT in place. From these sets of results both gains and noise figures are available for display or further processing via an interface bus. Gain measurement using the source noise as signal has the advantage that the bandwidth is automatically the noise bandwidth B_N.

Automatic measurements have the following advantages:

- Source ENR can be stored at a number of frequencies over the operating band and interpolated between these points.
- Room temperature can be monitored to determine the cold source temperature accurately.
- Calibration of the instrument, with the device under test removed, allows storage of noise figure and noise level for second-stage corrections and DUT gain determination.
- Signal averaging improves measurement accuracy.
- Both gain and noise figure are available in virtually real time, thus allowing rapid adjustments of source matching, e.g. by means of a slide tuner.

9.16 Other noise sources

Noise sources have been classified under the following headings:[18]

- Thermal
- Diode
- Gas discharge
- Solid state

A good description and comparison of the first three types can be found in the literature.[22] Thermal sources we shall deal with later, since they are the basis of the fundamental standards. Diode sources rely on shot noise in a temperature-limited thermionic diode, but have little or no application at microwave frequencies. The solid-state avalanche diode has already been discussed, but we must still consider gas-discharge sources, which, until recently, were the most accurate and widely used of microwave sources.[19]

There are two common types of gas-discharge tube: one for coaxial systems with a helix wound tightly around the tube; the other simply mounted across a waveguide at a sloping angle for optimum match, as shown in Fig. 9.23. The

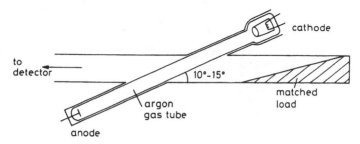

Fig. 9.23 *Gas-discharge noise source in waveguide*

former type, in Fig. 9.24, has been found to generate transient spikes in the noise output, which can damage small-geometry solid-state devices. One end of the helix is terminated and the other provides an output. The ENR of these devices depends on the effective electron temperature of the discharge plasma. Tube radius, wall thickness, concentricity, gas pressure and current all affect plasma stability and resultant ENR. SWR is less than 1·05 in both ON and OFF states, giving an excellent noise source for high ENRs (~ 15·5 dB). The stability of gas sources has proved to be < 0·005 dB over a two-year period with 0·02 dB correlation with NBS standards. The tubes become physically hot, operating at 86 V and 150 mA, which causes a problem of heat removal if a cold, OFF, immediately follows a hot measurement.

Instrumentation provided with these sources relies on measuring the hot/cold ratio, for insertion in a Y-factor calculation. Prior to microprocessor control a degree of automation was achieved by controlling the instrument IF gain in relation to one of the two noise measurements, to give a single output reading.[18]

Thermal noise sources usually refer to the hot and cold terminations that are at the centre of primary standards. They form a reference for all secondary standards, and are available in standards laboratories and other places where high-quality noise measurements are made, particularly for low-noise amplifiers with noise figures less than 1 dB. Fig. 9.25 illustrates the essential features of a hot-and-cold-load thermal standard.[19] The load, a thin-film resistive element at low frequencies and a conventional waveguide load at higher frequencies, is isolated from the interface adaptor by a uniform transmission line made of

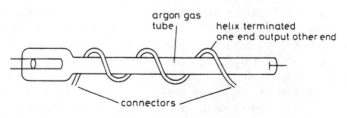

Fig. 9.24 *Gas-discharge noise source in helix for coaxial connection*

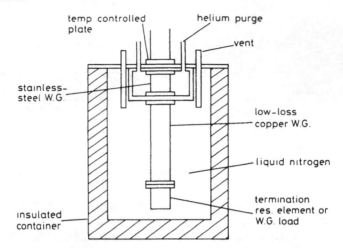

temp controlled plate

helium purge

vent

stainless-steel W.G.

low-loss copper W.G.

liquid nitrogen

termination res. element or W.G. load

insulated container

Fig. 9.25 *Low-temperature noise standard*

copper-plated stainless steel with distributed temperature and transmission-line factors. Construction techniques are similar for hot and cold loads, the former having accurate thermostatically controlled heaters at $373 \cdot 1°K$ and the latter being cooled to $77 \cdot 36°K$ in liquid nitrogen. Moisture condensation in the isolation section can cause problems with the cold load, but this is overcome by carefully adjusted heating at the output end of the section.

Computer modelling of temperature distributions and waveguide losses are necessary, and losses have to be measured to better than $0 \cdot 002$ dB. Parametric-pressure and ambient-temperature changes must also be corrected if extreme accuracy is essential. A good thermal source has an accuracy better than $\pm 1 \cdot 5°K$, provided the previously outlined precautions are given proper attention.

9.17 Very large noise figures

When the DUT noise temperature is very large, change of temperature, be it from hot or cold source, makes little difference at the output, and the Y-factor becomes inaccurate. A better estimate can be obtained from the gain/frequency characteristic of the DUT, found by using a signal generator at the input and a suitable detector at the output. Integrating the gain as in eqn. 9.17, we have the gain/noise bandwidth $G_0 B_N$, and can then calculate the quantity $k G_0 B_N T_0$ when the input noise is due to a resistor at temperature T_0. Next, with the resistor connected at the input, we measure N_0, the output noise from the amplifier, to calculate the noise figure according to eqn. 9.15 from the ratio

$N_0/kG_0B_NT_0$. The same principle is used in radiometers to measure the temperatures of radiating bodies, such as a hot steel furnace or the cold sky.

9.18 Noise measurement by spectrum analyser

With due care it is possible to estimate noise figure with a spectrum analyser.[20] A typical test arrangement is shown in Fig. 9.26. Noise figure is

$$F = \frac{N_0}{G_D G_a k T_0 B_N} \tag{9.100}$$

where N_0 is the output noise from the pre-amplifier in the Figure, G_D is the DUT gain, G_a the pre-amplifier gain, B_N the analyser's noise bandwidth and T_0 is the standard temperature of an input matched load. The pre-amplifier is optional depending on whether the noise level output from the DUT is above the analyser sensitivity. In using it, care must be taken to eliminate second stage noise, or it may be necessary to make a separate calibration with the DUT removed to establish a figure for second-stage effect. Noise figures so far described, with the possible exception of radiometer-type devices, have been spot values, i.e. specified over a limited frequency band.[9] Such spot noise figures are easily selected with a spectrum analyser by adjustment of its IF bandwidth to the required B_N. Since noise power is usually proportional to bandwidth, the maximum IF bandwidth is limited by the onset of gain compression caused by too high a noisy-density bandwidth product (see Appendix 1 for further discussion of gain compression in spectrum analysers). In decibels the noise-figure equation is

$$F = [N - (G_D + G_a) - 101 \log B_N + 174 + C] \text{ decibels} \tag{9.101}$$

where N is the noise level observed on the analyser in bandwidth B_N, 174 is the kT_0 term, $G_D + G_a$ is the system gain and C is a correction constant, usually between 1 and 2 dB to allow for the analyser detection law, including the effects of a video filter to integrate the output for post-detection display, with typical bandwidth one-hundredth that of the IF. C is usually quoted by the manufacturer. The method is not very accurate and is not suitable for noise figures of less than 3 dB.

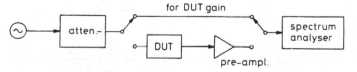

Fig. 9.26 *Noise figure with a spectrum analyser*

9.19 Noise in mixer image response

An important, and often troublesome, noise-figure calculation occurs in mixer applications. The IF channel is coupled by the down-conversion process to a signal and image channel, each of bandwidth B on either side of the local-oscillator frequency. Noise appears in both these channels, but the signal in only one. In Fig. 9.27, signal and image frequencies f_S and f_I, are separated from the local-oscillator frequency f_{LO}, by $\mp f_{IF}$, respectively. Both are heterodyned to f_{IF}. If G_S and G_I are the conversion gains for signal and image channels, the output signal/noise ratio is

$$\frac{S_0}{N_0} = \frac{S_0}{kB(T_0 + T_e)(G_S + G_I)} \tag{9.102}$$

where T_e is the effective noise temperature, assumed to be the same in both channels. The input signal/noise ratio is

$$\frac{S_i}{N_i} = \frac{S_i}{kT_0 B} \tag{9.103}$$

and the noise figure is

$$F = \frac{S_i/N_i}{S_0/N_0} = \frac{N_0}{G_S N_i}$$
$$= \frac{kB(T_0 + T_e)(G_S + G_I)}{kBT_0 G_S}$$

or

$$F = \frac{T_0 + T_e}{T_0}\left(1 + \frac{G_I}{G_S}\right) \tag{9.104}$$

This is sometimes called the single-sideband noise figure because the signal is restricted to f_S.

In practice, at microwave frequencies, the wide-band noise source in the noise meter is both 'signal' and noise in a gain/noise figure determination. The observed result is equivalent to putting $G_S + G_I$ instead of G_S in the denominator

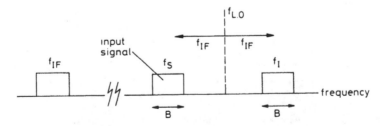

Fig. 9.27 *Showing image and signal noise channels*

in eqn. 9.104 because 'signal' and noise occupy the same total bandwidth and the observed result is referred to as the double-sideband noise figure.[8] If $G_I = G_S$, the measured noise figure is 3 dB lower than the correct value under normal operating conditions, that is, when the signal appears in only the signal channel.

9.20 Conclusion

We began by reviewing early work on fluctuation noise, defining noise figure and temperature and applying these ideas to understanding the behaviour of two-port networks. We continued with a description of noise sources, and concluded by showing how noise figure is measured in a variety of ways. Noise has been a desired output from sources intended for noise measurement, but signal oscillators are also sources of undesired noise, as we shall see in the next Chapter.

9.21 References

1 JOHNSON, J. B.: 'Thermal agitation of electricity in conductors', *Phys. Rev.*, July 1928, pp. 97–109

2 SCHOTTKY, W.: *Ann. der Phys.*, 1918, **57**, p. 541

3 NYQUIST, H.: 'Thermal agitation of electric charge in conductors', *Phys. Rev.*, July 1928, pp. 110–113

4 PARKER, P.: 'Electronics' (Edward Arnold, 1950) Chap. 18

5 RACK, A. J.: *Bell Syst. Tech. J.*, 1938, **17**, p. 592

6 ADIR BAR-LEV, : 'Semiconductors and electronic devices' (Prentice-Hall, 1984) pp. 309–315, 148

7 GRIFFIN, E. J.: 'Detectors and detection for measurement', *in* BAILEY, A. E. (Ed.): 'Microwave measurement' (Peter Peregrinus, 1985) p. 91

8 PASTORI, W. E.: 'Image and second stage corrections resolve noise figure measurement confusion', *Microwave Syst. News*, May 1983, pp. 67–86

9 HAUS, H. A.: 'Representation of noise in linear two-ports', *Proc. IRE.*, Jan. 1960, pp. 69–74

10 LIAO, S. Y.: 'Microwave devices and circuits' (Prentice-Hall, 1980) Chap. 6

11 MEYS, R. P.: 'A wave approach to the noise properties of linear microwave devices', *IEEE Trans.*, 1978, **MTT-26**, pp. 34–37

12 FUKUI, H.: 'Available power gain, noise figure and noise measure of two-ports and their graphical representation', *IEEE Trans.*, 1966, **CT-13**, pp. 137–142

13 KUHN, N. J.: 'Curing a subtle but significant cause of noise figure error', *Microwave J.*, June 1984, pp. 85–98

14 'S-parameter design'. Hewlett-Packard Application Note 154, April 1972

15 SWAIN, H. L., and COX, R. M.: 'Noise figure meter sets record for accuracy, repeatability and convenience', *Hewlett–Packard J.*, April 1983, pp. 23–34

16 HAITZ, R. H.: 'Noise in self-sustaining avalanche discharge in silicon: Low frequency noise studies', *J. Appl. Phys.* 1967, **38**, pp. 2935–2946

17 HAITZ, R. H., and VOLTMER, F. W.: 'Noise in self-sustaining avalanche discharge in silicon: Studies at microwave frequencies', *J. Appl. Phys.*, 1968, **39**, pp. 3379–3384

18 LAVEGHETTA, T. S.: 'Handbook of microwave testing' (Artech House, 1981) pp. 142 and 148

19 ARNOLD, J.: 'Amplifier evaluation techniques' *in* MORGAN, D. V., and HOWES, M. J. (Eds.): 'Microwave solid state devices and applications' (Peter Peregrinus, 1980) p. 153

20 'Spectrum analysis ... noise measurements'. Hewlett–Packard Application Note 150-4, April 1974
21 'Fundamentals of RF and microwave noise figure measurements'. Hewlett–Packard Application Note 57-1, July 1983
22 MILLER, C. K. S., DAYWITT, W. C., and ARTHER, M. G.: 'Noise standards measurements and receiver noise definitions', *Proc. IEEE*, 1967, **55**, pp. 865–877

9.22 Examples

1   A noisy amplifier has input resistance R_i and an equivalent noise resistance $R_N = R_i/2$. A signal source with output resistance R_S is connected at the input. Show, by means of a graph, how the noise at the amplifier output varies with the source resistance. Explain why this is different from the graph in Fig. 9.1*b*.

2   A receiving antenna has an effective noise temperature of 20°K. The first stage of the receiver has a gain of 15 dB and the second stage a gain of 10 dB with an effective noise temperature of 3000°K. If the total (equivalent plus antenna) input noise should not exceed 300°K, calculate the effective noise temperature and noise figure of the first amplifier stage.

3   A 50 Ω line connects a resistive load, with VSWR of 1·2 and a temperature of 290°K to the matched input of an amplifier with a noise figure of 3 dB. What is the noise temperature at the input to the receiver? What kind of load would give the minimum noise temperature?

4   A microwave transistor amplifier has an input reflection coefficient of $0.45 \underline{/-142°}$ at 2 GHz. The minimum noise figure is 3·5 dB for a source reflection coefficient of $0.4 \underline{/170°}$. The noise circle for $F = 4.5$ dB has a radius of 0·1. What is the radius when $F = 6.5$ dB?

5   For the amplifier of the previous question, estimate the value of the noise figure for the condition of maximum gain.

6   The minimum noise figure is 3·5 dB when the source reflection coefficient at the input to an amplifier is $0.4 \underline{/164°}$. If the source reflection coefficient is changed to $0.3 \underline{/164°}$, the noise figure is 4·8 dB. Write down the lumped-constant noise-figure equation for the amplifier.

7   The mean SWR of a noise source is 1·1 when ON and 1·15 when OFF. If it is used in a Y-factor measurement of an amplifier with an input SWR of 2, what is the maximum uncertainty in Y and in the values of the hot and cold temperatures? Comment on the effect of these uncertainties on the derived noise figure.

8   The equivalent noise bandwidth of an amplifier is 1 MHz and its midband gain is 60 dB. With no signal input, the output noise at room temperature is 1 mW. What is the noise figure of the amplifier?

9   The ENR of a noise source is 15·5 dB. Find the noise figure if the measured output noise from a DUT is 2 mW with the noise source on and 0·1 mW when off. What assumptions have you made in your calculation?

Chapter 10

Frequency stability and measurement

10.1 Noise in signal sources

An ideal source provides a single spectral line with zero width and finite energy. This strange mathematical world does not exist in practice, because real oscillators produce lines of finite width and have random amplitude, frequency and phase fluctuations. The spectral line, sometimes called the carrier, may also be accompanied by harmonics, usually generated by the non-linear behaviour of active components in the oscillator circuit. These are deterministic and not the subject of this discussion. Random variations modulate the carrier to produce noise sidebands in a bandwidth determined by the frequency range of the fluctuations. These noise sidebands rise above the general thermal background noise and tend to peak as the frequency approaches the carrier. Amplitude fluctuations can be minimised in a high-quality oscillator, leaving frequency and phase fluctuations as the main causes of source noise. Low-frequency fluctuations cause sidebands with very small separation from the carrier. Indeed, for very low-frequency variations the main observed effect is a long-term drifting of the carrier frequency, usually about a mean level, but also uniformly increasing or decreasing, particularly during a warm-up period. Short-term fluctuations show sidebands with increasing spectral spread with decreasing fluctuation times. A rough division between long- and short-term noise is a mean fluctuation rate of 1 Hz. These ideas are represented in Fig. 10.1, in which f_c is the carrier or nominal frequency of the oscillator. It is drifting slowly about its mean position or in a unidirectional manner, but it is also accompanied by the noise spectrum generated by short-term fluctuations. The amplitude of the spectrum is symmetrical and structured, with the predominant feature being a decrease to the thermal background level with increasing separation from the carrier frequency. Line spectra in the sidebands are not due to phase noise, but are typically a form of spectral leakage observed in real oscillators whose environment includes control circuits, perhaps for frequency selection

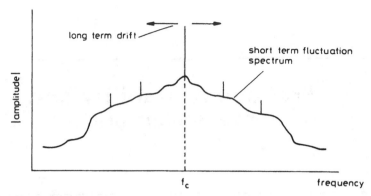

Fig. 10.1 *Close-to-carrier noise*

in phase-lock loops, capable of generating harmonic spikes near the source frequency.

We will begin with a fuller discussion of the types of short-term noise before comparing these with long-term drift.

10.2 Short-term noise

Short-term noise processes can be recognised by the frequency dependence of their spectral densities.[1] These in turn are related to the statistical form of their variations, and distinguished by their frequency dependency. In the following discussion we define each type in a composite idealised symmetrical spectrum illustrated by the upper-sideband noise levels and gradients in Fig. 10.2.

Random Walk FM ($1/f^4$): Environmental effects, such as mechanical shock, vibration or temperature changes cause low-frequency changes in carrier frequency, seen as a spectral distribution very close to the carrier. At the lowest frequencies this becomes long-term drift and is very difficult to separate from the carrier during measurement. It varies as the fourth power of the frequency separation from the carrier.

Flicker FM ($1/f^3$): This is not well understood physically, but appears to be related to the physical properties of active resonant circuits, or other parts of the electronics, or even to environmental factors. It is common in high-quality oscillators often mixed with other types, and varies inversely as the third power of the frequency from the carrier.

White FM ($1/f^2$): This appears in passive-resonator frequency standards with slave oscillators.

Flicker phase modulation ($1/f$): Usually related to a physical resonance mechanism, but also to noisy electrons. It is the most common noise in high-quality oscillators, chiefly because amplification involving electronics components is nearly always necessary.

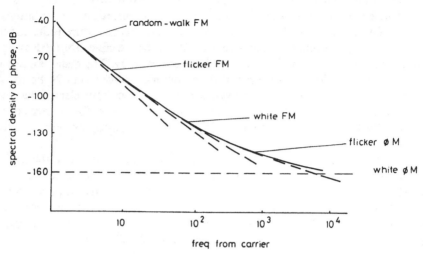

Fig. 10.2 *Types of carrier phase noise*

White phase modulation (f^0): This is broad-band noise, usually of similar origin to $1/f$ noise, but it can be kept to minimal levels by careful amplifier design and narrow-band filtering at the output.

Practical oscillators do not usually have all these noise types, neither are they so well distinguished by their spectral separation in the sidebands. Even though noise types are a mixture of frequency- and phase-modulation sidebands, they can be expressed in terms of either a frequency or a phase variance. For instance, short-term frequency accuracy refers to phase or frequency variations occurring in time intervals significantly less than 1 s. A variance, found by averaging frequency fluctuations from the carrier over no more than 1 s, gives a short-term fractional frequency error $\sqrt{(\Delta f/f_0)^2}$. A good crystal-controlled oscillator has typical short-term variance in the range 2×10^{-9} to 1×10^{-11}.

10.3 Long-term noise

This is sometimes called long-term frequency drift, and is due to temperature changes and line-voltage variations in all oscillators, and to crystal aging in quartz oscillators. The three basic quartz-oscillator types are distinguished by the temperature control of their operating environments. These are room temperature, temperature compensated and oven controlled.[2]

Room temperature crystal oscillators are manufactured to have a minimum frequency change over a given temperature range by proper choice of crystal cut. Temperature compensation is achieved by adding external components to the internal equivalent lumped L, C and R of a crystal with temperature co-efficient selected to compensate for the temperature variations in the internal elements. Oven control seeks to maintain a constant temperature independent of the external environment of the oscillator. It can be a simple ON/OFF system or an analogue proportional control for greater accuracy. The disadvantage of oven control is the long warm-up period, sometimes longer than 24 hours.

Line-voltage variations, both on oven control and oscillator electronics, are, through design, not greater in their effects than other factors for changes of up to 10%. Crystal aging is an unavoidable occurrence causing an accumulative drift in frequency with time.

Since the time scale for long-term effect is hours, days or months, the drift is too low for spectral effects, but it can still be expressed as a frequency variance normalised to the carrier frequency, provided the averaging time is appropriately chosen. In the case of accumulative drift it is always necessary to express it over a time interval. Table 10.1 shows a comparison of the long- and short-term accuracies given as $\sqrt{(\Delta f/f_0)^2}$ for each type of oscillator.

High-quality instruments, such as synthesisers or frequency counters, have oven-controlled crystal standards, particularly when used to measure phase-noise performance of highly stable microwave sources, such as high-power-Doppler radar transmitters or local oscillators in radars and multi-channel communications receivers.

Table 10.1   *Long- and short-term accuracies expressed as* $\sqrt{(\Delta f/f_0)^2}$

Type of instability	Room temperature oscillators	Temperature compensated	ON/OFF ovens	Proportional ovens
Temperature $(0°—50°C)$	$< 2{\cdot}5 \times 10^{-6}$	$< 5 \times 10^{-7}$	$< 1 \times 10^{-7}$	$< 7 \times 10^{-9}$
Line-voltage 10% change	$< 1 \times 10^{-7}$	$< 5 \times 10^{-8}$	$< 1 \times 10^{-8}$	$< 1 \times 10^{-10}$
Aging	$< 3 \times 10^{-7}$ per month	$< 1 \times 10^{-7}$ per month	$< 1 \times 10^{-7}$ per month	$< 1{\cdot}5 \times 10^{-8}$ per month $< 5 \times 10^{-10}$ per day
Short-term 1 s average	$< 2 \times 10^{-9}$ RMS	$< 1 \times 10^{-9}$ RMS	$< 5 \times 10^{-10}$ RMS	$< 1 \times 10^{-11}$ RMS

Long-term instabilities have an effect on system performance, which can be controlled by careful environmental design, but the spectral spread of short-term fluctuations can be very damaging to radar detection or to inter-channel noise in communications. It is therefore essential that supplier and user be able to measure short-term phase noise as accurately and consistently as possible. To explain how this can be done, we begin with some general theory of phase noise.

10.4 Theory of phase-noise measurement

Because of the great difference in power between the carrier and sideband noise levels, typically $\sim 100\,\mathrm{dB}$, special techniques are necessary to separate them when making noise measurements. There is an equivalence between frequency and phase modulation that allows the source noise to be expressed as a spectral density, either of fractional frequency variation or of phase fluctuations. In the former case we first measure the average fractional frequency fluctuation by mixing the carrier with a reference oscillator offset by a frequency f_b from the carrier, to get a fluctuating difference, or IF, output centred on f_b. If the output is fed to a frequency counter and the fluctuations δf are averaged over an interval much greater than $1/f$, where f is a selected upper frequency limit in the sideband, repeated measurement gives the frequency fluctuation in sidebands between the carrier and f. By varying the averaging time interval, the fluctuation can be found as a function of time. A special form of this relationship is called the Allan variance, and we shall see in Section 10.6 that it has a Fourier transform equal to the spectral density of frequency fluctuation $S_y(f)$.

Shorter averaging times extend the observations to fluctuations with greater offsets in the noise sidebands, and can lead to difficulties due to the tendency for decreasing noise levels with increasing offsets. For averaging times of less than about 1 s, greater accuracy is obtained by measuring phase rather than frequency fluctuations. We will show that, for small phase fluctuations, phase-fluctuation density can be derived from the noise-power spectral density; and that the difficulty of measuring relatively very small noise powers near to a high-power carrier is overcome by mixing with a reference oscillator in phase quadrature to suppress the carrier. Finally, as we shall see later, there is also a simple relationship between the spectral densities of fractional frequency fluctuation and phase fluctuation. The remainder of this Section is concerned with the theoretical relationship between noise-power density and phase-noise density.

To derive a measurable expression for phase noise we begin with a sine function with amplitude and phase errors as[4]

$$V(t) = [V_0 + \delta V(t)] \sin [\omega_0 + \phi(t)] \tag{10.1}$$

where

V_0 = nominal peak amplitude of carrier
$\delta V(t)$ = fluctuations in peak amplitudes
ω_0 = nominal fundamental frequency
$\phi(t)$ = fluctuation of phase from nominal

Since amplitude fluctuations are negligible in high quality oscillators, $\delta V(t)$ will be neglected in the following expansion of eqn. 10.1. Therefore

$$V(t) = V_0 \sin(\omega_0 t) \cos \phi(t) + V_0 \sin \phi(t) \cos(\omega_0 t) \tag{10.2}$$

This can be further simplified, if $\phi(t)$ is sufficiently small, to give

$$V(t) \approx V_0 \sin(\omega_0 t) + V_0 \phi(t) \cos(\omega_0 t) \tag{10.3}$$

Certainly, if the phase fluctuations arise from a frequency-modulation process with modulation index $\delta f/f$, where δf is the equivalent frequency fluctuation at offset frequency f in the sideband, then the phase variations, which are also $\delta f/f$, decrease with increasing f. In general, phase variations may be assumed small to within 1 Hz of the carrier, or alternatively for averaging times[3,6,7] less than 1 s.

Eqn. 10.3 is a time-domain representation of the carrier and phase noise. The equivalent frequency-domain expression can be found through Fourier transformation, or more directly by first considering the phase fluctuation to be deterministic and of the form

$$\phi(t) = \phi \sin \omega t \tag{10.4}$$

and by expanding eqn. 10.1 to give

$$V(t) = V_0\{J_0(\phi) \sin \omega_0 t + J_1(\phi)[\sin(\omega_0 + \omega)t - \sin(\omega_0 - \omega)t] + \cdots\} \tag{10.5}$$

where $J_0, J_1, \ldots$ are Bessel functions which, for small arguments, are

$$J_0(\phi) \approx 1, \qquad J_1(\phi) \approx \phi/2$$

to give, for eqn. 10.5,

$$V(t) \approx V_0 \sin \omega_0 t + V_0 \frac{\phi}{2} \sin(\omega_0 + \omega)t - V_0 \frac{\phi}{2} \sin(\omega_0 - \omega)t \tag{10.6}$$

In fact, $\phi(t)$ is not deterministic if it represents random phase variations at all frequencies in the sidebands. The sum-and-difference pairs in eqn. 10.6 at each frequency appear as upper and lower sideband components, each with amplitude $V_0\phi(\omega)/2$. Carrier frequency fluctuation spreads power into the sidebands, with a corresponding decrease in carrier power. The ratio of sideband to carrier power at a single frequency is

$$\frac{[J_1(\phi)]^2}{[J_0(\phi)]^2} \approx \frac{\phi^2}{4}$$

with phase-noise sideband levels at -100 dBc, this gives $\phi \approx 2 \times 10^{-5}$, thus justifying our first-order approximations in eqns. 10.3 and 10.6.

An idealised representation of phase noise consisting of sideband pairs due to random phase fluctuations is shown in Fig. 10.3. In a high-quality oscillator, source noise reaches the thermal noise floor typically at offsets from the carrier of $\pm\omega_0\, 5 \times 10^{-4}$ radi/s. The carrier has been removed from the Figure because it is normally eliminated in the noise-detection process.

Since the sideband power in eqn. 10.6 is $|V_0\phi/2|^2$, phase-noise determination is reduced to the measurement of noise power in spot-frequency ranges in the noise sidebands, with just one additional measurement of carrier power to find $|V_0|^2$. The carrier is first suppressed by mixing it with a reference signal of the same frequency ω_0, but in phase quadrature and feeding the output to a spectrum analyser via a low-pass filter. In Fig. 10.4, the reference signal is at the same frequency as, and in phase quadrature with, the carrier whose phase noise is to be measured:[3,4]

$$v_R = V_R \sin\left(\omega_0 t + \frac{\pi}{2}\right) \tag{10.7}$$

and the second-order product term at the mixer output is

$$V(t)v_R = bV_0V_R \sin(\omega_0 t) \sin\left(\omega_0 t + \frac{\pi}{2}\right)$$

$$+ bV_0V_R\frac{\phi}{2}[\sin(\omega_0 + \omega)t - \sin(\omega_0 - \omega)t]\sin\left(\omega_0 t + \frac{\pi}{2}\right) \tag{10.8}$$

where $V(t)$ is from eqn. 10.6 and b is the second-order conversion constant in a polynomial expansion whose form is given in eqn. 1.32. In reality $V(t)$ is a

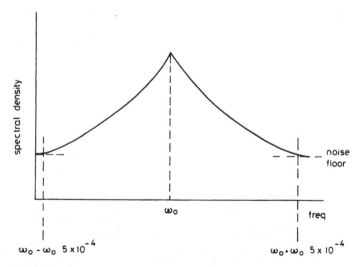

Fig. 10.3 *Idealised representation of carrier phase noise*

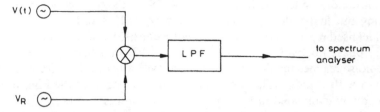

Fig. 10.4 *Phase noise by reference comparison and spectrum analyser*

noise voltage retained here in a deterministic form only to show how phase-quadrature mixing can select sidebands whilst rejecting the carrier. Expansion of eqn. 10.8 gives

$$\frac{V(t)v_R}{b} = \phi \frac{V_0 V_R}{2} \sin(\omega t) + \frac{V_0 V_R}{2} \sin(2\omega_0 t)$$

$$+ \phi \frac{V_0 V_R}{4} [\sin(2\omega_0 + \omega)t + \sin(2\omega_0 - \omega)t] \qquad (10.9)$$

Noise at frequency ω from the carrier has been translated to a zero-amplitude carrier at zero frequency. Upper and lower sidewaves have been folded into a base-band region which can be separated from other components in eqn. 10.9 by means of the low-pass filter. This is illustrated in Fig. 10.5 for the case of a source with a typical range of noise sideband frequencies. There is no carrier at ω_0, only a harmonic at $2\omega_0$ with sidebands symbolised by the single lines at $(2\omega_0 - \omega)$ and $(2\omega_0 + \omega)$. The amplitude of the base-band line at ω is $b\phi V_0 V_R/2$.

Noise-power distribution in the baseband can be examined with a spectrum analyser with bandwidth B, adjusted as in Fig. 10.6 to suit the required or practically possible resolution. The statistical nature of the phase fluctuations

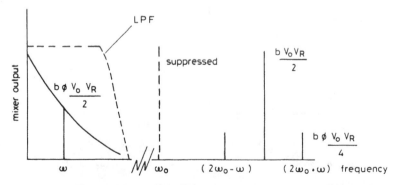

Fig. 10.5 *Mixer products and base-band noise sideband*

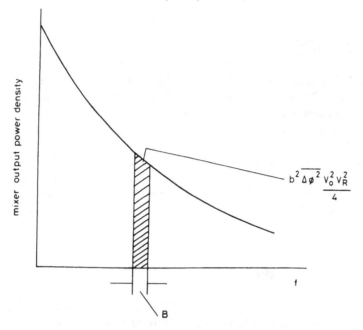

Fig. 10.6 *Examination of base-band noise distribution*

can be accounted for by writing $\sqrt{\overline{(\Delta\phi)^2}}$ for ϕ in the base-band region to give the noise power at frequency ω and in bandwidth B as

$$N_B = b^2\overline{(\Delta\phi)^2}\,\frac{V_0^2 V_R^2}{4} \tag{10.10}$$

From which the spot-noise power density is

$$\frac{N_B}{B} = b^2\,\frac{\overline{(\Delta\phi)^2}}{B}\,\frac{V_0^2 V_R^2}{4} \tag{10.11}$$

The quantity $\overline{(\Delta\phi)^2}/B$, the mean-square phase fluctuation per hertz, is called the phase spectral density, or

$$S_{\Delta\phi}(f) = \frac{\overline{(\Delta\phi)^2}}{B}\quad \text{radian}^2/\text{Hz} \tag{10.12}$$

$S_{\Delta\phi}(f)$ is often called the one-sided spectral density of phase fluctuation because both sidebands are folded into a single baseband.

The measured quantity N_B/B in eqn. 10.11 can yield $S_{\Delta\phi}(f)$ only if $b^2 V_0^2 V_R^2/4$ is known. A simple calibration procedure to find this quantity is slightly to de-tune the reference oscillator to ω_0', to produce sum and difference products in the mixer output at $\omega_0' \pm \omega_0$ with amplitude equal to $bV_0 V_R/2$. Thus by measuring the amplitude of the beat frequency $(\omega_0' - \omega_0)$ the calibration factor is given as

$$K_d = b\,\frac{V_0 V_R}{2} \tag{10.13}$$

An implicit assumption in this theory has been that the reference oscillator is noise free. Provided the noise level of the reference is at least 10 dB lower than the source under test, this assumption is tenable, but if, for instance, both oscillators have equal noise, the measured $S_{\Delta\phi}(f)$ should be reduced by 3 dB.

A second important omission is that no account has been taken of the effects of long-term drift on the short-term measurements. Each oscillator drifts independently and may therefore spoil the long-term frequency coincidence between them. This can be overcome by phase-locking the reference to the test source.

10.5 Phase-lock detection

The noise base-bandwidth is from 0 to $10^5 \pi f_0$ in a high-quality oscillator but, as we will see in Section 10.6, long-term drift in the range 0—1 Hz is best measured in the time domain through the Allan variance, because phase fluctuations are too large for the approximations to hold in quadrature phase-detection methods. Both the carrier and the large long-term phase variations should be suppressed before measuring the noise-power equivalent of short-term phase noise. The time-domain representation in Fig. 10.7 shows the carrier with small amplitude fluctuations due to short-term phase noise, whilst the whole pattern moves in time over a longer term. Corresponding frequency bounds are shown on the

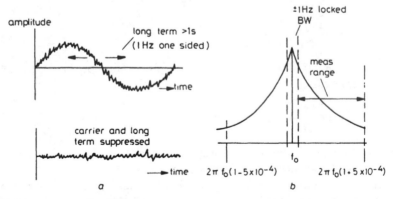

Fig. 10.7 *Short- and long-term phase noise*
a Time
b Frequency

frequency domain, in which the long-term effects are confined to a ± 1 Hz bandwidth on the carrier and the measured short-term range lies between this and the onset of the thermal-noise floor. Carrier suppression has already been achieved by quadrature mixing and low-pass filtering, but long-term drift remains.

Phase-locking the test to the reference oscillator causes it to track within the locking bandwidth of the loop. If the bandwidth is set to 1 Hz, only short-term fluctuations are observable at the mixer output. Fig. 10.8 shows how phase lock is used in conjunction with the quadrature mixer by the addition of a voltage-controlled oscillator as a reference. Short-term phase remains as a difference between the test and reference, but long-term phase excursions are decreased by the loop gain of the feedback loop.

If ϕ_T and ϕ_R are the instantaneous phases of the test and reference oscillators, the output voltage is proportional to their difference according to

$$V_d = \Delta V = V(t)v_R = b\frac{V_0 V_R}{2}(\phi_T - \phi_R) = K_d\phi \tag{10.14}$$

where V_d is measured in the folded baseband of the detector as the first term of eqn. 10.9. V_d is supplied to the VCO to give a rate of change of phase

$$\frac{d\phi_R}{dt} = K_0 A V_d \tag{10.15}$$

where A is the amplifier gain and K_0 is the sensitivity of the VCO. If we take the Laplace transform, with s as the complex frequency, eqn. 10.15 gives[8]

$$s\phi_R(s) = K_0 A V_d(s)$$

which, on substitution in eqn. 10.14, gives

$$V_d(s) = b\frac{V_0 V_R}{2}\left[\phi_T(s) - \frac{K_0 A}{s}V_d(s)\right]$$

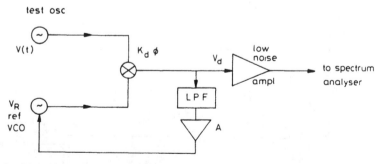

Fig. 10.8 *Phase lock with quadrature mixer*

or

$$V_d(s) = \frac{sK_d\phi_T(s)}{s + K_dK_0A} \tag{10.16}$$

A frequency-domain response is found by setting the complex frequency s to $j\omega$ to give

$$V_d = \frac{\omega K_d\phi_T}{\omega - jK_dK_0A} \tag{10.17}$$

The response is zero at $\omega = 0$ with cut off frequency at

$$\omega_c = K_dK_0A \tag{10.18}$$

For $\omega \gg \omega_c$

$$V_d = K_d\phi_T = b\frac{V_0V_R}{2}\phi_T$$

When ϕ_T is a statistical quantity, it is replaced with $\sqrt{\overline{(\Delta\phi)^2}}$ and the output power is measured as

$$|V_d|^2 = b^2\frac{V_0^2V_R^2}{4}\overline{(\Delta\phi_T)^2} \tag{10.19}$$

which is identical with eqn. 10.10 and leads directly to the spectral density of phase fluctuation $S_{\Delta\phi}(f)$.

Fig. 10.9 shows the frequency response of the phase-lock loop increasing from zero to the 3 dB cut off at 1 Hz, and reaching K_d at large frequencies. A cut off at 1 Hz corresponds to an integration time of $1/2\pi$ seconds, and allows measurement of $S_{\Delta\phi}(f)$ to within 1 Hz of the carrier.

If V_d is measured in a spectrum analyser of bandwidth B and $|V_{RMS}|^2$ is the mean square voltage per Hz,

$$|V_{RMS}|^2 = b^2\frac{V_0^2V_R^2}{4}\frac{\overline{(\Delta\phi_T)^2}}{B} \tag{10.20}$$

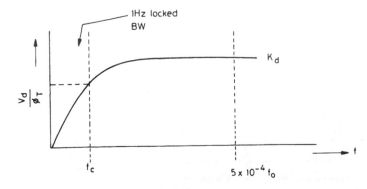

Fig. 10.9 *Frequency response with phase lock*

and the spectral density of phase fluctuation is

$$S_{\Delta\phi}(f) = \left(\frac{V_{RMS}}{K_d}\right)^2 \qquad (10.21)$$

With $K_d = bV_0 V_R/2$ in volts per radian and V_{RMS} in volts per $(\text{Hz})^{1/2}$; $S_{\Delta\phi}(f)$ is in radian^2 per hertz.

10.6 Frequency fluctuations and Allan variance

Short-term phase fluctuations are expressed as fractional frequency errors in Table 10.1, since this is often a more practical means of comparing different sources. These errors are the mean-square summation of a distribution of frequency fluctuations in the noise sidebands of the oscillator, that can be derived from the spectral density of phase fluctuation by first examining the deterministic relation between frequency and phase modulation. Thus, if $\delta\phi$ is the phase change occurring in a carrier of frequency f_0, the frequency change equals the rate of change of fractional phase, or

$$\delta f = \frac{1}{2\pi} \frac{d(\delta\phi)}{dt}$$

If, for instance, $\delta\phi = \Delta\phi \sin \omega t$, $\delta\phi$ is determinate and

$$\delta f = f \Delta\phi \cos \omega t$$

Expressed as a fractional change in the carrier frequency, this is

$$\frac{\delta f}{f_0} = \frac{f}{f_0} \Delta\phi \cos \omega t = y(t) \qquad (10.22)$$

The fluctuation δf is at offset frequency f in the sideband of the carrier at f_0, and is a function of time. When $\delta\phi$ is indeterminate it becomes a random fluctuating quantity $\sqrt{\overline{(\Delta\phi)^2}}$ with an equivalent random frequency fluctuation $\sqrt{\overline{(\delta f)^2}}$ such that

$$\frac{\overline{(\delta f)^2}}{f_0^2} = \left(\frac{f}{f_0}\right)^2 \overline{(\Delta\phi)^2} = \overline{y^2(t)} \qquad (10.23)$$

Any instrument used to measure $\overline{(\Delta\phi)^2}$ at frequency f has bandwidth B. If we divide both sides of eqn. 10.23 by B, the fractional square frequency fluctuation per hertz at frequency f is

$$S_y(f) = \frac{1}{B} \frac{\overline{(\delta f)^2}}{f_0^2} = \left(\frac{f}{f_0}\right)^2 \frac{\overline{(\Delta\phi)^2}}{B} = \frac{\overline{y^2(t)}}{B} \qquad (10.24)$$

But

$$\frac{\overline{(\Delta\phi)^2}}{B} = S_{\Delta\phi}(f)$$

therefore

$$S_y(f) = \left(\frac{f}{f_0}\right)^2 S_{\Delta\phi}(f) \qquad (10.25)$$

$S_y(f)$ is the spectral density of fractional frequency fluctuation, and it is related to the spectral density of phase fluctuation through eqn. 10.25. The total frequency error as expressed in Table 10.1 can be found by integrating $S_y(f)$ over all f in the phase-noise sidebands of the oscillator.

We have already seen that $S_y(f)$ is best derived from measured $S_{\Delta\phi}(f)$ at frequencies separated from the carrier by more than 1 Hz, whereas timed averages of frequency fluctuations give more accurate figures close to the carrier. Long-term frequency variations up to f hertz will be measured if samples of $y(t)$ are averaged in time periods $\tau = 1/f$, because all higher rates of change will average to zero. Mean-square $y(t)$ is directly related to $S_y(f)$ in eqn. 10.24, but it is not easily derived from the linear averages directly. Instead, we use an alternative function called a pair variance or Allan variance, defined as[5,6]

$$\sigma_y^2(\tau) = \frac{1}{2(M-1)} \sum_{k=1}^{M-1} (\overline{y}_{k+1} - \overline{y}_k)^2 \qquad (10.26)$$

where M is the number of data values of $\overline{y}_k$, and $\overline{y}_k$ is the average over time τ of the kth data point.

The spectral density of fractional frequency fluctuations can be found from the Fourier transformation of the Allan variance with parametric time variation τ. $\sigma_y^2(\tau)$ is therefore measured over a range of τ from zero to $1/f$, where $f \approx 1$ Hz, so that the truncated Fourier components remain approximately valid up to 1 Hz from the carrier. In practice, the Allan variance is often used up to $\tau \gtrsim 10$ ms or $f \lesssim 100$ Hz.

A heterodyne method, illustrated in Fig. 10.10, is used to measure the Allan variance. The signal under test is down-converted to an intermediate frequency f_b, which is counted for a fixed period τ to measure the average

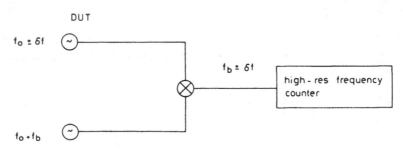

Fig. 10.10 *Heterodyne measurement of Allan variance*

frequency. Division by the measured carrier frequency gives $\bar{y}_k$. Repeated measured sequences give $\sigma_y^2(\tau)$ as a function of τ. The sensitivity of the method decreases at offsets $> 100\,\text{Hz}$, or where the noise is flat or decreasing at $1/f$ or less.

10.7 Direct measurement of two-sided power spectral density

If we return to eqn. 10.3, namely

$$V(t) \simeq V_0 \sin(\omega_0 t) + V_0\phi(t)\cos(\omega_0 t) \qquad (10.3)$$

we see that, for small modulation angle, phase modulation in the time domain becomes the quadrature addition of two carriers, one of which is modulated by a voltage equal to the product of the original carrier amplitude and the phase variation $\phi(t)$. When the phase variations are random, $\phi(t)$ becomes $\Delta\phi(t)$ and there is an equivalent noise power $V_0^2(\Delta\phi)^2/2$ associated with the random phase variations $\Delta\phi(t)$. The spectrum of $V_0^2(\Delta\phi)^2/2$ is the one-sided per hertz distribution with respect to zero carrier frequency, normally generated by quadrature mixing and measured as V_{RMS} in eqn. 10.20. If the noise-power density in the spectrum is N_0 volts per hertz, it follows from eqn. 10.20 that

$$N_0 = V_{RMS}\frac{2}{b^2 V_R^2} = \frac{V_0^2}{2}\frac{\overline{(\Delta\phi)^2}}{B} = \frac{V_0^2}{2}S_{\Delta\phi}(f) \qquad (10.27)$$

N_0 is therefore[4] the one-sided mean-square amplitude spectral density that is equivalent to the phase spectral density $S_{\Delta\phi}(f)$.

Before down-conversion to a one-sided baseband, the spectrum is two-sided about the carrier at ω_0. The ratio of the power density in one of the sidebands at frequency f from the carrier to the total carrier power is

$$\mathscr{L}(f) = \frac{1}{2}\frac{N_0}{V_0^2/2} = \frac{S_{\Delta\phi}(f)}{2}$$

or

$$S_{\Delta\phi}(f) = 2\mathscr{L}(f) \qquad (10.28)$$

$\mathscr{L}(f)$ is called the two-sided power-spectral-density ratio at offset f from f_0. It is a particularly useful form when a direct measurement of oscillator noise is made with a spectrum analyser, since the measurement can be conveniently confined to one sideband of the noise spectrum. The spectrum analyser is connected directly to the source output, and the noise power is measured in a bandwidth and offset determined by the analyser setting.[9]

Unlike the base-band method, direct measurement suffers from local-oscillator noise, and, because $\mathscr{L}(f)$ is essentially the carrier/noise-power ratio, there

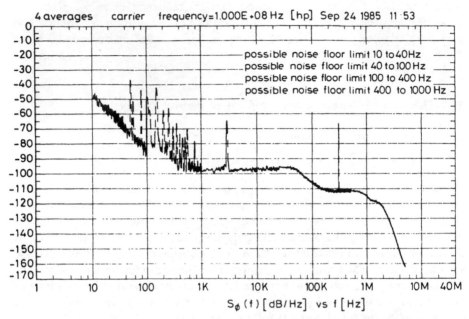

Fig. 10.11 $S_\phi(f)$, $[S_{\Delta\phi}(f)]$, for Hewlett–Packard 8341A synthesised signal generator

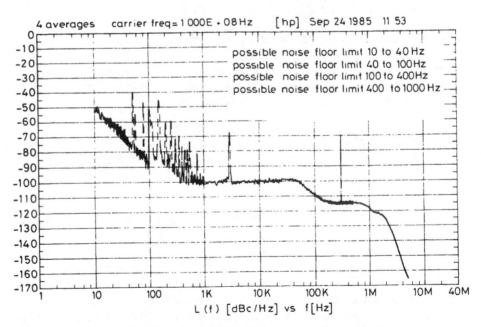

Fig. 10.12 $\mathscr{L}(f)$ for Hewlett–Packard 8341A synthesised signal generator

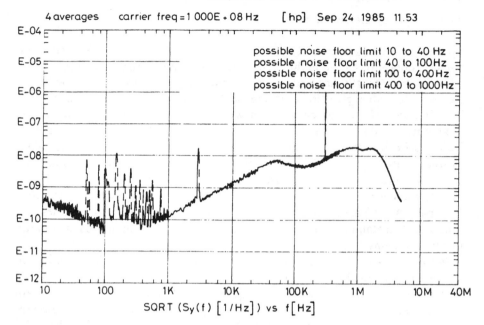

4 averages   carrier freq = 1 000E + 08 Hz   [hp] Sep 24 1985 11.53

possible noise floor limit 10 to 40 Hz
possible noise floor limit 40 to 100Hz
possible noise floor limit 100 to 400Hz
possible noise floor limit 400 to 1000Hz

SQRT $(S_y(f)$ $[1/Hz])$ vs $f[Hz]$

Fig. 10.13 $S_y(f)$ *for Hewlett–Packard 8341A synthesised signal generator*

is the further disadvantage of a limited dynamic range. Direct measurement is therefore of little value with high-quality sources where characteristic ratios exceed 100 dBc.

$S_{\Delta\phi}(f)$ and $\mathscr{L}(f)$ are usually plotted in dBc with $\mathscr{L}(f)$ just 3 dB lower than $S_{\Delta\phi}(f)$. Modern automated equipment incorporates software conversion to provide $\mathscr{L}(f), S_y(f)$ etc. from measurement of $S_{\Delta\phi}(f)$. For instance, the HP11740S automatic equipment prints out these quantities to an accuracy of ± 2 dB to 1 MHz offsets, and ± 4 dB to 40 MHz offsets; these are in a 1 Hz bandwidth. A 0·4 dB error occurs when the reference noise is 10 dB below the test noise, though it is possible, using three measured sources, to find the absolute noise of the reference and eliminate it from future measurements. Examples of print outs of measurements on the HP8341A synthesised signal generator are given in Figs. 10.11—10.13. $\mathscr{L}(f)$ and $S_{\Delta\phi}(f)$ are different by only 3 dB, but $S_y(f)$, the spectral density of frequency fluctuation, is not so simply related and has been calculated using eqn. 10.25. The total frequency fluctuation of the carrier, in the form stated in Table 10.1, can be found by integrating $S_y(f)$ over the full spectral range from carrier to noise floor level. Many of the narrow peaks are determinate, owing to environmental factors or spurious responses in the phase-lock chains.

10.8 Frequency measurement

Spectral noise is measured at spot frequencies offset from a carrier, and so far we have concentrated on finding the mean phase and frequency variations. But now we turn to the accuracy of spot-frequency measurements. Microwave frequency counters give direct readings with accuracies better than 1 part in 10^7. This is accomplished by comparison with a high-precision quartz oscillator, temperature stabilised in an oven and protected against line-voltage variations. Table 10.1 shows the long- and short-term error ranges from the different kinds of quartz oscillator found as comparison standards in laboratory counters,[11] and serves as a reminder that phase noise and long-term drift offset the measuring instrument as well as the source it may be measuring. Greater accuracies (~ 1 part in 10^{12}) are available from atomic-frequency standards, but these are found only in standards laboratories or as time-reference standards in communication systems.

Before the arrival of general-purpose microwave counters, frequency measurements depended on the calculable resonances of short-circuited or capacitively loaded transmission lines; e.g., quarter- and half-wave coaxial resonant cavities like those in Fig. 10.14. In Fig. 10.14a when $l = \lambda/2$ the line resonant frequency is calculable from the length. In the case of the quarter-wave line with $l \approx \lambda/4$ in Fig. 10.14b, there is a capacitive effect due to the open circuit at one end which must be taken into account in determining frequency from resonant length. Further modification of the calculation is necessary owing to the loading of a coupling loop, often directly connected to an external diode as in Fig. 10.15. When the cavity or resonator is tuned by varying the insertion depth of the centre conductor, it becomes a wave meter. A calibrated scale, engraved on the barrel of the meter, can be read directly in frequency. At resonance the detector output peaks in the transmission cavity of Fig. 10.15. An absorption cavity, illustrated in a waveguide version in Fig. 10.16, causes a dip in the main waveguide response owing to absorption losses at resonance. Accuracy is limited by cavity losses. In the microwave region to 18 GHz, coaxial-cavity

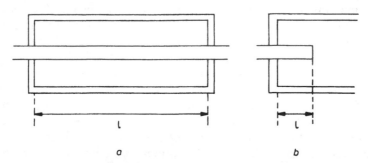

a   b

Fig. 10.14   *Short- and open-circuit coaxial resonators*

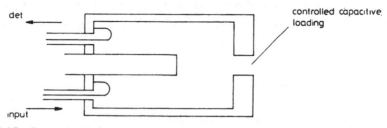

Fig. 10.15 *Transmission frequency meter*

Q-factors are from 1200 to 4000, decreasing with frequency. Waveguide wavemeter Q-factors are greater than 8000. If f_0 is the centre frequency and Δf is the span between the 3 dB points on the resonance curve,

$$Q = \frac{f_0}{\Delta f} \qquad (10.29)$$

Thus for a Q-factor of 2000, the 3 dB resolution accuracy is 1 part in 5×10^4.

Except in the case of very accurately made standards the Q-factor limit is rarely approached, because other errors, such as mechanical backlash in the plunger, scale resolution, humidity and temperature effects, are overwhelmingly greater. Coaxial wave meters can thus be accurate to $< 0.2\%$ and waveguide types to better than 0.1%. Generally speaking, the greater the precision the smaller the tuning range: with a maximum of an octave for the former and half this for the latter. A 10% bandwidth is a realistic maximum as accuracies approach the Q-factor limit.

Wave meters are less popular in laboratories now that high-frequency electronic counters are readily available at reasonable cost. Not only are the latter potentially more accurate, if used properly, but the measurement is easier to

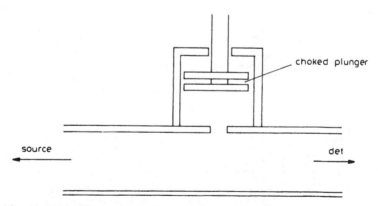

Fig. 10.16 *Absorption frequency meter*

acquire either as a digital read out or via an interface to a computer. Frequency counters are available with microprocessor control for remote or automatic operation.

10.9 Frequency counters

Microwave frequency counters generally rely on frequency translation to a lower frequency, since digital-counting circuits have a current maximum rate of 3 GHz with 500 MHz to 1 GHz being more typical of most commercially available examples.[11,12] The translated lower frequency is measured using standard low-frequency methods and displayed after applying a suitable scaling factor. Before describing the different down-conversion options, we will review the operating principles of low-frequency counters starting with the block diagram in Fig. 10.17. The signal is converted to a train of pulses, each one corresponding to one cycle of the input frequency, before entering the input circuits of the gate. To control the time interval for which the gate is open, the time-base divider counts down from a precision quartz oscillator. During the open period t, pulses pass through the gate and are counted in the register. From the gate time t and the number of counts n, the frequency is

$$f = \frac{n}{t} \tag{10.30}$$

Gate time is a decade multiple of 1 s to simplify calculation and display. Accuracy is dependent on time measurement, and for this reason time-base-oscillator frequencies are usually between 1 and 10 MHz with long-term stability of 1×10^{-7}.

Sometimes the reciprocal of f, or period P, is measured by finding t for n counts, to give

$$P = \frac{t}{n} \tag{10.31}$$

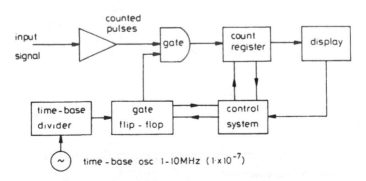

Fig. 10.17 *Low-frequency counter circuit*

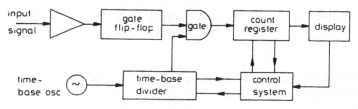

Fig. 10.18 *Counter based on period measurement*

In this case, the time-base oscillator is counted whilst the gate is held open during several periods n of the signal, as illustrated in Fig. 10.18. If N is the number of pulses counted in n periods, t is given by the divided time-base frequency f_{TB} and

$$P = \frac{N}{nf_{TB}} \tag{10.32}$$

The upper frequency limit of such direct counters can be extended by pre-scaling the input frequency. Pre-scalers divide the input frequency so that only every Nth pulse is counted. Resolution is reduced by N, but this can be recovered by extending the gate time by N to trade measurement time against resolution for increasing frequency. Fast digital dividers do not extend the range much above 2 GHz, so they are not widely used in microwave counters. Instead, frequency translation by down-conversion is preferred before digitisation and input to a direct low-frequency counter. The earliest methods used a transfer oscillator with a comb generator, but heterodyne techniques are now more common. Modern counters with built-in microprocessor control use elements of both methods. These different methods will now be discussed.

10.9.1 Transfer oscillators
In the simplified form of transfer oscillator shown in Fig. 10.19, suppose the input signal is 10 GHz and that it mixes with the Nth harmonic of the VCO

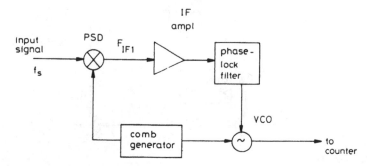

Fig. 10.19 *Transfer-oscillator principle*

from the comb generator. The intermediate frequency is fixed by the IF amplifier pass band, so that, for lower sideband mixing,

$$F_{IF1} = Nf_{VCO} - f_S \tag{10.33}$$

If $F_{IF1} = 10$ MHz and $N = 100$ in eqn. 10.33 f_{VCO} becomes 100·1 MHz. The input microwave frequency f_S is therefore converted to 10 MHz in a phase-lock loop using the Nth harmonic of a voltage-controlled oscillator, and the output of the VCO is measured directly on a low-frequency counter. Provided F_{IF1} and N are known, the signal frequency is determined. N can be found by extending the circuit as shown in Fig. 10.20. A power divider splits the input signal between two mixers with their respective VCOs mutually offset by frequency f_0. Thus

$$F_{IF2} = N(f_{VCO} \pm f_0) - f_S$$
$$= F_{IF1} \pm Nf_0 \tag{10.34}$$

F_{IF2} is now mixed with a sample of F_{IF1} to generate Nf_0. N is finally determined by passing Nf_0 and f_0 to a ratio counter, and the time-base divider of the main frequency counter can then be set to give the correct reading on the display.

A ratio counter finds the ratio of two frequencies by using the lower frequency for gate control whilst counting the higher frequency. The principle is illustrated in Fig. 10.21, where, if the count time t is equal to $1/f_0$, the number of pulses counted is N.

Transfer oscillators have two major disadvantages. Resolution is reduced by harmonic number N, or, for equivalent resolution, the measurement time increases by N. Single measurements at high frequencies (~ 18 GHz) can take 1 min or more if 1 Hz resolution is required. A more serious disadvantage is

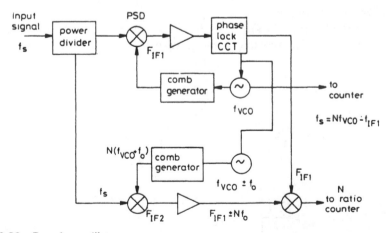

Fig. 10.20 *Transfer oscillator*

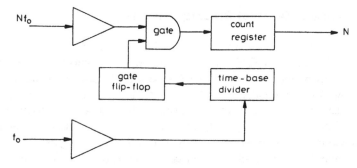

Fig. 10.21 *Ratio counter*

that the continuous phase locking of transfer oscillators limits the amount of frequency modulation the input may carry. Advanced forms of heterodyne down-convertors have overcome this problem.

10.9.2 Heterodyne down-convertors

The restrictions of continuous phase locking are avoided by computer selection of the Kth harmonic of a comb generator through a YIG/PIN switch filter. In Fig. 10.22, the frequency input to the comb generator f_i is known to the accuracy of the multiplied time-base oscillator. The IF frequency f_{IF} ($=f_s - Kf_i$) for the Kth harmonic is measured in a low-frequency counter and the signal frequency is displayed as

$$f_S = Kf_i + f_{IF} \tag{10.35}$$

Counters are often hybrids of these two types and significant advances are still being made in their design.[2,10,11,12]

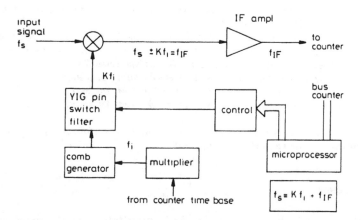

Fig. 10.22 *Heterodyne down-convertor*

10.10 Performance factors of frequency counters

Counters currently operate from several hertz to 26·5 GHz, some using an external head to reach 40 GHz and even higher. Input sensitivities are typically -25 to -30 dBm and dynamic range 30—40 dB. A major consideration is FM tolerance, which is typically 10—50 MHz. Counters usually select the largest signal for display, but require amplitude separations in excess of 6 dB and several hundred megahertz frequency separation for successful acquisition. Accuracy is chiefly dependent on time-base accuracy, though gate-tuning errors of the order of picoseconds are also possible. A ± 1 count error, which is fundamental to the method, can be reduced by having longer counts. Acquisition times are < 150 ms mean in counters from 10 Hz to 18 GHz, but, for pulses of low PRF, acquisition can be difficult and may require manual operation. With increasing on-board intelligence, counters are still rapidly developing towards improved acquisition of complex signals, though fundamental frequency accuracy is unlikely to improve except by means of special time-base reference oscillators.

10.11 References

1 HOWE, D. A.: 'Frequency domain stability measurements: A tutorial introduction'. US National Bureau of Standards, Technical Note 679, 1976
2 'Fundamentals of electronic counters'. Hewlett–Packard Application Note 200, July 1978
3 'Understanding and measuring phase noise in the frequency domain'. Hewlett–Packard Application Note 207, October 1976
4 BAGHDADY, E. J., LINCOLN, R. N., and NELIN, B. D.: 'Short-term frequency stability: Characterisation, theory and measurement'. IEEE–NASA Symposium, NASA SP-80, Goddard Space Flight Center, Greenbelt, MD, Nov. 1964, pp. 65–87
5 ALLAN, D. W.: 'The measurement of frequency and frequency stability of precision oscillators'. NBS Technical Note 669, July 1975
6 'Phase noise characterisation of microwave oscillators: Phase detector method'. Hewlett–Packard Product Note 11729B-1, Aug. 1983
7 SHEPHERD, P. R., FAULKNER, N., VILAR, E., and BRADSELL, P.: 'Relation between power spectrum and phase spectrum of oscillators', *Electron. Lett.*, 1982, **18**, pp. 614–615
8 GARDNER, F. M.: 'Phaselock techniques (Wiley, 1966) Chap. 2
9 'Spectrum analysis: Noise measurements'. Hewlett–Packard Application Note 150-4, April 1974
10 MILLER, L.: 'Trends in microwave counter technology', *Microwave J.*, April 1983, pp. 105–112
11 'Fundamentals of microwave frequency counters'. Hewlett–Packard Application Note 200-1, Oct. 1977
12 SCOTT, R. G.: 'Gallium arsenide lowers cost and improves performance of microwave counters'. *Hewlett–Packard J.*, Feb., 1986, pp. 4–10

10.12 Examples

1   At the mixer output of a phase-lock quadrature phase detector, $K_d = 1$ V/rad, V_{RMS} (45 Hz) $= 100$ nV per root hertz. Solve for $S_{\Delta\phi}$ (45 Hz) in decibels.

2   Calculate the beat-note calibration factor in a quadrature mixer, and show how this relates to the phase-detector constant for small phase differences.

3   A transfer-oscillator counter measures frequencies in the range 4—18 GHz. If the IF is at 100 MHz and the 60th VCO harmonic is used at 18 GHz, calculate the tuning range of the VCO. What is the measurement time for an accuracy of 1 Hz at 18 GHz?

Chapter 11

Time-domain reflectometry

11.1 General principles

It is possible to find the position of an obstacle in a transmission system by sending a pulsed RF wave into the system and measuring the return time of the pulse after reflection at the obstacle. It is more convenient to launch a step function as in Fig. 11.1, when the measurement is confined to the time domain.[5,6] The waveforms have an equivalent spectral description which can be simulated as a set of discrete spectral lines with appropriate amplitudes and phases, and the reflected spectra then contain the same information as the reflected time functions. This information is found by taking the Fourier transform of reflected spectra measured by vector or scalar methods, according to whether phase is available directly or through wave interference in the test set.

Modern time-domain measurements are made, not by launching the step function at an instant of time, but by sequentially launching its discretely

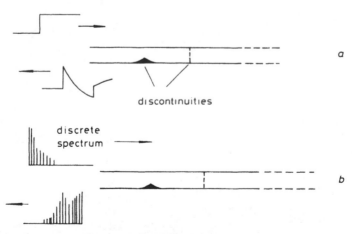

Fig. 11.1 *Time (a) and frequency (b) reflectometry*

sampled spectral components and noting their modified amplitudes and phases on return from the discontinuity. Time information is generated by taking a discrete Fourier transform (DFT) of the spectral returns, thus yielding the same information on the discontinuity as was available from the original step-function method. For this reason, the term frequency-domain reflectometry (FDR) is sometimes applied to the spectral technique. The equivalence of time and frequency domains is illustrated in Fig. 11.1, in which the forward step function either contains all frequencies simultaneously or can be sent as a set of single frequencies one after the other over a period of time. On return, the timed arrival of the step-function leading edge is determined by the propagation velocity of the waveform and the distance to the discontinuity; or alternatively, the spectral amplitudes and phases may be transformed to find this same time delay.

The advantages of FDR are:[2]

- An improvement in signal/noise and jitter
- Other waveforms than a step function can be simulated; e.g., an impulse or RF pulse
- Zero-level drift and zero ambiguity are reduced
- Spectral shaping by windowing gives greater control of responses
- Time gating allows separation of multiple reflection

The generation of the spectrum and subsequent recovery of the time-domain information depends on vector analysis and extensive computer software. Direct measurement of returns is not possible with scalar analysis, though, as we have previously demonstrated, a physical interaction in the test set can sometimes be arranged to occur before square-law detection, and therefore yield vector type results. An important example of this is fault location in cables, in which a Fourier transform of a square-law detected standing wave, caused by a discontinuity, is able to reveal its magnitude and location. We will begin with an investigation of this technique before returning to the general principles of vector methods.

11.2 Fault location in transmission lines

Consider a cable of a few metres length with a matched load connected at its far end through a high-quality connector with a return loss of -35 dB. At distance z from the load the cable has been damaged such that the reflection is equivalent to a return loss of -10 dB. The flowgraph for the cable is shown in Fig. 11.2. At distance z' from the measurement reference plane the input reflection coefficient due to the fault and load connector is

$$S_{11} = \rho_L = \rho_1 + \frac{\rho_2 \exp(-2j\beta z)}{1 - \rho_1\rho_2 \exp(-2j\beta z)} \tag{11.1}$$

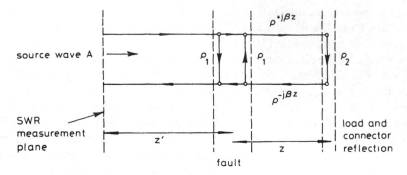

Fig. 11.2 *Flowgraph for scalar fault location*

From the result in eqn. 3.32 the standing-wave voltage measured at the reference plane is

$$V = A \exp(-j\beta l') \left[1 + \left(\rho_1 + \frac{\rho_2 \exp(-2j\beta z)}{1 - \rho_1\rho_2 \exp(-2j\beta z)} \right) \exp(-2j\beta z') \right] \quad (11.2)$$

where l' is the distance from the source to the reference plane. For a square-law detector we take the modulus of eqn. 11.2 to give

$$|V|^2 = |A|^2 \left\{ \frac{\begin{aligned} & 1 + \rho_1^2 + \rho_2^2 + \rho_1^2\rho_2^2 + \rho_1^4\rho_2^2 \\ & - \rho_1\rho_2(1 + \rho_1^2)\cos 2\beta z + \rho_1(2 + \rho_1^2\rho_2^2 - \rho_2^2)\cos 2\beta z' \\ & + \rho_2(2 - 2\rho_1^2 - \rho_1^2\rho_2)\cos 2\beta(z + z') - 2\rho_1^2\rho_2 \cos 2\beta(z - z') \end{aligned}}{1 - 2\rho_1\rho_2 \cos 2\beta z + \rho_1^2\rho_2^2} \right\}$$

$$(11.3)$$

When the denominator is expanded binomially to first order the only significant effect is to add a term $-\rho_1^2\rho_2^2 + 2\rho_1\rho_2 \cos 2\beta z$, giving, to fourth order,

$$|V|^2 = |A|^2 \{ 1 + \rho_1^2 + \rho_2^2 + \rho_1\rho_2(1 - \rho_1^2)\cos 2\beta z + \rho_1(2 - \rho_2^2)\cos 2\beta z'$$

$$+ \rho_2(2 - 2\rho_1^2 - \rho_1^2\rho_2)\cos 2\beta(z + z') - 2\rho_1^2\rho_2 \cos 2\beta(z - z') \} \quad (11.4)$$

Frequency components are evident in the β values, and cause the scalar voltage $|V|$ to vary as frequency changes. A sweeper and scalar network analyser provides the means of finding the frequency dependence of $|V|^2$. Thus in Fig. 11.3 the sweeper output is split in the power divider, half routed to the detector and the other half to the transmission line under test. Reflections in the line cause some of the energy to be reflected back into the detector, where it interferes with the forward fraction of the input signal. At the detector/divider reference plane the voltage is

$$V = A'\rho_L \exp(-2j\beta z') + A \quad (11.5)$$

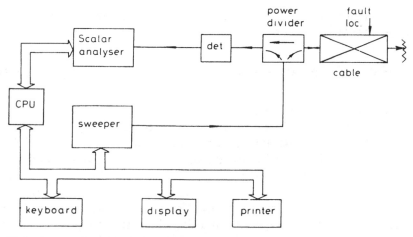

Fig. 11.3   *Test set for scalar fault location*

where A is the incident amplitude after splitting, $\rho_L \exp(-2j\beta z')$ is the reflection coefficient of the transmission line at the reference plane and A' is the effective amplitude of the returning wave before reflection at the discontinuities. A' is generally less than A because of the two-way transmission loss of the cable. In a long cable this difference could be quite large. It is therefore first necessary to estimate the transmission loss of the cable, by the standard methods described in Chapter 6, but we will assume that $A' = A$ in the following discussion.

If ρ_L in eqn. 11.5 is substituted using eqn. 11.1, we have

$$V = A\left[1 + \left(\rho_1 + \frac{\rho_2 \exp(-2j\beta z)}{1 - \rho_1\rho_2 \exp(-2j\beta z)}\right) \exp(-2j\beta z')\right] \qquad (11.6)$$

indicating that $l' = 0$, because the reference plane coincides with the source and detection plane. The square of the modulus remains as in eqn. 11.4.

In practice, the square of the effective standing-wave magnitude $|V|^2$ might be measured at 256 discrete frequencies in the chosen sweeper scan range to give the frequency spectrum of $|V|^2$. A discrete Fourier transform (DFT) of the data converts the response to the time domain according to the following expression:[11]

$$F(t_k) = \sum_{n=0}^{N-1} f(n\omega_1) \exp(-jn\omega_1 t_k) \qquad (11.7)$$

where

$$f(n\omega_1) = |V|^2_{sampled}, \qquad t_k = \frac{k}{N\omega_1}, \qquad k = 0, 1, \ldots, N-1 \qquad (11.8)$$

and $n\omega_1 = 2\pi n f_1$, where f_1 is the lowest scan frequency.

It is convenient to use the integral form of the transform to investigate the time response of $|V|^2$. Thus

$$F(t) = \int_{-\infty}^{+\infty} f(\omega) \exp(-j\omega t) \, d\omega \tag{11.9}$$

where we now have $f(\omega) = |V|^2$.

There are two types of term in eqn. 11.4. A constant-amplitude spectrum, $1 + \rho_1^2 + \rho_2^2$, gives a single impulse $\delta(0)$ at the time origin in Fig. 11.4; and cosine terms of the form $\cos 2\beta z$ give impulses $\delta(t \pm t_z)$ at advanced and delayed times, depending on propagation distance to reflecting obstacles in the transmission line. We can examine this further by considering the term, $\rho_1(2 - \rho_2^2) \cos 2\beta z'$, that is most strongly affected by the line fault at z' from the reference plane. It can be expressed in a time dimension by writing

$$2\beta z' = \omega \frac{2z'}{v_p} = \omega t_{z'} \tag{11.10}$$

where $t_{z'}$ is the time for a round trip to the discontinuity at z' and back to the reference plane, and v_p is the phase velocity in the transmission line. In a dispersive medium, such as a waveguide, v_p is a function of frequency which has to be properly included in the Fourier transform. The time-domain response of $\cos \omega t_{z'}$ is shown in Fig. 11.4 as two delta functions at $t = \pm t_{z'}$. But in reality the number of lines N in a discrete spectrum is finite with total scan width $S = N\omega_1/2\pi$. Consequently, the δ functions are convoluted with a shaping factor approximately given by $(\sin 2\pi St)/2\pi St$. This is indicated by dotted curves on each of the δ functions in Fig. 11.4. Finite scan width therefore causes spectral broadening of the ideal line spectrum and sets a limit on the resolution of closely spaced lines. The time resolution of two functions of this shape corresponds to the -6 dB width, since any closer overlap does not adequately separate the peaks. The -6 dB time width is twice the time given by

$$2\pi St = 1\cdot 9$$

or

$$\Delta t = \frac{1\cdot 9}{\pi S} \tag{11.11}$$

But the time width relates to a distance separation Δz through the phase velocity of the wave, so that, by extension from eqn. 11.10, we have

$$2\Delta z = v_p \, \Delta t$$

or

$$\Delta z \approx \frac{v_p}{\pi S} \tag{11.12}$$

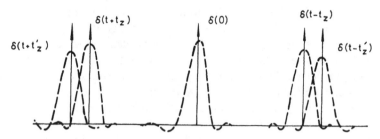

Fig. 11.4 *Time resolution*

Other terms in eqn. 11.4 might overlap that at $t_{z'}$ depending on their relative delays. For instance, if the fault is approximately halfway between the reference plane and load connector, $z' \sim z$, so that the multiple reflection term $\rho_1 \rho_2 (1 - \rho_1^2) \cos 2\beta z$ and the fault will both give returns close to $t_{z'}$. This is illustrated by the overlapping responses in Fig. 11.4.

It is now possible to write down by inspection the Fourier transform of each term in eqn. 11.4. This is listed in Table 11.1 with the calculated magnitudes of the responses in decibels.

Table 11.1 *Fourier transforms of each term in eqn. 11.4*

Frequency domain	Time domain	Response, dB
$1 + \rho_1^2 + \rho_2^2$	$(1 + \rho_1^2 + \rho_2^2)\delta(0)$	0·83
$\rho_1 \rho_2 (1 - \rho_1^2) \cos 2\beta z$	$\rho_1 \rho_2 (1 - \rho_1^2)\delta(t \pm t_z)$	$-45\cdot80$
$\rho_1 (2 - \rho_2^2) \cos 2\beta z'$	$\rho_1 (2 - \rho_2^2)\delta(t \pm t_{z'})$	$-3\cdot99$
$\rho_2 (2 - \rho_1^2 - \rho_1^2 \rho_2) \cos 2\beta(z + z')$	$\rho_2 (2 - \rho_1^2 - \rho_1^2 \rho_2)\delta(t \pm t_{z+z'})$	$-29\cdot80$
$2\rho_1^2 \rho_2 \cos 2\beta(z - z')$	$2\rho_1^2 \rho_2 (t \pm t_{z-z'})$	$-48\cdot89$

Even if there is no overlapping in time responses, as would be the case if $z' = 2$ m and $z = 1$ m, Table 11.1 shows that response levels are still affected by multiple reflections. The figures for ρ_1 and ρ_2, already given, are 0·316 and 0·0018, respectively, and the amplitudes are shown in a decibel ratio to the forward-wave amplitude A. Each term is separated from the others by the propagation time differences between reflections. For instance, the third term $\rho_1 (2 - \rho_2^2)\delta(t \pm t_{z'})$, predominantly due to the fault in the line, and the fourth $\rho_2 (2 - \rho_1^2 - \rho_1^2 \rho_2)\delta(t \pm t_{z+z'})$, due to the load connector, are separated by $t_{z'} - t_z$, or, from eqn. 11.10, by

$$(z' - z) = \frac{v_p(t_{z'} - t_z)}{2}$$

ρ_1 is a -10 dB return loss, but it is doubled in the Table to give an expected -4 dB. An error of 0.01 dB occurs due to the effects of $\rho_1\rho_2^2$. Similarly ρ_2, at -35 dB return loss becomes -29.8 dB, when doubled and the extra error terms added. These are the expected quantities that are observed in the presence of the remaining spurious responses caused by multiple reflections and the processes of the Fourier transformaton. If these extra responses are not separated from the true ones by at least the resolution distance, the measurement becomes impossible. The limit is set by the scan range. For instance, if S is from 72.94 MHz to 18.6 GHz, the resolution distance is 0.5 cm. In practice as the line length to the discontinuity increases, the resolution deteriorates due to imperfections in the line, which are in turn the cause of attenuation and dispersion. Typically, at 30 m the resolution is 30 cm.

Spectral spreading due to finite scan width is sometimes referred to as spectral leakage. It can be reduced by shaping the amplitudes of the input spectrum in a process called windowing, always at the expense of time resolution, but often with a marked improvement against the effects of spurious responses. A full discussion follows in the next Section on vector time-domain reflectometry.

11.3 Time-domain from frequency-domain measurements

The essential property of time-domain techniques is the ability to distinguish events, such as reflections, by their time separations. Discontinuous time functions, such as pulses or step functions, suffer distortions owing to the different propagation times of their frequency components and to the spectral sensitivity of reflecting objects; but provided these distortions are not too great, time-resolved reflections can identify the positions of obstacles in transmission media. Perhaps the best known example of this is pulsed radar, where the medium is relatively unbounded and the reflecting objects take a variety of forms. The frequency spectrum of discontinuous time functions consists of continuous waves over an infinite frequency range, each wave behaving differently in propagating and reflecting from objects in the medium. If we wish to achieve a time discrimination from spectral measurements, the incident spectra are restricted to those that generate appropriate time functions. We must therefore first review the principles of time-domain measurements to show how they relate to the more powerful techniques of frequency-domain analysis.

Pulses of suitable shape can be sent into a network and the locations of propagation discontinuities determined from their return delays. Step functions, with spectra extending to zero frequency, have been most extensively used for such measurements. A schematic illustration of the components of a time-domain test system is shown in Fig. 11.5. The step generator triggers a sampling oscilloscope as a step function enters the DUT. In this example, both the reflected and transmitted returns are taken to the sampled display. The step rise time is typically 150 ps with a constant level at $t > 0$ for a period in excess of

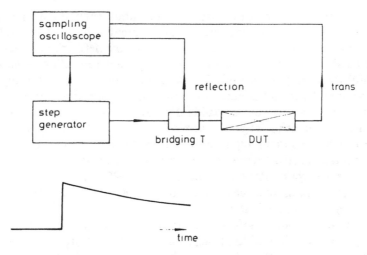

Fig. 11.5 *Time domain with sampling oscilloscope*

$2\,\mu$s. Resolution depends on the rise time, which is inversely related to the bandwidth of the step function. In a good non-dispersive cable the distance resolution is about 2 cm. 4000 separate samples are taken every 20 ms, corresponding to a pulse-repetition frequency of $5\,\mu$s or an equivalent of 700 ms in free space.[6]

Time-domain measurements of this form suffer a number of disadvantages, many of which can be overcome by working in the frequency domain. But the shape of the waveforms depends on recovery of phase as well as amplitude; hence a vector network analyser is essential if the accuracy of the original time-generated time function is to be retained. Discrete frequency-domain test sets, illustrated in Fig. 11.6, are similar to conventional reflectometer arrangements. The sweeper is synthesised for accurate phase determination, but phase lock and roll can also be used if greater speed is essential. Unlike the scalar analyser for cable-fault location, the vector-analyser error-correction procedures eliminate reflection and transmission errors to the level of the standards,

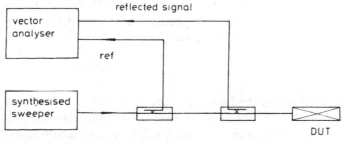

Fig. 11.6 *Frequency domain with vector analyser*

but at the expense of increased measurement time. Input time functions, no longer restricted to a step function, are approximated by their equivalent frequency spectra within the available bandwidth of the network analyser and the natural bandwidth of the device under test. High time resolution demands wide bandwidths, typically 0—20 GHz for about 1 cm resolution, but many devices have operating bandwidths much less than this, giving reflection or transmission returns of unwanted out-of-band information that can obscure the in-band components, or causing large numbers of low-level out-of-band samples near the noise level. The main application of time-domain reflectometry from frequency-domain measurements is therefore a diagnostic study of wide-band transmission networks. Reflection coefficient, VSWR, return loss etc. can also be displayed, with, in many cases, the extra advantage of separating the causes and showing their individual effects.

Common input time waveforms are an impulse, simulated by a sequence of repeating pulses of finite width; a step, simulated by a square wave; and an RF pulse, simulated by a selected range of frequency spectral lines. System bandwidth determines the limits to pulse and line widths, whilst line separation in the frequency spectrum, the reciprocal of time between pulses, determines the aliasing time or distance range. Periodic simulations can be expressed as infinite Fourier series expansions of the time waveform.[2,3] Thus, if $V_i(t)$ is the input-simulated time function with repetition period T,

$$V_i(t) = \frac{V_{c0}}{2} + \sum_{k=1}^{\infty} \left[V_{ck} \cos \frac{2\pi kt}{T} + V_{sk} \sin \frac{2\pi kt}{T} \right] \tag{11.13}$$

with spectral components

$$V_{ck} = \frac{2}{T} \int_{-T/2}^{+T/2} V_i(t) \cos \frac{2\pi kt}{T} \, dt \tag{11.14}$$

$$V_{sk} = \frac{2}{T} \int_{-T/2}^{+T/2} V_i(t) \sin \frac{2\pi kt}{T} \, dt \tag{11.15}$$

The infinite summation of frequency components in eqn. 11.13 is not possible in practice, since it is truncated by the available bandwidth of the source. For example, if $V_i(t)$ is a single impulse repeating every T seconds, it can be represented as an infinite summation of δ functions

$$\sum_{n=0}^{\infty} \delta(t - nT) \tag{11.16}$$

This can be simulated in the frequency domain by a finite number, $k_n + 1$, of equal amplitude carriers, one at zero frequency and the others equally spaced by the fundamental frequency f_1, where $f_1 = 1/T$. The highest-frequency line is

$k_n f_1$, if k in eqn. 11.13 runs from 0 to k_n. Substitution of the δ functions in eqns. 11.14 and 11.15 shows that $V_{sk} = 0$ and V_{ck} consists of the equally spaced spectral lines. Writing the summation for $V_i(t)$ in exponental form allows the use of a two-sided rectangular base-band spectrum indicated by the dotted lines in Fig. 11.7. The equivalent time function in two-sided form is

$$V_i(t) = \sum_{-k_n}^{+k_n} V_{ck} \exp j \frac{2\pi k}{T} t \tag{11.17}$$

For a set of equal amplitude components V_c, the summation is

$$V_i(t) = V_c \frac{\sin 2k_n(\pi/T)t}{\sin (\pi/T)t} \tag{11.18}$$

In the region of the main lobe and if $t \ll T$, $V_i(t)$ approximates to a $(\sin X)/X$ function, or

$$V_i(t) = 2k_n V_c \frac{\sin 2k_n(\pi/T)t}{2k_n(\pi/T)t} \tag{11.19}$$

Time responses repeat in intervals T, called the aliasing time, given by the reciprocal of the frequency spectral line separation. They are also broadened by spectral truncation that could cause overlap if the response width is comparable with the aliasing or pulse-repetition time. An estimate of the width can be found

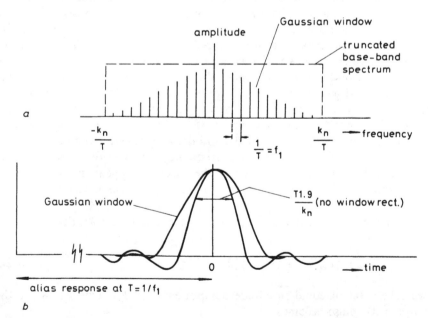

Fig. 11.7 *Frequency sampling, time aliasing and windowing at base-band*

as in eqn. 11.11 by calculating twice the time to the $-6\,dB$ point in eqn. 11.18. Thus, for the $-6\,dB$ point,

$$2\pi \frac{k_n t}{T} = 1\cdot 9 \tag{11.20}$$

giving the width as

$$\Delta t = \frac{T}{k_n} \frac{1\cdot 9}{\pi} \tag{11.21}$$

The narrower time response in Fig. 11.7b is a $(\sin X)/X$ function, corresponding to the rectangular spectrum of Fig. 11.7a. Thus to avoid overlap, after reflection or transmission, between preceding and following pulses, they should be separated by no less than Δt. Accompanying side lobes spread in theory to infinity, and can cause spectral leakage between separated time responses if the separation times allow overlap of close-in side lobes. For instance the first side-lobe level is at $-13\,dB$ for a $(\sin X)/X$ function, and this may sometimes appear as a spurious response after interaction with side lobes from other wanted components. Leakage can be dammed at the cost of increased main-lobe width (or decreased resolution), by application of a window function to the frequency spectrum, as illustrated by the Gaussian tapering of the amplitudes in Fig. 11.7a. The input spectrum is first multiplied by the window function and then Fourier transformed to reveal the broader main lobe in Fig. 11.7b. Since the Fourier transform of a Gaussian is another Gaussian function; the resultant time pulse has no side lobes, but its increased main-lobe width reduces the overlap distance to the next pulse, thus degrading time resolution.

When equivalent time waveforms have DC levels they generate base-band spectra with a line at zero frequency. Radio-frequency pulses, on the other hand, have band-pass spectra since there is no average DC level. They can be simulated by a finite number k_n of carriers, equally spaced in the designated pass band of the radio-frequency carrier. If the line separation is Δf, the aliasing, or pulse-repetition, time T is $1/\Delta f$, and the lines spread from $f_B - k_n/2T$ to $f_B + k_n/2T$, where f_B is the RF carrier frequency. In two-sided form again by writing eqn. 11.13 in exponential terms, the spectrum appears in two bands at $\pm f_B$, as shown in Fig. 11.8, and eqn. 11.17 for $V_i(t)$ becomes

$$V_i(t) = \sum_{-k/2}^{+k/2} V_{ck}[\exp j2\pi(-f_B + kt/T) + \exp j2\pi(f_B + kt/T)]$$

$$= \sum_{-k/2}^{+k/2} V_{ck} \exp\left(j\frac{2\pi kt}{T}\right)[2\cos(2\pi f_B t)] \tag{11.22}$$

If we take a set of equal-amplitude components for V_{ck}, eqn. 11.18 for the band-pass RF pulse becomes

$$V_i(t) = V_c \frac{\sin k_n(\pi/T)t}{\sin(\pi/T)t}[2\cos(2\pi f_B t)] \tag{11.23}$$

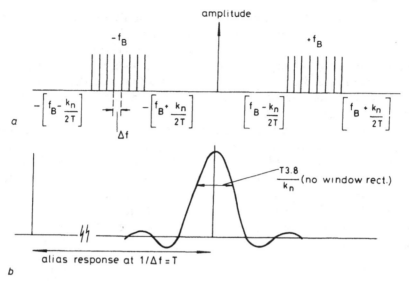

Fig. 11.8 *Frequency sampling and time aliasing for RF pass-band*

with a $-6\,\text{dB}$ width of

$$\Delta t = 2\frac{T}{k_n}\frac{1\cdot 9}{\pi} \tag{11.24}$$

These transformations are illustrated in Fig. 11.8, where the width of the spectrum and the time resolution are effectively halved compared to the base-band impulse or step function.

Choice of window shape is a matter of experience and judgment about the particular discontinuity to be investigated. A comprehensive review can be found in the literature.[1] Among the most popular are the Gaussian, because of its great power in reducing side lobes, and the Kaiser–Bessel for its high resolution between closely spaced carriers, as in two-tone measurements. With no windowing and for equal-amplitude carriers we can see that, for a base-band simulation, with scan width $S = k_n/T$, eqn. 11.21 gives a time resolution of

$$\Delta t \approx \frac{0\cdot 6}{S} \tag{11.25}$$

and for band pass we have from eqn. 11.24,

$$\Delta t \approx \frac{1\cdot 2}{S} \tag{11.26}$$

Windowing increases these by a factor from 1 to 2·4, depending on the chosen carrier weightings.

In the analysis of the cable-fault location there is an implicit assumption of base-band simulation, even though the lowest frequency was stated as 72·9 MHz. In practice, zero frequency is not available from swept-frequency sources and there would be difficulties in detecting a DC response. This can be overcome by extrapolating the low-frequency response to zero frequency, though that may lead to trace bounce, e.g. on a low-pass step response with low-frequency terms near the noise level. Signal averaging reduces the effect but does increase measurement time.[4]

So far, we have discussed only the input spectrum and its Fourier transform to a specified time waveform, whereas it is, in fact, the reflected or transmitted spectra that are transformed to reveal the nature and locations of obstacles we wish to isolate for, or eliminate from, measurement. Thus we begin with input spectral components V_{ck}, V_{sk}, weighted according to the required equivalent time and window functions, and successively find the response of the DUT to each one. Thus in complex form, if S_{ok} is the kth reflection or transmission coefficient,

$$S_{ok} = S_k \exp j\theta_k \tag{11.27}$$

where S_k is the magnitude and θ_k the phase. The two kth terms of eqn. 11.13 are modified after reflection or transmission in the DUT to

$$V_{ok} = S_k \left\{ V_{ck} \cos\left(\frac{2\pi}{NT} kt + \theta_k\right) + V_{sk} \sin\left(\frac{2\pi kt}{NT} + \theta_k\right) \right\} \tag{11.28}$$

When the DC response, found from extrapolation of the low k responses, is included, a truncated version of the time-domain reflection follows from eqns. 11.13 and 11.28 as

$$V_0(t) = V_{DC} + \sum_{k=1}^{N-1} S_k \left\{ V_{ck} \cos\left(\frac{2\pi kt}{NT} + \theta_k\right) + V_{sk} \sin\left(\frac{2\pi kt}{NT} + \theta_k\right) \right\} \tag{11.29}$$

If t_0 is the delay time to a discontinuity and back, the terms in eqn. 11.29 are measured complex samples with phase shifts

$$\theta_k = -2\pi k f_1 t_0 \tag{11.30}$$

Each term contributes to produce a time-response peak in $V_0(t)$ at $t = t_0$ with side lobes dependent on the applied window, and a range given by the aliasing time

$$t_R = \frac{1}{f_1} \qquad \text{for base band} \tag{11.31}$$

or

$$t_R = \frac{1}{\Delta f} \qquad \text{for pass band} \tag{11.32}$$

and a corresponding range $v_p t_R / 2$.

11.4 Interpretation of time-domain-reflectrometry displays

Before looking at some frequency-domain applications, it is essential that we briefly review earlier work on step functions, because, even though these were originally directly generated time functions, the interpretations we are to examine remain equally valid for the parametric time functions found from discrete spectral measurements. We will consider just three aspects of time-domain reflectometry (TDR), related to propagation in cables with unmatched load terminations, distributed losses and internal multiple reflections.[5-8]

11.4.1 Reflections due to unmatched terminations
The reflection coefficient at a load Z_L in a transmission line with characteristic impedance Z_0 is

$$\rho = \frac{V^-}{V^+} = \frac{Z_L - Z_0}{Z_L + Z_0} \tag{11.33}$$

where V^+ and V^- are the incident and reflected voltages, respectively. If the load is measured by TDR, a step voltage of amplitude V^+, launched at the cable input, takes time $t_0/2$ to reach the load, where a fraction V^- is returned to the input in a further time $t_0/2$ and is added to the original step function with a delay t_0. Some examples of the resulting waveforms at the input, measured in a sampling oscilloscope as in Fig. 11.5, are given in Figs. 11.9 and 11.10. Such Figures provide a catalogue of expected results from known terminations against which experimentally determined results can be compared to estimate the nature of unknown terminations. For instance, in Fig. 1.9a the incident voltage V^+ is reflected at an open circuit, so that V^- returns and adds to the input step after interval t_0. Thus $V^- = V^+$ because $\rho = 1$ when $Z_L \to \infty$ in eqn. 11.33. Similarly in Fig. 11.9b $V^- = -V^+$, since $Z_L = 0$, and the step returns to zero after time t_0. Partial reflections occur in Figs. 11.9c and d, since

$$V^- = \tfrac{1}{3}V^+, \qquad \text{with } \rho = \tfrac{1}{3} \qquad \text{if } Z_L = 2Z_0$$

and

$$V^- = -\tfrac{1}{3}V^+, \quad \text{with } \rho = -\tfrac{1}{3} \quad \text{if } Z_L = Z_0/2$$

In the remaining diagrams of Figs. 11.10a—d, the components of complex load impedances can be determined from the time constants, and the asymptotic amplitudes of exponential returns.

11.4.2 Lossy cable
Depending on whether a cable has series or shunt losses, the reflected edge of the step function has either a rising or falling exponential shape. This can be

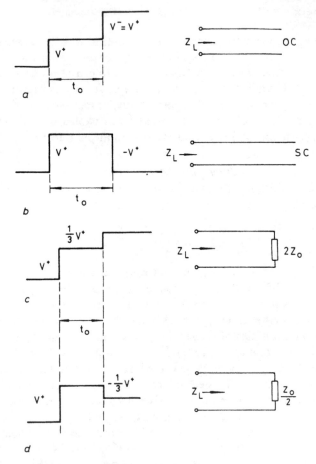

Fig. 11.9 *Characteristic responses to time step functions*
(a) Open circuit $V^- = V^+$
(b) Short circuit $V^- = -V^+$
(c) $Z_L = 2Z_0$
(d) $Z_L = \frac{1}{2}Z_0$

seen by considering an infinitely long lossy line with characteristic impedance given by

$$Z_0 = \sqrt{\frac{R + j\omega L}{G + j\omega C}}$$

When series losses predominate $G \ll \omega C$ and

$$Z_0 = \sqrt{\frac{L}{C}} \left(1 + \frac{R}{j\omega L}\right)^{1/2}$$

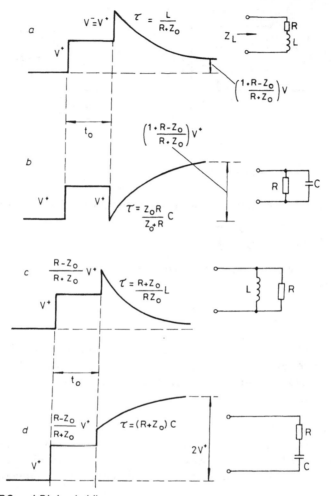

Fig. 11.10 *RC and RL loaded line responses*
a Series *RL*
b Shunt *RC*
c Shunt *RL*
d Series *RC*

which for small $R/\omega L$ is

$$Z_0 \approx \sqrt{\frac{L}{C}} \left(1 + \frac{R}{j2\omega L}\right)$$

The characteristic impedance therefore approximates to a series circuit with equivalent components $R' = \sqrt{L/C}$ and $C' = 2\sqrt{LC}/R$, and an equivalent circuit as shown in Fig. 11.11*a* with the step function response of Fig. 11.11*b*.

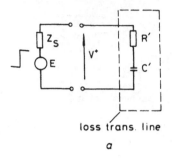
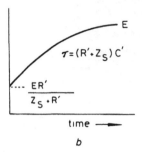

Fig. 11.11 *Lossy-line response—series loss*

When, on the other hand, shunt losses predominate $R \ll \omega L$ and

$$Y_0 = \sqrt{\frac{C}{L}}\left(1 + \frac{G}{j2\omega C}\right)$$

to give the equivalent shunt circuit and step-function response of Fig. 11.12*a* and *b*, in which

$$G' = \sqrt{\frac{C}{L}} \quad \text{and} \quad L' = 2\sqrt{LC}/G$$

The responses of real cables are likely to be much less easily related to particular properties either of the cable or of its termination, especially in the presence of multiple reflections between faulty or inappropriate connectors and damaged parts of the cable.

11.4.3 Multiple discontinuities

Time-pulsed TDR can handle situations involving more than one discontinuity. When the mismatched components are stretched sequentially along the line, their effects can be individually observed to determine their equivalent reactive

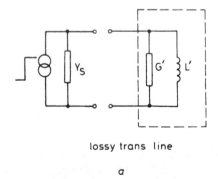
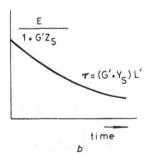

Fig. 11.12 *Lossy-line response—parallel loss*

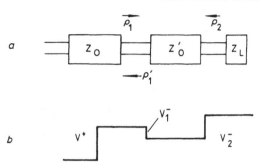

Fig. 11.13 *Multiple discontinuities—time domain*

or resistive values. Because of the time-separated observations, the Smith chart can be applied to each mismatch to eliminate or reduce its effects.[5] A simplified presentation of a multiple-reflection discontinuity appears in Fig. 11.13*a* with its TDR display in Fig. 11.13*b*. It shows two transmission lines of differing characteristic impedance cascaded and with a terminating load. Each reflection is time separated in TDR, thus allowing individual observation. However, if the multiple reflections are large, it becomes difficult to make useful measurements and the situation rapidly worsens as the number of discontinuities increases.

We have already seen that error correction for each frequency line in a frequency-domain measurement can eliminate the effects of at least the multiple reflections in the test set, but we have yet to examine how time gating can also be used with the Fourier transform to overcome the overlapping of frequency responses.

11.5 Frequency response from time gating

Vector analysis of discrete frequency responses, followed by Fourier transform to find the time responses, has significant advantages over direct generation of time functions. Most notable are the improved signal/noise ratio, due to the inherently narrower bandwidth and the availability of the total source power for each line. Though this is at the cost of increased measurement time compared to the broad-band time-generated waveforms, with modern computer-controlled vector network analysers, spectral methods are proving both more accurate and versatile. For example, the discrete Fourier transforms of frequency-domain reflectometry form a two-way link between the frequency and time domains, because of the equivalence of frequency and time samples (see Appendix 3). We can see from Fig. 11.14 how this can be exploited to find the reflection properties of a number of cascaded components in a network consisting of an air line, a 10 dB pad and a short circuit joined through a number of

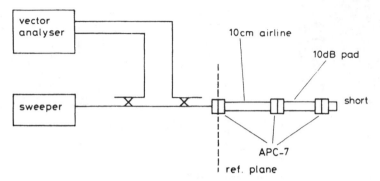

Fig. 11.14 *Multiple discontinuities—frequency domain*

APC–7 connectors.[2] From the Fourier transform of the discrete frequency responses in the range 0—17 GHz, the time response in Fig. 11.15 shows four separate reflection ranges, corresponding to the short circuit at a distance between 16 and 20 cm, the pad and centre APC–7 between 9 and 12 cm, and the test-set internal reflections up to the first APC–7. This nearest region, between 0 and 3 cm, is labelled CAL in the diagram to indicate that it could normally be eliminated by calibration at the test port. It is possible to find the frequency response for a single component, by taking a Fourier transform in a range restricted to the period of its time response. For example, in Fig. 11.15*b*, the VSWR of the short circuit follows from the transform of the peak occurring in the range 16—20 cm in Fig. 11.15*a*. Similarly, the two dotted lines are the

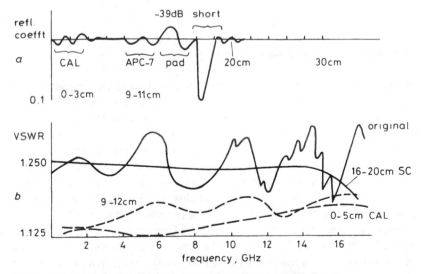

Fig. 11.15 *VSWR from time gating (After Hines and Stirehelfer, ref. 2)*

frequency dependence of VSWR for the CAL and centre APC–7, respectively. The original variation of VSWR with frequency, also shown in Fig. 11.15*b*, which includes all these effects simultaneously, can therefore be time-gated to reveal a corrected VSWR, provided the time-domain version of it sufficiently differentiates between the interfering reflections. Spectral leakage and time resolution are major factors affecting time separations, thus emphasising the importance of the initial windowing of the input spectral lines.

11.6 Time gating and scattering

Time-domain reflectometry, incorporating the error correction and computing power of a modern network analyser, is finding increasing application in antenna test-range measurements. The time-domain response often has a useful separation between the wanted signal, from the antenna or scatterer under test, and the unwanted, usually reflected, energy from the walls of an anechoic chamber or site objects on an open range. Provided path differences between returns are greater than the resolution length, the time response can be gated to select out any obscuring effects, before taking a Fourier transform for the corrected frequency display.

Radar targets are defined by their scattering cross-sections or equivalent scattering areas with reference to an isotropic radiator placed in the same incident field. Cross-sections are found in an anechoic chamber by illuminating the target from a transmitting antenna and measuring the reflected energy at a given angle at discrete frequencies in the operating bandwidth of the radar. Many targets with very small cross-sections, having reflections comparable to the background scattering from the walls of the chamber, can be measured only by applying the methods of time-gated frequency-domain reflectometry. One experimental layout is illustrated in Fig. 11.16, in which the swept source is in synthesised step mode to ensure accurate phase control.[9] The reference and test channels enter the test set, where they are down-converted by harmonic mixing before passing to the analyser. Transmit and receive antennas are located together and remote from the scattering DUT. Band-pass mode selects a range of frequencies to give an RF pulse, with, for instance a span from 600 MHz to 4·6 GHz, giving a site range (to aliasing) of 30 m if there are 401 sample points, and a resolution of 10—20 cm, depending on the window selected. These are well matched to the dimensions of typical anechoic chambers.

Error correction requires a number of known standards, depending on the complexity of the error-calibration model. Some simple objects are available, e.g. cylinders or spheres, whose scattering cross-sections are reliably determined from theory. A one-port error model in Fig. 11.17 requires only a single standard, and has the merit that it can take advantage of error-correcting software in the HP8510 analyser system by selecting parts of the resident S_{11} one-port error model. $E_{isolation}$ represents residual reflections from the chamber and direct

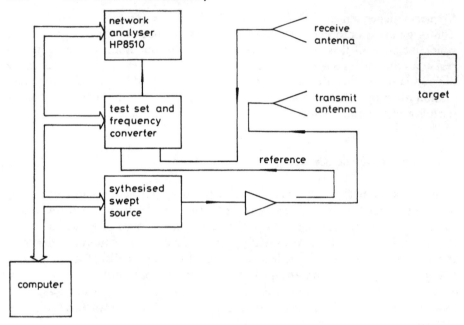

Fig. 11.16 *Scattering patterns from time gating*

leakage between the transmit and receive antennas; $E_{response}$ is the tracking error between reference and test channel; and S_{21} refers to the chamber transmission loss. There are two further reasons for this choice; namely the increased speed of a simple procedure and the ability to use the time-gating routine, available only with the internal error-correcting procedures[9] of the HP8510. These considerations may indicate a weakness in the instrumentation, but they do also illustrate the importance of fully appreciating the limitations of general-purpose instruments and the need sometimes to extend their use, either by extensive software development or by judicious choice of the standard software.

Fig. 11.8 illustrates the effects of time gating of the frequency-domain scattering cross-section of a metal sphere. In Fig. 11.18a we have the uncalibrated radar cross-section (RCS) frequency and time-domain responses of a 12 in-diameter sphere. The result of a one-port calibration procedure, in Fig. 11.18b,

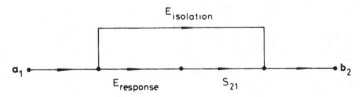

Fig. 11.17 *Reduced flowgraph approximation for range reflection and transmission*

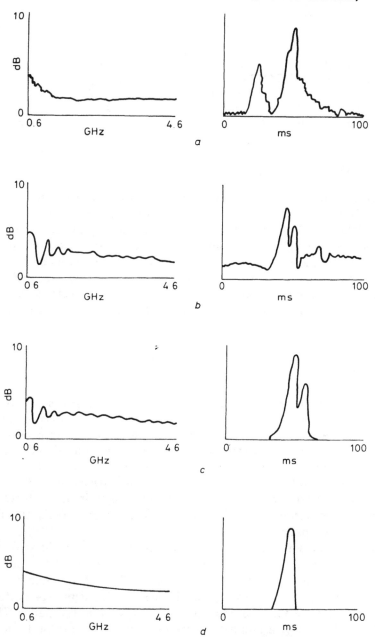

Fig. 11.18   *Scattering cross-section of a sphere from time-gated measurements*

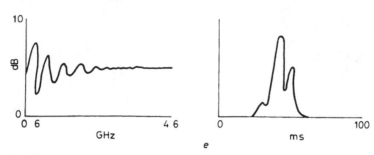

Fig. 11.18 *Continued*

is a reduction in chamber reflections and the clear emergence of the peaks of a sphere reflection in the centre of the time period. Time gating is next adopted in Fig. 11.18c to select the central time portion and display its spectrum. Fig. 11.18d is meant to highlight the importance of gate shape and width, by showing a smooth frequency response when only one of the reflection peaks is gated. The remaining slight slope with frequency was due to the characteristics of the reference, or calibration target. When that is subtracted vectorially, the final version in Fig. 11.18e closely resembles the theoretical RCS response of the sphere.

11.7 Time gating in transmission

All the examples given so far have been concerned with reflection, but there are important applications of time- and frequency-domain methods to transmission systems. Antenna radiation patterns are measured by transmission between a test and source antenna on ranges or in anechoic chambers, where significant site reflections may sometimes be unavoidable. If the site reflection paths are sufficiently greater than the direct path between the two antennas, time gating can again provide a means of separating their effects. Since bandwidth determines resolution and antennas are often narrow band, the method has only limited application.

As an example we consider the antenna measurement illustrated in Fig. 11.19, in which the receiver and transmitter are separated by 170 ft of cable, and the computer is extended between the site ends.[10] The test antenna is a dipole tunable in the range 50—150 MHz, though the swept-frequency span extends to 300 MHz to increase the time resolution to 2·5 ns or 1 m in distance. The test antenna can be rotated azimuthally in the far field of the transmitter to allow measurements in the frequency domain at 5° intervals, from which the equivalent time-domain responses can be found by Fourier transformation. Error correction is not necessary, since it is possible to eliminate residual errors after gating, by normalisation of the angular pattern at each frequency to the main bore-sight trace.

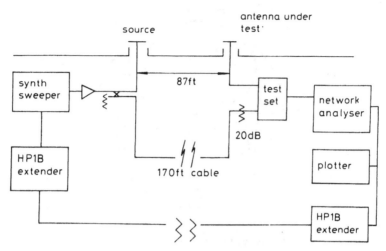

Fig. 11.19 *Antenna range for time-gated pattern measurement*

The traces in Fig. 11.20 are for 201 frequencies in the range 50—300 MHz and from 0 to 100 ns in parametric time. In Fig. 11.20*a* the antenna is at bore sight, showing a resonance at 115 MHz and a time response with two main regions: one due to the antenna, the other largely due to the effect of site reflection. This is confirmed in Fig. 11.20*b*, where the gated portion gives the smoother frequency peak typical of a theoretical dipole resonance. A narrower

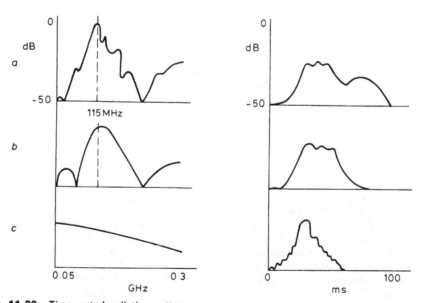

Fig. 11.20 *Time-gated radiation patterns*

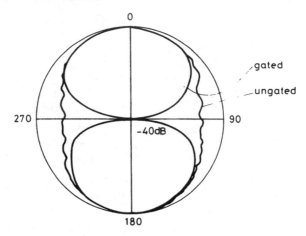

Fig. 11.21 *Comparison of gated and ungated polar patterns*

gate, as in Fig. 11.20c, clearly does not capture enough of the time trace to adequately recover the frequency response. A conventional polar plot for the test antenna at 125 MHz in Fig. 11.21 shows the improvement achieved by time gating. Apart from the limitations of narrow-band antennas in an essentially broad-band system, there is also a time penalty. 20 min are required for a full 360° polar scan at 5° intervals and 1 h to obtain a full frequency trace at one angular position.

11.8 Conclusion

Time-domain reflectometry requires large bandwidths for high resolution. Many devices are narrow band with high Q-factors that cause time delays and ringing, thereby preventing adequate time separation of the events at network discontinuities. It must therefore be considered carefully as one possible option among the methods available for network analysis.

11.9 References

1 HARRIS, F. J.: 'On the use of windows for harmonic analysis with the discrete Fourier transform', *Proc. IEEE*, 1978, **66**, pp. 51–83
2 HINES, M. E., and STIREHELFER, H. E.: 'Time oscillographic microwave network analysis using frequency domain data', *IEEE Trans.*, 1974, **MTT-22**, pp. 276–282
3 ULRIKSSON, B.: 'A time domain reflectometer using a semi-automatic network analyser and the fast Fourier transform', *IEEE Trans.*, 1981, **MTT-29**, pp. 172–174
4 'Operating and programming manual', HP8510 Network Analyser, March 1984
5 ADAM, S. F.: 'Microwave theory and application' (Prentice-Hall, 1969) p. 148
6 'Time domain reflectometry'. Hewlett–Packard Application Note 62

7 'Cable testing with time domain reflectometry'. Hewlett–Packard Application Note 67, 1965
8 'Selected articles on time domain reflectometry'. Hewlett–Packard Application Note 75, 1964
9 BOYLES, J. W.: 'Radar cross-section measurements with the HP8510 network analyser'. Hewlett–Packard Product Note 8510-2, April 1985
10 BOYLES, J.W.: 'Using a network analyser to measure the radiation pattern of a dipole antenna using time domain and gating'. ICAP 85, IEE Conference Publication 248, 1985, pp. 218–222
11 LYNN, P.A.: 'An introduction to the analysis and processing of signals' (Macmillan, 1982) p. 62

11.10 Examples

1   Show by integration of real functions that the Fourier response to a flat base-band spectrum from 0 to B has a 3 dB width of $1/2B$. By a similar method, prove for a band of frequencies from B_1 to B_2, where $B_2 - B_1 = B$ and all B are positive, that the 3 dB width of the Fourier transform is $1/B$. (Let the spectral components be $\cos 2\pi ft$.)

2   What is the site range and resolution length in an anechoic-chamber measurement using TDR and gating, if the frequency span is from 600 MHz to 4·6 GHz, and the number of frequency samples is 401? Describe briefly how gating can improve wanted or unwanted ratios in the measurements.

3   A matched 50 Ω coaxial line is connected to a length of 70 Ω coaxial line. Each is filled with a material of dielectric constant 2·54. The displayed delay from a step input to the 70 Ω line is 2·5 ns. Sketch the shape of the waveform at the input and calculate the length of the 70 Ω line.

Chapter 12

Antenna measurements

12.1 Introduction

Communication between remote sites by radio-frequency radiation depends on the proper design and operation of antennas. Many design models allow accurate and reliable prediction of performance, based on gain, beam shape, input reflection, polarisation, side-lobe level and noise temperature; but the complex and demanding specifications of satellite and radar applications have led to intensified developments and heightened the importance of antenna testing.[1,2,3] The end point of antenna testing is an assessment of far-field performance, and the most comprehensive way of recording the information is by means of far-field pattern and main-beam gain.

For a receiving antenna, the former is obtained directly by rotating the test antenna in the far field of a transmitting source, and the latter by substitution and comparison with a standard antenna of known gain. With reciprocal radiators it is possible to obtain the same results irrespective of which transmits or receives. Patterns are three-dimensional, whereas rotation about a single axis produces a two-dimensional 'pattern cut'. For example, in the spherical co-ordinate system of Fig. 12.1, with the test antenna at the origin, a pattern cut may be obtained by fixing one of the angles, θ, ϕ, and varying the other. Fixed θ gives a conical scan about z, sometimes called an azimuthal pattern, since ϕ is in an azimuthal plane with respect to the z-direction. It follows that elevation patterns are found by fixing ϕ and varying θ. By common consent an antenna's own spherical co-ordinates are referenced to the peak of its main beam, so that, for instance, azimuthal patterns, with the main beam and the distant transmitting source in the horizontal plane, are first found by tilting the antenna to $\theta = 90°$. Therefore if α is the elevation angle to the azimuthal x, y plane, $\alpha = 90 - \theta$ degrees. A clear distinction should therefore be made between the antenna-under-test co-ordinate system, whose z-axis now lies in the original x, y plane, and that of the turning gear on the test mount, whose axis remains along the original z-direction. Since it is common to tilt the test antenna by $-\alpha$ to obtain its azimuthal pattern at $+\alpha$, it follows that the mount, in these

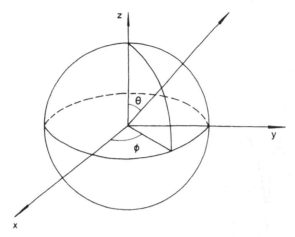

Fig. 12.1 *Spherical polar co-ordinates*

circumstances, must be turned to $\theta = 90 + \alpha$. An alternative method, sometimes adopted, points the z-axis of the turning gear along the horizontal towards the source and tilts the main beam in θ after first pointing it horizontally at $\theta = 0$, $\phi = 0$, so that the z-direction becomes a roll axis for ϕ variations. A complete pattern test for all polarisation entails a large programme of spatially orthogonal measurements over all values of θ, ϕ.

It is not necessary to go to the far field for information on complete patterns. Indeed it is often not practical to do so, particularly for some orientations, when, for instance, main-beam reflection from the ground may be greater than, say, the required level in a pattern null, or, again, large antennas may be too heavy, as for fixed radar systems, or too flimsy, as for satellite dishes, to be moved about every axis. Near-field testing aims to provide sufficient details of the fields very close to a radiator for calculating the far-field patterns. It has the advantage that all the patterns at any angles and for all polarisations can be calculated from one complete set of near-field results. Disadvantages include the high accuracies required, the cost and size of the test chamber, dependence on expensive software for field transformations and measurement times of hours or even days.

In the following Sections we will consider the different methods for testing antennas against their design specifications. These performance factors are illustrated in Fig. 12.2, which should be regarded as a compendium of features rather than specific to one example. There are theoretical relations between all these features that, when violated in the test results, should cause the experienced experimenter to question either the test set or the original design. For instance, beam width and gain are related, as are side-lobe level and beam shape. It is unavoidable that we should cover a number of basic theoretical topics, but the underlying theory of antenna measurements is rightly to be

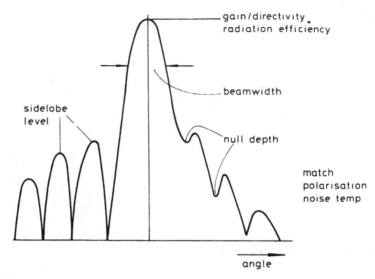

Fig. 12.2 *Antenna pattern features*

found only in the general literature[2,5,6,9] and forms too large a subject to be treated fully here. Our discussion uses aperture theory to review the fundamentals of far-field patterns, gain and effective aperture, before treating an antenna pair as a network and proceeding to an examination of the near, intermediate and far-field boundaries. With this minimal background we can then look at far-, intermediate- and near-field testing; polarisation measurements; and a selection of techniques relating to antenna noise temperature and to null accuracy in tracking antennas.

12.2 Radiation from apertures

Fields radiated at great distance from an antenna are often approximated as a plane wave. For example, an isolated source in free space radiates equal quantities of energy in every direction, so that it spreads evenly over the surface of a large sphere. Its amplitude decreases inversely as, and its phase delays proportionally to, distance from the source. Closer to the source, where the antenna dimensions become comparable to the radiation distance, neither amplitude nor phase are simple functions of distance. Moving away from the radiation source, there are three identifiable regions without sharp boundaries, dividing the radiation space into reactive near field, a radiating near field (Fresnel region) and the far field (Fraunhofer region). Far-field measurements are often based on several implicit approximations, such as orthogonal separability of field distributions in radiating-aperture planes. Because of the simple scalar relationship that exists between aperture and far fields via the Fourier transform we will base our theoretical discussion on aperture antennas, whilst noting that the fundamental concepts of far-field gain, beam width, side lobes etc., and the equally fundamental ones of near and intermediate fields, are similar for all antennas.

An ideal radiating aperture is one for which the total field can be calculated. Real apertures will generally fall short of the ideal, and may even behave so differently as to render useless the ideal predictions. If that is so a better ideal model is sought. One ideal form, giving a good model for radiation from horns and parabolas, consists of an opening in an infinite conducting plane behind which is a region containing radiating sources. If all the sources are enclosed in an infinitely conducting surface, one part of which fills the original aperture, currents can be placed on the surface to generate the original fields external to the enclosure, whilst inside the source region all fields are zero. In the half-space to the right in Fig. 12.3, J_M, an equivalent magnetic current density over the aperture, radiates the fields due originally to the sources. J_M causes radiation inwards and outwards, but the conducting surface, shown dotted in the aperture, reflects the inward wave so that, if E_a is the aperture field,

$$J_M = 2E_a \times n \tag{12.1}$$

An integral solution to a vector wave equation for the electric field E, generated by J_M, can be found by writing

$$E = -\frac{1}{\varepsilon} \nabla \times F \tag{12.2}$$

where ε is the permittivity of the propagating region and F is the electric field vector given by

$$F = \frac{\varepsilon}{4\pi} \int_S \frac{J_M}{R} \exp\left(-jk \cdot R\right) ds \tag{12.3}$$

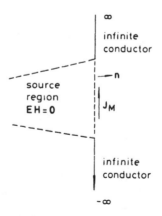

Fig. 12.3 *Radiating aperture in an infinite conducting plane*

where k is the propagation constant or vector wave number. The integration has to be carried out over the aperture. J_M is defined in terms of aperture fields in eqn. 12.2, and the dependency of R on the aperture co-ordinates can be determined by referring to Fig. 12.4, where R is the distance from a point P' in an aperture in the $(x, y, 0)$ plane to a point P in the right half-space of Fig. 12.3; r and r' are the distances of P and P', respectively, from an arbitrary origin in the aperture. a_R, a_r and $a_{r'}$ are unit vectors, so that with θ' defined in the triangle OPP', such that

$$r' \cos \theta' = a_r \cdot r'$$

$$r' = x' a_x + y' a_y \tag{12.4}$$

it is possible from the cosine rule to find

$$R^2 = r^2 + r'^2 - 2rr' \cos \theta' \tag{12.5}$$

and to expand this binomially as

$$R = r - r' \cos \theta' + \frac{1}{r} \left(\frac{r'^2}{2} \sin \theta' \right) + \frac{1}{r^2} \left(\frac{r'^3}{2} \cos \theta' \sin^2 \theta' \right) + \cdots \tag{12.6}$$

The phase-path difference between r and R is

$$k \cdot (R - r) = -kr' \cos \theta' + \frac{k}{r} \left(\frac{r'^2}{2} \sin \theta' \right) + \frac{k}{r^2} \left(\frac{r'^3}{2} \cos \theta' \sin^2 \theta' \right) + \cdots \tag{12.7}$$

The total field at P is the sum of contributions from each point P' in the aperture, according to the integral of eqn. 12.3, with the current density from

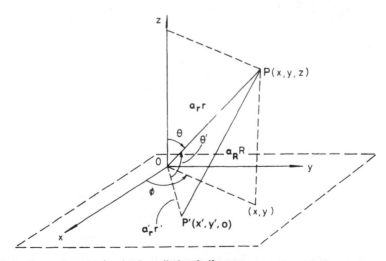

Fig. 12.4   *Co-ordinates of point in radiation half-space*

eqn. 12.1. For the far-field condition that all contributions should be as plane waves at P, terms in $1/r$ on the left of eqn. 12.7 must be zero. This plane-wave condition occurs only when $r \to \infty$, with the only significant term being $kr' \cos \theta'$. The far-field phase approximation is therefore

$$\boldsymbol{k} \cdot \boldsymbol{R} = k_0 r - k_0 r' \cos \theta' \tag{12.8}$$

where, if λ_0 is the free-space wavelength,

$$k_0 = \frac{2\pi}{\lambda_0}$$

Now in Fig. 12.4, $r' \cos \theta' = \boldsymbol{a_r} \cdot \boldsymbol{r'}$ and $R \approx r$ with $\boldsymbol{r'} = x'\boldsymbol{a_x} + y'\boldsymbol{a_y}$ in the aperture where $z = 0$; and the unit vector $\boldsymbol{a_r}$ is related to the rectangular unit vectors as

$$\boldsymbol{a_r} = \sin \theta \cos \phi \boldsymbol{a_x} + \sin \theta \sin \phi \boldsymbol{a_y} + \cos \theta \boldsymbol{a_z} \tag{12.9}$$

to give, on substituting in eqn. 12.8,

$$\boldsymbol{k} \cdot \boldsymbol{R} = k_0 r - (k_0 x' \sin \theta \cos \phi + k_0 y' \sin \theta \sin \phi) \tag{12.10}$$

On substituting for $\boldsymbol{k} \cdot \boldsymbol{R}$ in eqn. 12.3 and with $\boldsymbol{J_M} = 2\boldsymbol{E_a} \times \boldsymbol{a_z}$, the vector potential $\boldsymbol{F}$ at $P(x, y, z)$ is

$$\boldsymbol{F} = \frac{\varepsilon}{4\pi r} \exp(-jk_0 r) \int_x \int_y 2\boldsymbol{E_a} \times \boldsymbol{a_z} \exp j(k_0 x \sin \theta \cos \phi + k_0 y \sin \theta \sin \phi) \, dx \, dy$$

The primes are omitted on x and y, since the integration is confined to the aperture. But the aperture field,

$$\boldsymbol{E_a} = \boldsymbol{a_x} E_x + \boldsymbol{a_y} E_y$$

to give

$$\boldsymbol{E_a} \times \boldsymbol{a_z} = -\boldsymbol{a_y} E_x + \boldsymbol{a_x} E_y$$

Therefore

$$\boldsymbol{F} = \frac{\varepsilon}{4\pi r} \exp(-jk_0 r)[f_y(\theta\phi)\boldsymbol{a_x} - f_x(\theta\phi)\boldsymbol{a_y}] \tag{12.11}$$

where

$$f_x = \int_x \int_y E_x(x, y) \exp j(k_0 x \sin \theta \cos \phi + k_0 y \sin \theta \sin \phi) \, dx \, dy \tag{12.12}$$

and

$$f_x = \int_x \int_y E_y(x, y) \exp j(k_0 x \sin \theta \cos \phi + k_0 y \sin \theta \sin \phi) \, dx \, dy \tag{12.13}$$

For the ideal aperture of Fig. 12.3, we can now find the far-field electric components from the curl of F. But in the far-field limit only r variations of F are significant. This is most easily seen by dividing the aperture into m infinitesimal areas, each with elemental current J_M, and summing the curls to give the electric field from eqn. 12.3. Thus

$$E = -\frac{1}{4\pi} \int_S \frac{\nabla \times J_M}{R} \exp(-jk \cdot R)\, ds$$

and in the limit when $R = r$ and $k = k_0$, the electric-vector contribution from each element is

$$F_M = \frac{J_M}{r} \exp(-jk_0 r)$$

and is a function only of r, with a curl

$$\nabla \times F_M = a_r(0) - \frac{a_\theta}{r}\frac{\partial(rF_{M\phi})}{\partial r} + \frac{a_\phi}{r}\frac{\partial(rF_{M\theta})}{\partial r}$$

It follows from eqn. 12.2 that the electric field in spherical co-ordinates is

$$E_\theta = -jk_0 F \cdot a_\phi$$

$$E_\phi = jk_0 F \cdot a_\theta$$

$$E_r = 0 \tag{12.14}$$

Since $a_\phi = -\sin \phi a_x + \cos \phi a_y$ and $a_\theta = \cos \theta(\cos \phi a_x + \sin \phi a_y)$ in eqn. 12.11 for F, the far electric fields become

$$E_\theta = \frac{j}{\lambda r} \exp(-jk_0 r)[\, f_x(\theta\phi)\cos\phi + f_y(\theta\phi)\sin\phi] \tag{12.15}$$

$$E_\phi = \frac{j}{\lambda r} \exp(-jk_0 r)\cos\theta[\, f_y(\theta\phi)\cos\phi - f_x(\theta\phi)\sin\phi] \tag{12.16}$$

with f_x and f_y as in eqns. 12.12 and 12.13.

Simple forms of these expressions result if we consider the principal planes defined by $\phi = 0$ and $\phi = \pi/2$, respectively, in Figs. 12.5a and b. In both cases the aperture lies in the x, y plane, whereas r lies in the x, z plane in the former and the y, z in the latter. With $\phi = 0$, and at $\theta = 0$, E_θ and E_ϕ are parallel to x and y, respectively, whereas with $\phi = \pi/2$, E_θ and E_ϕ are reversed with respect to x and y. Corresponding expressions for the principal-plane far fields are:

with $\phi = 0$

$$E_\theta = \frac{j}{\lambda r} \exp(-jk_0 r)f_x(\theta\phi)$$

$$E_\phi = \frac{j}{\lambda r} \exp(-jk_0 r)f_y(\theta\phi)\cos\theta \tag{12.17}$$

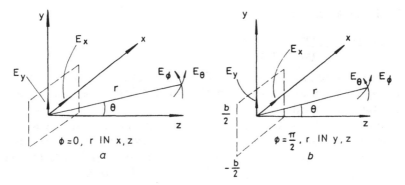

Fig. 12.5 *Far-field components from a radiating aperture*

with $\theta = \pi/2$

$$E_\theta = \frac{j}{\lambda r} \exp(-jk_0 r) f_y(\theta\phi)$$

$$E_\phi = \frac{j}{\lambda r} \exp(-jk_0 r) f_x(\theta\phi) \cos\theta$$

(12.18)

The expressions for eqns. 12.15 and 12.16 shows that E_θ and E_ϕ are functions of θ and ϕ, and can be used to calculate the radiation patterns, or the power density at specified angles in space. The two cuts at $\phi = 0$ and $\pi/2$ can be found by plotting E_θ and E_ϕ according to eqns. 12.17 and 12.18; and other cuts are possible by suitable selection of a fixed θ and ϕ. These, of course, give only the field pattern, but they can be converted to power patterns through the Poynting vector, $S = E \times H$, with the magnetic field found from

$$H = \frac{-a_r \times E}{\zeta}$$

(12.19)

where ζ is the intrinsic impedance of free space.

As a simple example we will take the case when $\phi = \pi/2$ with an aperture field E_y that is constant over x and y, so that $E_x = 0$, $f_x = 0$ and

$$f_y = \int_x \int_y E \exp(jk_0 y \sin\theta) \, dx \, dy$$

giving

$$E_\theta = \frac{j}{\lambda r} \exp(-jk_0 r) Ea \int_{-b/2}^{+b/2} \exp(jk_0 y \sin\theta) \, dy$$

(12.20)

where a is the x- and b the y-dimension of the aperture.

Since the phase change and spatial-attenuation dependency on r does not affect the pattern shape, it is sufficient to write the solution to eqn. 12.20 in the usual $(\sin X)/X$ form as

$$E_\theta = Eab \frac{\sin\left[(\pi b/\lambda)\sin\theta\right]}{(\pi b/\lambda)\sin\theta} \qquad (12.21)$$

This pattern shape has $-13\,\text{dB}$ first side lobes and zeros at $(\pi b/\lambda)\sin\theta = n\pi$. We will often use it as a standard pattern to aid description, for instance, of monopulse and conical scan.

This completes the essential theory of radiation from apertures, and provides a background for the measurement techniques described in later Sections. We also have a model for the derivation of gain and directivity and for detailed discussion of the effects of polarisation on far-field patterns.

12.3 Directivity, gain, loss and effective aperture

If $P(\theta, \phi)$ is the power radiated by an antenna at angle θ, ϕ per unit solid angle, we know from the previous discussion of radiation patterns that it is not uniformly, or isotropically, distributed, but shows peaks and nulls in different directions. Integrating $P(\theta, \phi)$ over a sphere gives the total power radiated as

$$P_0 = \int_0^{2\pi}\int_0^{\pi} P(\theta, \phi)\sin\theta\,d\theta\,d\phi \qquad (12.22)$$

If this were radiated isotropically, the power per unit solid angle would be

$$P_{iso} = \frac{P_0}{4\pi}$$

The directivity of the antenna at angle θ, ϕ is defined as the power density at θ, ϕ as a ratio of the power density from an isotropic source radiating the same total power. Thus if $D(\theta, \phi)$ is the directivity

$$D(\theta, \phi) = \frac{\text{Power per unit solid angle}}{\text{Isotropic power per unit solid angle}}$$

$$= \frac{P(\theta, \phi)}{P_0/4\pi}$$

and with P_0 calculated from eqn. 12.22, this becomes

$$D(\theta, \phi) = \frac{4\pi P(\theta, \phi)}{\iint P(\theta, \phi)\sin\theta\,d\theta\,d\phi} \qquad (12.23)$$

The majority of antennas have a main beam with a power density exceeding the isotropic density and a side-lobe region with power levels well below the isotropic level. An aperture illuminated with an equi-phase wave and axes as in

Fig. 12.5 has a main beam at $\theta = 0$ and a directivity

$$D(0, \phi) = \frac{4\pi P(0, \phi)}{\int\int P(\theta, \phi) \sin\theta \, d\theta \, d\phi} \qquad (12.24)$$

Since total power radiated requires accurate knowledge of a radiation pattern that may be difficult to calculate or measure, it may not always be possible to integrate the far field in eqn. 12.22, but, in principle, P_0 could be found from the power P_T at the input terminal of the antenna. However, this will result in error, because of losses and/or mismatch in the antenna feed and radiating parts.

The gain of an antenna is given by

$$G = \frac{4\pi P(\theta, \phi)}{P_T} \qquad (12.25)$$

and is similar to the directivity in eqn. 12.23, except that attenuation and mismatch losses are included. P_0 is less than P_T in a ratio η, called the radiation efficiency, defined in terms of gain and directivity by setting $P_T = P_0/\eta$ in eqn. 12.25 to give, from eqn. 12.23,

$$G = \eta D \qquad (12.26)$$

There is a further loss due to polarisation that occurs when, for instance, a receiving antenna is not of the same polarisation as that of an incoming wave. If the incident wave is polarised, there is a phase shift between the orthogonal components E_x, E_y, so that the electric vector E_i is

$$E_i = a_x E_x + a_y E_y \exp(j\alpha_i)$$

Similarly, the electric vector in the receiving antenna is

$$E_a = a_x E_x + a_y E_y \exp(j\alpha_a) \qquad (12.27)$$

The polarisation loss factor p is often expressed as

$$p = \cos^2\phi \qquad (12.28)$$

where ϕ is the mean of the phase angles α_i and α_a between the orthogonal polarisation pairs in the receiving antenna and the incoming wave. Polarisation loss factor is discussed in greater detail in Section 12.12.3.

Since directivity increases with decreasing beam width or, in the case of aperture antennas, with increasing aperture area, it should be possible to define an effective radiating area for any antenna, even when, as for a Yagi array, its physical structure has little resemblance to a radiating aperture. To illustrate the relationship between directivity and effective aperture, consider an aperture polarised with an electric field E_y. When $\theta = 0$ eqns. 12.17 and 12.18 reduce to a single form

$$E(0, \phi) = \frac{j}{\lambda r} \exp(-jk_0 r) \int_x \int_y E_y \, dx \, dy \qquad (12.29)$$

with corresponding magnetic field as in eqn. 12.19. Power flow in the far field at the peak of the main beam is

$$S = \tfrac{1}{2}\mathrm{Re}\,|E \times H| = \frac{\tfrac{1}{2}|E(\theta, \phi)|^2}{\zeta} \tag{12.30}$$

and the power per unit solid angle is

$$P(0, \phi) = r^2 S \tag{12.31}$$

which, on substitution from eqns. 12.29 and 12.30, becomes

$$P(0, \phi) = \frac{1}{2\zeta\lambda^2}\left|\int\limits_x \int\limits_y E_y \, dx \, dy\right|^2$$

The total power radiated is the power in the aperture fields found by integrating $|E_y|^2/2\zeta$ over the aperture to give

$$P_0 = \int\limits_x \int\limits_y \frac{|E_y|^2}{2\zeta} \, dx \, dy$$

and the directivity is

$$D(0, \phi) = \frac{4\pi P(0, \phi)}{P_0}$$

$$= \frac{4\pi}{\lambda^2} \frac{|\int\int E_y \, dx \, dy|^2}{\int\int |E_y|^2 \, dx \, dy} \tag{12.32}$$

The expression

$$\frac{|\int\int E_y \, dx \, dy|^2}{\int\int |E_y|^2 \, dx \, dy}$$

has the dimensions of area, as we can see by considering the case of a uniformly illuminated aperture with x-dimension from $-a/2$ to $a/2$ and y from $-b/2$ to $b/2$. Eqn. 12.32 is then

$$D(\theta, \phi) = \frac{4\pi}{\lambda^2} ab$$

$$= \frac{4\pi}{\lambda^2} A \tag{12.33}$$

and the effective area is simply the aperture area A. Effective area is a maximum for a uniformly illuminated aperture and is therefore typically less than the physical area, because aperture fields are rarely unifom. A is written as A_e and becomes the effective aperture for directivity D.

12.4 Antenna-pair transmission network

Measurement uncertainties in well bounded network analysis are closely related to multiple reflections between mismatches at connectors and other component discontinuities. A transmitting and receiving antenna pair, treated as a two-port network, has the same kind of component discontinuities in the input and output circuits, but the transmission coefficients in the radiation region are greatly complicated by reflections from other objects, including the antenna supporting structures and the ground. Nevertheless, such a pair, illustrated in Fig. 12.6, can be modelled as the two-port shown in Fig. 12.7. S_{11} and S_{22} are the transmitter and receiver input reflection coefficients, and S_{21}, S_{12} account for all the effects of the radiation region between the antennas. At the receiver, distance r from the transmitter, the power density is

$$P_T = \eta_T \frac{P_0 D_T}{4\pi r^2} \tag{12.34}$$

where η_T is the transmitter radiation efficiency, D_T is its directivity and P_0 is the total transmitted power. The power entering the receive antenna is

$$P_R = P_T A_R \tag{12.35}$$

where A_R is the effective aperture of the receive antenna, which is related to the receiver radiation efficiency η_R and directivity D_R as

$$A_R = \eta_R \frac{D_R \lambda^2}{4\pi}$$

Total transmission loss between the antennas can be found by first connecting the generator directly to the receiver load, when the power received is

$$P_{R1} = \frac{\frac{1}{2}b_g^2(1 - |\rho_L|^2)}{|1 - \rho_L \rho_g|^2} \tag{12.36}$$

But the power available from the generator, given by the conjugate match

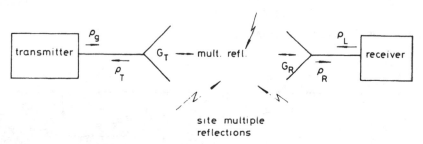

Fig. 12.6 *Transmitter–receiver pair*

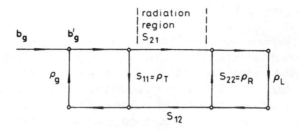

Fig. 12.7 *Flowgraph for transmitter–receiver pair*

condition $\rho_L = \rho_g^*$, is $P_{av} = \frac{1}{2}b_g^2/(1 - |\rho_g|^2)$, so that

$$P_{R1} = P_{av} \frac{(1 - |\rho_g|^2)(1 - |\rho_L|^2)}{|1 - \rho_L \rho_g|^2}$$

Next the two antennas are inserted and the power received becomes

$$P_{R2} = \frac{\frac{1}{2}b_g^2(1 - |\rho_L|^2)|S_{21}|^2}{|1 - \rho_g \rho_T - \rho_L \rho_R + \rho_g \rho_T \rho_L \rho_R - \rho_g \rho_L S_{21} S_{12}|^2} \tag{12.37}$$

If the antennas are well separated, the second-order term $\rho_g \rho_L S_{21} S_{12}$ is negligibly small, though in some measurements to be described, this will not always be the case. The ratio of eqns. 12.36 and 12.37 is then

$$\frac{P_{R2}}{P_{R1}} = \frac{|S_{21}|^2 |1 - \rho_L \rho_g|^2}{|1 - \rho_g \rho_T|^2 |1 - \rho_L \rho_R|^2}$$

Or in terms of power available

$$\frac{P_{R2}}{P_{av}} = \frac{|S_{21}|^2 (1 - |\rho_L|^2)(1 - |\rho_g|^2)}{|1 - \rho_g \rho_T|^2 |1 - \rho_L \rho_R|^2} \tag{12.38}$$

Ignoring the second-order term is equivalent to neglecting multiple reflections in the radiation region, and allows the transmitter to be considered in isolation according to the flowgraph of Fig. 12.8a. The radiated power, unaffected by external objects, is therefore[4]

$$P_0 = P_{av} \frac{(1 - |\rho_g|^2)(1 - |\rho_T|^2)}{|1 - \rho_g \rho_T|^2} \tag{12.39}$$

where P_{av} is the available power into a conjugate match.

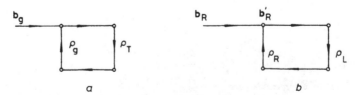

Fig. 12.8 *Transmitter and receiver flowgraphs*

The power density at the receiver is $\eta_T P_0 D_T / 4\pi r^2$, and if A_R is the effective aperture and η_R the radiation efficiency of the receiving antenna, the power entering the receiver is

$$\eta_R \eta_T \frac{P_0 D_T A_R}{4\pi r^2}$$

which gives, for the power into a matched receiver,

$$\tfrac{1}{2}|b_R|^2 = \eta_R \eta_T \frac{P_0 D_T A_R}{4\pi r^2}(1 - |\rho_R|^2) \tag{12.40}$$

b_R is the wave that would be received if $\rho_L = 0$ and becomes b'_R for an unmatched receiver as in Fig. 12.8b, to give for the power received at the load

$$P_{R2} = \frac{\tfrac{1}{2}|b_R|^2(1 - |\rho_L|^2)}{|1 - \rho_R \rho_L|^2}$$

On substituting for $\tfrac{1}{2}|b_R|^2$ from eqn. 12.40, this becomes

$$P_{R2} = \eta_R \eta_T P_0 \frac{D_T A_R(1 - |\rho_R|^2)(1 - |\rho_L|^2)}{4\pi r^2 |1 - \rho_R \rho_L|^2} \tag{12.41}$$

The addition of a polarisation loss factor p and substitution for P_0 from eqn. 12.39, gives for the power into the receiver load,

$$P_{R2} = P_{av} \left(\frac{\lambda}{4\pi r}\right)^2 p\eta_R \eta_T D_T D_R \frac{(1 - |\rho_g|^2)(1 - |\rho_L|^2)(1 - |\rho_R|^2)(1 - |\rho_T^2|)}{|1 - \rho_g \rho_T|^2 |1 - \rho_R \rho_L|^2} \tag{12.42}$$

in which A_R has been replaced by $\lambda^2 D_R / 4\pi$.

If we compare eqns. 12.38 and 12.42, the spatial transmission loss $|S_{21}|^2$ is

$$|S_{21}|^2 = \left(\frac{\lambda}{4\pi r}\right)^2 p\eta_R \eta_T D_T D_R(1 - |\rho_R|^2)(1 - |\rho_T|^2) \tag{12.43}$$

It is essential to emphasise that this expression accounts for all sources of loss, with the exception of the effects of multiple reflections between the antennas. It shows antenna match and polarisation loss directly affect the transmission efficiency, in a way that can be calculated from measured values. Overall received power in eqn. 12.42 is subject to the uncertainty of multiple reflections between generator and transmit-antenna input, and between receiver load and receive-antenna output. In the simplest case the antennas are co-polar with no polarisation loss and all components are matched to the transmission line. Thus $\rho_g = \rho_L = \rho_T = \rho_R = 0$ and P_{av} equals the measured power P_M from the generator into a matched load. With these assumptions, eqn. 12.42 reduces to the Friis transmission equation

$$\frac{P_{R2}}{P_M} = G_T G_R \left(\frac{\lambda}{4\pi r}\right)^2 \tag{12.44}$$

where the relation $\eta D = G$ has also been used. $(\lambda/4\pi r)^2$ is called the free-space loss factor.

12.5 Primary and secondary gain standards

The Friis formula is the basis of absolute gain measurements, for the provision of primary gain standards. Two antennas, with matched polarisation directions, are separated by distance r and Z_0-matched to their connecting transmission lines, as in Fig. 12.9. P_T is the measured power from the generator and P_R the power in the receiver load. If G_A and G_B are the antenna gains, the Friis equation in décibels is

$$G_A + G_B = 20 \log \left(\frac{4\pi r}{\lambda}\right) - 10 \log \frac{P_T}{P_R} \tag{12.45}$$

An absolute gain measurement can be made if the antennas are identical primary standards, when

$$G_A = G_B = \frac{1}{2}\left[20 \log \left(\frac{4\pi r}{\lambda}\right) - 10 \log \frac{P_T}{P_R}\right] \tag{12.46}$$

Only three quantities are required to be measured: one length, the frequency and a power ratio. If identical antennas are not available, three can be measured in pair combinations to give three simultaneous equations of the form

$$G_A + G_B = 20 \log \left(\frac{4\pi r}{\lambda}\right) - 10 \log \left(\frac{P_T}{P_R}\right)_{AB}$$

$$G_A + G_C = 20 \log \left(\frac{4\pi r}{\lambda}\right) - 10 \log \left(\frac{P_T}{P_R}\right)_{AC}$$

$$G_B + G_C = 20 \log \left(\frac{4\pi r}{\lambda}\right) - 10 \log \left(\frac{P_T}{P_R}\right)_{BC} \tag{12.47}$$

which can be solved for the gains on substituting for three measured power ratios.

Secondary gain standards are calibrated against primaries in substitution measurements to be described later, but Schelkunoff, using Fresnel integrals, has calculated the gain of pyramidal horns carefully manufactured as secondary standards. The aperture fields have a quadratic phase error, due to the spherical wave originating from a phase centre close to the apex of the slant heights l_a, l_b,

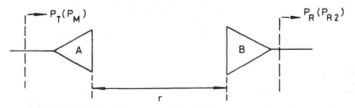

Fig. 12.9 *Showing separation of apertures and positions of input and output reference planes*

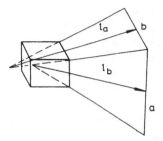

Fig. 12.10 *Standard-gain horn*

shown in Fig. 12.10. Silver[9] gives normalised gains as a function of a/λ with l_a/λ or l_b/λ as a running parameter, from which standard-gain horns may be calculated. Horns are available commercially in fixed frequency bands with gains from about 14—25 dB. Calculated calibration curves are accurate to $\pm 5\%$ above, and $\pm 12\%$ below, 2·6 GHz. A typical measured gain, for a secondary horn, shown against the smooth theoretical prediction in Fig. 12.11, has undulations that are probably due to multiple reflections between the aperture and the transition from waveguide near the throat.[4]

Horns are used as secondary standards from 0·35 to 90 GHz and dipoles[1] from 77 MHz to 1 GHz. Dipoles, normally tuned to half-wave resonance where the gain is 2·15 dB, are linearly polarised and may be affected by the feed lines and nearby objects. Horns are less susceptible to external effects, but have an inherent axial ratio of about 40 dB.

Gains of standards are transferred on far-field test ranges by finding the ratio of the responses P_S and P_t of the standard and test antennas at the same distance from a transmitting or receiving antenna. First, the test-antenna input reflection coefficient is measured before connecting to a matched receiver. It is then replaced with the matched standard to find the ratio of received powers as

$$G_t = (P_t/P_S)G_S \qquad\qquad (12.48)$$

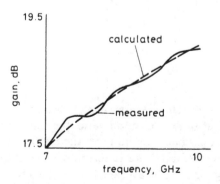

Fig. 12.11 *Comparison of measured and calculated standard gain*

The corrected gain of the test antenna, after allowing for input reflection co-efficient ρ_t, is $G_t/(1 - |\rho_t|^2)$. Site reflections are averaged by moving the horn over an area with principal dimensions sufficiently large to include any amplitude variations and making repeated measurements. When the antennas are of similar size, both may require movement. For best results G_S and G_t should be within 10 dB of one another.

Site reflections can be minimised by reducing the separation of the antenna pair, but the price may be that far-field conditions no longer apply. An early attempt to avoid interference from the ground was made by Purcell,[9] who used a reflecting surface as a mirror so that the antenna's image became a second identical antenna in the Friis formula, as illustrated in Fig. 12.12. However, that also suffers from multiple reflections, demonstrated by amplitude ripples that occur with axial movement of the horn. If ρ is the reflection coefficient of the mirror,

$$\frac{P_R}{P_T} = |\rho|^2 = \left(\frac{S-1}{S+1}\right)^2 = \frac{G^2\lambda^2}{4\pi r^2} \tag{12.49}$$

where S is the standing-wave ratio in the antenna feeder.

These range multipath reflections, also present between two antennas in close proximity, have been turned to good effect by Newell *et al.*,[8] by reducing the separation to near- and intermediate-field dimensions. The simple $1/r$ dependency in the Fraunhofer region is replaced by a series expansion in powers of $1/r$, and multiple reflections between the antennas are accounted for by odd multiples of the separate phase additions of the reflected components. The received-wave amplitude is written as a double series

$$b_R = C \sum_{p=0}^{\infty} \frac{\exp -j(2p+1)k_0 r}{r^{2p+1}} \sum_{q=0}^{\infty} \frac{A_{pq}}{R_q} \tag{12.50}$$

in which C is a constant depending on the transmitted amplitude and reflection coefficient in the receive antenna. The series is easier to understand if written in the following form:

$$b_R = \frac{\exp(-jk_0 r)}{r} \left[A_{00} + \frac{A_{01}}{r} + \frac{A_{02}}{r^2} + \cdots \right]$$
$$+ \frac{\exp(-j3k_0 r)}{r} \left[A_{10} + \frac{A_{11}}{r} + \frac{A_{12}}{r^2} + \cdots \right]$$
$$+ \cdots \tag{12.51}$$

Coefficients of $\exp(-jk_0 r)/r$ are the direct transmission terms; those of $\exp(-j3k_0 r)/r$ are first-order reflection; and the remaining are higher-order reflections. A_{00} is the only one remaining in the far-field region.

If relative intensity is measured as a function of separation r, cyclic variations are observed as in Fig. 12.13. The average line of the variations corresponds to A_{00}, thus enabling an accurate gain determination by extrapolation to large

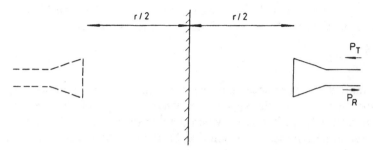

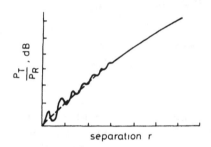

Fig. 12.12 *Purcell method for gain measurement*

Fig. 12.13 *Relative intensity for increasing antenna-pair separation*

distances. Reported measurement uncertainties[8] were within ± 0.05 dB, obtained at distances less than $2D^2/\lambda$, thus avoiding the site multipaths of longer ranges. To achieve these results, there were at the National Bureau of Standards two ranges, one outdoors at 60 m and another indoors at 9 m. The antennas moved apart on rails, with only 0.01 dB error attributable to rail misalignment. The method was extended to a three-antenna system for absolute gain measurement, one of which could be circularly polarised.

12.6 Fresnel region

Transition from intermediate- to far-field conditions is an arbitrary choice[6] dependent on the relative insignificance of the second compared to the first term in eqn. 12.7. Certainly, in the limit, the condition

$$R - r = -r' \cos \theta' \tag{12.52}$$

is necessary, but that implies all other terms, particularly the second, are zero. Since that may be a practical impossibility, there has to be a criterion

$$\frac{kr'^2}{2r} \sin^2 \theta' < \phi_{min} \tag{12.53}$$

where ϕ_{min} is to be decided according to the nature of the experiments in hand. $\sin \theta'$ is a maximum when $\theta' = \pi/2$; so the criterion is written

$$\frac{kr'^2}{2r} < \phi_{min} \tag{12.54}$$

It is a matter of experience that the patterns of many antennas appear to approximate to a far-field appearance when $\phi_{min} = \pi/8$. We met this same problem in time/frequency terms when the transition from a Fresnel to Fraunhofer integral determined the resolution bandwidth of a spectrum analyser. In this case, if $2r'$ is a characteristic maximum dimension in the aperture, often denoted by D for a circular diameter or the larger side of a rectangular aperture, the far-field boundary is given by

$$\frac{kD^2}{2 \cdot 4r} \leqslant \frac{\pi}{8}$$

or

$$r \approx R_{min} \geqslant \frac{2D^2}{\lambda} \tag{12.55}$$

Suppose a receiving antenna, with principal dimension D, is in the spherical wave from a transmitter. The second term is assumed large enough for the quadratic phase error to increase from zero at the centre, lying on an axis between transmitter and receiver, to $\pi/8$ at the aperture edge at $\pm D/2$. This situation, illustrated in Fig. 12.14, leads to a corresponding distance error of $\lambda/16$. According to this definition, the Fresnel boundary between the far and intermediate fields of an antenna, is at $2D^2/\lambda$ from its equivalent radiating surface.

The inner boundary of the Fresnel region depends on an arbitrary criterion of significance on the third term in eqn. 12.7. If this is again chosen as a minimum phase increase of $\pi/8$ from the central to extreme propagation line between two antennas, the third-term condition is

$$\frac{k_0 r'^3}{2r^2} \cos \theta' \sin^2 \theta' \leqslant \frac{\pi}{8} \tag{12.56}$$

with $\cos \theta' \sin \theta'$ set to its maximum value. On differentiating eqn. 12.56, we have for a maximum

$$\frac{\partial}{\partial \theta'} \left[\frac{1}{r^2} \left(\frac{r'^3}{2} \cos \theta' \sin^2 \theta' \right) \right] = \frac{r'^3}{2r^2} \sin \theta' [-\sin^2 \theta' + 2 \cos^{-2} \theta'] \tag{12.57}$$

or

$$\tan \theta'_M = \pm \sqrt{2}$$

which on substituting back in eqn. 12.56 gives for the lower boundary

$$R_F \geqslant 0 \cdot 62 \sqrt{\frac{D^3}{\lambda}} \tag{12.58}$$

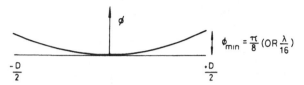

Fig. 12.14 *Phase error for Fresnel distance*

Similar calculations show that, if D_R is the receive and D_T the transmit principal dimension, far-field conditions require a separation of $2(D_T + D_R)^2/\lambda$. However, for antennas with significantly different dimensions, only the larger one need be considered.

12.7 Far-field ranges

Far-field ranges approximate to free-space and plane-wave conditions at the test position, by having an end-to-end distance in excess of the Fresnel range to ensure a quadratic phase error of less than $\pi/8$ in the test aperture. Test antennas are commonly rotated on two-axis positioners, though as many as six axes are sometimes used. The best positioners have azimuthal accuracies of the order of $0\cdot04°$, even when carrying several tonnes. Positioner and antenna co-ordinate systems do not always coincide, and it is then necessary to transform pattern and positioner references to obtain transferable specification data. Fig. 12.15 shows the structure and co-ordinate system of an azimuth over elevation positioner, and Fig. 12.16 shows how the addition of a second lower azimuth allows the use of a roll axis in the same positioner.

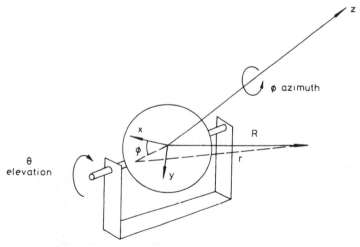

Fig. 12.15 *Azimuth over elevation positioner*

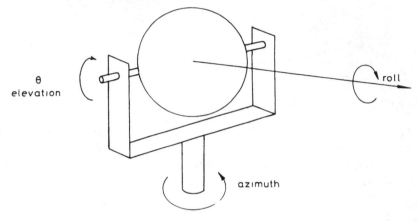

Fig. 12.16 *Roll axis with elevation and azimuth*

At microwave frequencies, far-field ranges include elevated ranges,[10] compact ranges[11] and anechoic chambers.[12] Reflection ranges, only briefly considered here, are common at lower frequencies, where ground reflections are difficult to eliminate. The principle of operation, outlined in Fig. 12.17, depends on inter-ference between the ground and direct wave to produce a cosine pattern with a peak at the centre of the test antenna and a null at the base of the test tower. Radiative coupling between the source and test position is 45 dB below the required signal level if a $\frac{1}{4}$ dB taper is arranged across the receive aperture. These considerations put height restrictions on the source and test antenna as follows:[13]

$$h_t \geqslant 3 \cdot 3D$$

$$h_s = \frac{\lambda R}{4h_t}$$

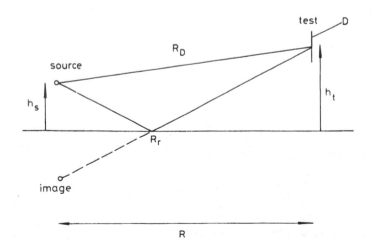

Fig. 12.17 *Reflection range*

Smoothness and reflectivity of the ground are important factors and require careful control, but at microwave frequencies ground reflections are usually minimised; so we will not prolong the discussion any further.

12.7.1 Elevated ranges
In this arrangement (Fig. 12.18) both antennas are mounted on towers or buildings to reduce or eliminate contributions from the surrounding environment. Among the factors affecting performance are

- Directivity and side-lobe level of the source
- Obstacles reflecting energy towards the test position
- Interference from other radiating sources
- Phase curvature of the illuminating wave front
- Amplitude taper across the test aperture
- Radiative coupling between the antennas
- Horizontal misalignment of the source and test antennas

The range length R should be large enough to ensure a good far-field approximation. If side-lobe level and main-beam shape are the prime consideration, $R = 2D^2/\lambda$ is usually long enough. But for low-side-lobe antennas, where null depth may be important, two or three times that distance may be required. For instance, on a -30 dB side lobe, the first null is virtually filled at $2D^2/\lambda$, and at $4D^2/\lambda$ its level is only a few decibels below the side-lobe peak.

12.7.1.1 Amplitude taper across the test antenna
Amplitude tapering, even up to 0·5 dB from centre to edge of the illuminating wave at the test antenna, has little effect on side-lobe peaks and null depths, but it does increase radiation coupling, owing to multiple reflections between the source and test antennas.[13] Such mutual coupling does not occur if the antennas are matched, whilst otherwise decreasing with source beam width θ_s

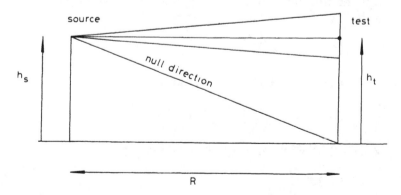

Fig. 12.18 *Elevated range*

and increasing with the angle subtended by the test antenna at the source, α_t. Thus, for a uniformly illuminated source aperture,

$$\theta_s = \frac{1 \cdot 22 \lambda}{d} \quad \text{and} \quad \alpha_t = \frac{D}{R}$$

where d is the source, D is the test antenna diameter and

$$R = \frac{KD^2}{\lambda}$$

with K chosen $\geqslant 2$, then

$$d = 1 \cdot 22 KD \frac{\alpha_t}{\theta_s} \tag{12.59}$$

The power in the test antenna is

$$P_t = \frac{P_0 G_s}{4 \pi R^2} \eta_t A_t \tag{12.60}$$

where P_0 is the total power radiated and G_s, the source gain, is

$$G_s = 4 \pi \eta_s \frac{A_s}{\lambda^2}$$

and η_t, A_t are the test-antenna radiation efficiency and effective aperture. Since $A_s = \pi d^2/4$ and $A_t = \pi D^2/4$, eqn. 12.60 becomes

$$P_t = P_0 \left[0 \cdot 92 \eta_s \eta_t \left(\frac{\alpha_t}{\theta_s} \right)^2 \right]$$

If K_t is the fraction scattered back by the test antenna, the power recaptured by the source is

$$P_s = K_t P_0 \left[0 \cdot 92 \eta_s \eta_t \left(\frac{\alpha_t}{\theta_s} \right)^2 \right]$$

and if K_s is the fraction of this re-transmitted, the received power after two reflections is

$$P_t' = P_0 K_s K_t (0 \cdot 92 \eta_s \eta_t)^3 \left(\frac{\alpha_t}{\theta_s} \right)^6$$

As a ratio of the original received power this is

$$\frac{P_t'}{P_t} = K_s K_t (0 \cdot 92 \eta_s \eta_t)^2 \left(\frac{\alpha_t}{\theta_s} \right)^4$$

Choosing typical values of $K_s = K_t = 0 \cdot 25$ and $\eta_s = \eta_t = 0 \cdot 5$,

$$\frac{P_t'}{P_t} = 3 \cdot 3 \times 10^{-3} \left(\frac{\alpha_t}{\theta_s} \right)^2 \tag{12.61}$$

Using eqn. 12.61 with a specified main-beam pattern for the source, the amplitude taper across the test antenna can be found as a function of α_t/θ_s. Thus, for a $(\sin X)/X$ beam shape, the coupling level P'_t/P_t is at least 45 dB below the original received amplitude if the amplitude taper is $\leqslant 0.25$ dB. Since these results depend on the assumed values of K_s, K_t and the radiation efficiencies, care should be exercised in their estimation.

12.7.1.2 Test-tower height

To avoid excessive ground reflection, no part of the main beam should illuminate the ground between the towers. To ensure this, the first null is directed to the base of the receive tower as shown in Fig. 12.18. For a typical source antenna this null might be at $1.5\lambda/d$. Thus

$$\theta_{null} = \frac{3}{2}\frac{\lambda}{d} < \frac{h_t}{R}$$

from which

$$d \geqslant \frac{3}{2}\frac{\lambda R}{h_t}$$

or

$$d \geqslant 1.5\frac{KD^2}{h_t}$$

But for the 0.25 dB amplitude illumination taper $\alpha_t/\theta_s = 0.3$ in eqn. 12.59 to give

$$d \leqslant 0.37 KD$$

Therefore, to keep the whole of the main beam above the bottom of the receive tower,

$$h_t > 4D \tag{12.62}$$

12.7.1.3 Source-tower height

Both antennas should normally be at the same horizontal level on a flat range, but it may sometimes by advantageous to vary this. For instance, if side-lobe levels are the most critical characteristics, it is better to have the source higher than the test antenna to reduce the effects of reflections on them; but when back lobes are of greater importance, the test antenna is usually placed higher. Again, it is often the case that a flat range is not available and relative heights are then a matter of site-reflection evaluation and topology. The resulting optimum positions may well lead to unwanted bore-sight misalignment from the horizontal. With vertical azimuthal axes on fixed-axis positioners, measured patterns will not lie in great-circle planes nor will θ or ϕ be held independently constant. The situation is illustrated in Fig. 12.19, where the test antenna has to

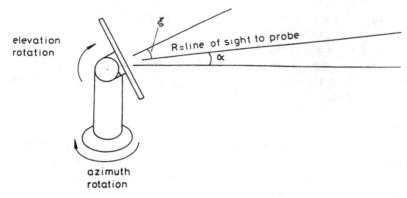

Fig. 12.19 *Elevated line of sight to source antenna*

be elevated by α to align the bore-sight with the source, and then by another angle ξ for azimuthal cuts at different elevations. Clearly, both θ and ϕ in the test antenna co-ordinate system are varied during an azimuthal rotation. We see in Fig. 12.20 that the test antenna is at the origin and that its main-beam bore-sight initially points at the source by tilting the positioner about its elevation axis x by an angle α. The antenna co-ordinate system, x, y, z begins with

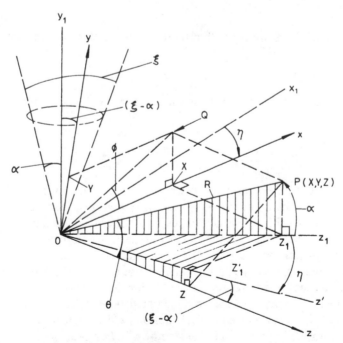

Fig. 12.20 *Co-ordinate rotations for elevated line of sight*

y vertical along y_1, and z horizontal along z_1. It is then tilted back in the plane containing OPZ_1, by α to point the main-beam bore-sight at the source, along a line from the origin to $P(X, Y, Z)$ in antenna co-ordinates. To take a positive cut at elevation angle ξ, the y-axis is now tilted forward about the x-directed elevation axis by ξ. A pattern cut can now be taken for a resultant tilt angle from the horizontal of $(\xi - \alpha)$, by rotating the x-axis in the horizontal plane about the y_1 axis. In Fig. 12.20 the azimuthal angle is η, ϕ is the angle in the x, y plane between the x axis and the projection of OP on to that plane and θ is the angle between OP and the z-axis. x_1, x, z_1, and z' are all in the horizontal plane. It follows from the Figure that

$$OZ = OZ_1 \cos \eta \cos (\xi - \alpha)$$

$$OP = \frac{OZ_1}{\cos \alpha}$$

Therefore

$$\cos \theta = \frac{OZ}{OP} = \cos \alpha \cos \eta \cos (\xi - \alpha) \qquad (12.63)$$

Also

$$OX = Z_1'Z_1 = OZ_1 \sin \eta = OP \cos \alpha \sin \eta$$

$$OQ = OP \sin \theta$$

Therefore

$$\cos \phi = \frac{OX}{OQ}$$

$$= \frac{\cos \alpha \sin \eta}{\sin \theta} \qquad (12.64)$$

The angles in the antenna co-ordinate system are therefore

$$\theta = \cos^{-1} [\cos \alpha \cos \eta \cos (\xi - \alpha)]$$

and

$$\phi = \cos^{-1} \left[\frac{\cos \alpha \sin \eta}{\sqrt{1 - \cos^2 \alpha \cos^2 \eta \cos^2 (\xi - \alpha)}} \right] \qquad (12.65)$$

In practice, α is usually small and its effects can be ignored, at least for measurement near the main beam. For polarisation measurements on vertical azimuthal axes, especially at large angles, errors may become excessive if the source and receiver are not horizontally aligned, unless co-ordinate transformations using eqns. 12.65 are made to build complete patterns in true antenna spherical co-ordinates.

12.7.2 Slant ranges

Slant ranges, with the source at ground level and test antenna above it on a tower as in Fig. 12.21, are more economical of ground space than elevated ranges. The centre of the source beam illuminates the test antenna, and its first null is directed at the spectacular reflection point on the ground. A null condition is difficult to achieve and requires careful adjustment. The reverse procedure, with source on elevated tower, is often used for testing satellite antennas, because their fragile structures are best secured at ground level.

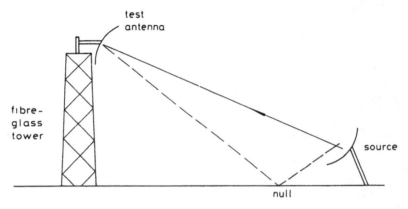

Fig. 12.21 *Slant range*

12.7.3 Diffraction fences

Diffraction fences are used on elevated and slant ranges to deflect energy skywards from the specular region. However, conditions can be made worse if the source main beam interacts with them, and their placing can then become very critical. Fences are constructed of wooden frames with a metal screen or mesh surface, and their tops are sometimes shaped to cause efficient scattering in desired directions.

12.7.4 Compact ranges

Many of the difficulties experienced on far-field ranges can be overcome by taking advantage of collimation in the near field of a paraboloid reflector. Fig. 12.22 shows a typical arrangement with an offset feed to avoid interference from blockage or scattering. Collimation in the radiating near field between 10λ and D^2/λ from the paraboloid provides plane waves, if the linear dimensions of the test antenna are three or four times less than those of the source. The main problem is diffraction from the edge of the source dish, which can be ameliorated by serrating the edge to increase the randomness of scattered energy.

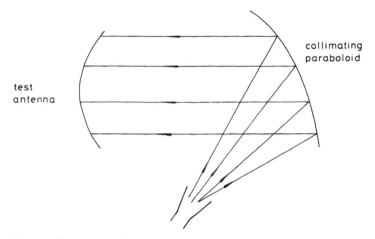

Fig. 12.22 *Plane-wave region in compact range*

Problem areas in compact ranges are:[11]

- Direct radiation from the feed
- Diffraction from feed and supports
- Diffraction from reflector edges
- Depolarisation
- Space attenuation from primary feed
- Interaction between range and test antenna
- Stray radiation within room
- Reflector surface tolerances

Direct radiation from the feed can be reduced by designing a feed with low radiation in the direction of the test antenna and by careful use of high-quality absorbing material, as illustrated in Fig. 12.23. Diffraction from feed supports is eliminated by offsetting the feed from a paraboloid section, leaving only the edge diffraction to be reduced by empirically designed serrations similar to those illustrated in Fig. 12.24.

Depolarisation arises in the primary feed and in the doubly curved paraboloidal reflector. For the former it can be minimised by employing high-quality circular or corrugated horns; and for the latter it decreases with focal length, being least near the reflector axis. A long focal length is also desirable to minimise the effects of feed space attenuation, caused by the longer path lengths for rays furthest from the axis in the collimated beam. If the feed is aimed at the top of the aperture, its pattern taper tends to cancel space attenuation. Amplitude and phase data, taken on a typical compact range[1] at 10 GHz, are shown in Fig. 12.25. Amplitude variation is <0·8 dB over the operating

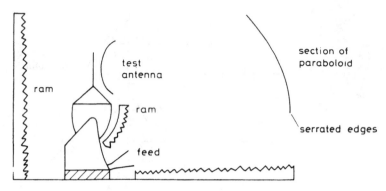

Fig. 12.23 *Compact range*

range, sometimes called the plane-wave zone, and phase shows only a random variation as distinct from the quadrature radial dependence of a far-field range.

There are potentially serious problems from interaction between range and test antenna, which can be avoided by minimising standing waves due to reflections among the feed, reflector and test antenna. A well-matched feed and minimum energy interception by the test antenna are therefore both essential design aims. For low-side-lobe measurements stray radiation should be below -50 dB relative to collimated levels, because stray components may enter the main beam when a null is aligned with the collimation axis, as illustrated in Fig. 12.26.

Surface tolerance has restricted doubly curved reflector surfaces for compact ranges to an upper frequency of 30 GHz, whilst the lower limit of about 4 GHz is set by the difficulties of installing large indoor apertures. Surface-tolerance effects not only depend on actual departure from true position, but also on the

Fig. 12.24 *Serrated edge to reduce scatterring*

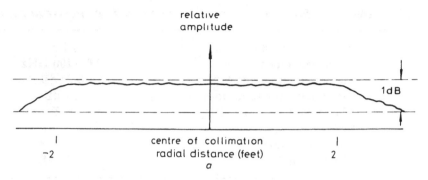

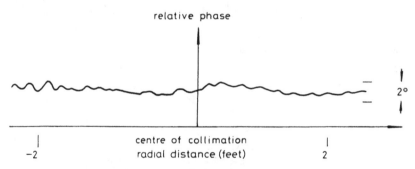

Fig. 12.25 *Amplitude and phase variation in the test region*

correlation area, or the area over which a single distortion occurs. If correlation areas are $< \lambda^2$, random additions from them at the test position tend to average out, and for large distortion areas, comparable to the reflector dimensions, deterministic additions cause only slight defocusing with little amplitude non-uniformity. The greatest difficulty occurs for correlation areas between these extremes, with predicted amplitude error ~ 0.5 dB for a surface deviation of only 0.007λ for correlation over one Fresnel zone.[11]

Further developments in compact ranges involve two orthogonally placed cylindrical reflectors.[14] One acts as a sub-reflector to the other, and is illuminated with an offset feed, with input SWR of $1.05 : 1$ to ensure minimum mutual radiation coupling. The maximum size of the dual-reflector system for a given size of test antenna is 50% less than for single reflector geometry; and, because it is easier to control surface errors on cylindrical reflectors, the operating frequency extends to 100 GHz. The physical arrangement is shown in Fig. 12.27 and some measured phase and amplitude variations in the plane-wave zone are listed in Table 12.1

Table 12.1   *Phase and amplitude variations in plane-wave zone*

Plane-wave zone diameter	1·2 m
Frequency range	4·0—100 GHz
Maximum-amplitude taper	0·25 dB
Maximum-amplitude ripple	0·4 dB
Maximum phase variation	$\pm 2°$
Cross-polarisation	-30 dB
Stray signal level	-55 dB
Mutual coupling	-50 dB

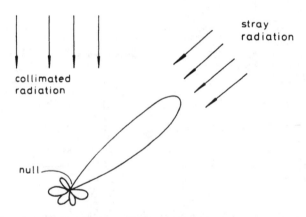

Fig. 12.26   *Showing effects of stray radiation*

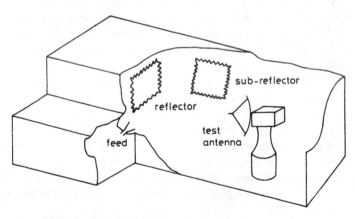

Fig. 12.27   *Compact range with cylindrical reflectors*

12.7.5 Anechoic chambers

Far-field measurements can be made on small antennas, such as feeds, in chambers lined with RF absorbing material. In common with compact ranges, it has the advantage of indoor working conditions in a controlled temperature and reflection environment, thus allowing all-year working irrespective of weather conditions. Free-space conditions are simulated by reducing wall reflections to minimal proportions. There is a long history of developing absorbing materials,[15] but modern absorbers consist of polyurethane foam loaded with carbon, or, more expensively, thin ferrite wedges that are also more frequency sensitive. A smooth impedance variation from free space to the backing metal plate is achieved by fixing pyramid-shaped absorbers to the metal plate, or by having electrically graded alternating layers. Absorbers operate from 100 MHz to 100 GHz. Pyramid length decreases with frequency, being a few metres at the lower frequencies. At normal incidence they are capable of -50 dB reflection for a material depth of four free-space wavelengths. As the angle of incidence increases from the normal, reflection increases; being of the order of -25 dB at 60°.

12.7.5.1 Rectangular chambers

Free-space conditions are simulated in rectangular chambers by covering all the walls with absorbing material, paying particular attention to areas of specular reflection, and restricting angles of incidence to less than 70°. Thus, in Fig. 12.28 the width W and range length R are related by

$$W \geqslant \frac{R}{2 \cdot 75} \tag{12.66}$$

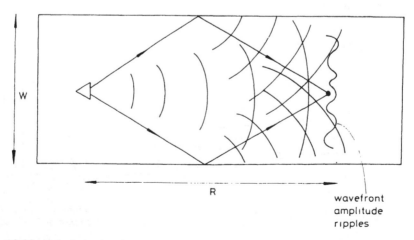

wavefront
amplitude
ripples

Fig. 12.28 *Anechoic chamber*

Rectangular-chamber design usually follows a geometrical-optics approach in which specular reflections are assumed at the walls. Phase changes at reflections are ignored, and worst-case reflection amplitude additions at the test point are compared to direct amplitude with source pattern and frequency range included. The method is subject to error, since it cannot properly account for polarisation and wide-angle and/or non-specular reflections.[16]

12.7.5.2 Tapered chambers

These are shaped liked a pyramidal horn with the source at the apex end. At low frequencies the horn is placed at the apex, so that the specular reflections, outlined in Fig. 12.29, are at small angles of incidence close to the source to ensure small phase differences between direct and reflected rays. The resulting amplitude in the quiet zone around the test region is a smooth taper rather than the ripple, typical of rectangular chambers and open-air free-space ranges. Positioning the source is a matter of experimentation, though the evident similarity to a ground-reflection range allows reasonable estimation of a start position. Thus in Fig. 12.30 the path difference between a reflected and direct wave can be found by first calculating R_D and R_R, the direct and reflected path lengths. Source and test positions are at perpendicular heights h_s and h_t from the 'ground' plane, which in this case, is the anechoic chamber wall. If R is the distance along the wall between the perpendiculars,

$$R_D^2 = (h_t - h_s)^2 + R^2$$
$$R_R^2 = (h_t + h_s)^2 + R^2$$

or

$$R_R^2 - R_D^2 = 4h_t h_s$$

which if

$$R_R \sim R_D \sim R$$

gives

$$2R(R_R - R_D) = 4h_t h_s \tag{12.67}$$

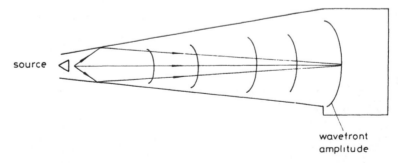

source

wavefront
amplitude

Fig. 12.29   *Tapered anechoic chamber*

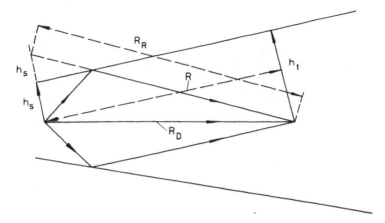

Fig. 12.30 *Source positioning in tapered chambers*

If we assume a 180° phase change on reflection $(R_R - R_D)$ must equal $\lambda/2$ for in-phase addition at the test position. Therefore

$$h_s \approx \frac{\lambda R}{4h_t} \qquad (12.68)$$

If the source is placed forward in the chamber, path differences rapidly increase and deep interference nulls occur. For the same reason, as frequency increases, the source must be placed nearer to the apex. Eventually the height condition becomes impossible to meet and it is necessary to suppress reflection by using a higher-gain source antenna. The chamber then behaves more like a free-space range or rectangular chamber.

Anechoic chambers and free-space ranges are evaluated from field measurements in the 'quiet zone' around the test point. This is an area perpendicular to the axis between source and test antennas, whose diameter may be specified in terms of an amplitude and/or phase variation along linear tracks, usually vertical or horizontal, though, in fact, any radial direction will do. Similar measurement tracks parallel to the axis are also used to evaluate stray radiation from all parts of the chamber. The signal received, say, by a horn on a track is the sum of a direct and reflected wave, and if the horn is directional, reflections in the main-beam direction predominate, thus making it necessary to repeat measurements on the same track but with the horn reorientated angularly. The presence of interference is indicated by a ripple along a track path as shown in Fig. 12.31. If $f(\theta)$ is the voltage wave pattern of the test horn, and θ is the angle along a reflected wave, peak-to-peak ripple is given by

$$2\left|\frac{E_{ripple}}{E_D}\right| = \left|\frac{1 + (E_R/E_D)f(\theta)}{1 - (E_R/E_D)f(\theta)}\right| \qquad (12.69)$$

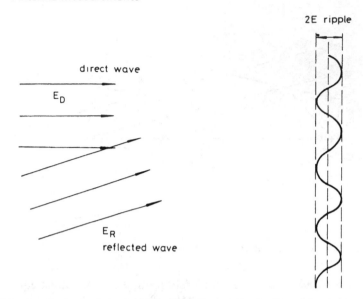

Fig. 12.31   *Amplitude ripple in the quiet zone due to reflection from chamber wall*

The horn can be replaced by three orthogonal dipoles to provide an omnidirectional probe requiring only three orthogonal tracks to characterise the chamber. There is no standardised figure of merit for chambers, though in comparison with ~0·5 dB for conventional far-field ranges, a typically good ripple might be 0·1—0·2 dB peak-to-peak.

Another method for evaluating anechoic chambers relies on comparison of the patterns from a standard-gain horn on a mount moved to several locations in the quiet zone. Analysis of normalised and superimposed patterns gives reflection levels and angles from observed changes in side-lobe levels, but the method is less satisfactory than the previous one because of difficulty in determining maximum reflection levels.

12.8 Intermediate-field ranges

Linear and planar apertures are normally focused at infinity. Far-field measurements can only approximate to the resulting pattern at infinity, but if the antenna is re-focused to shorter range, we may expect measured and predicted patterns to coincide. Intermediate ranges have source and test antennas at separations less than $2D^2/\lambda$. If the measured patterns are as expected for that distance, the test antenna is then re-focused to infinity and the far-field pattern is assumed to be determined within some estimated range of error.[2]

Paraboloids and lenses can be re-focused at short distances by axial positioning of the feed and linear arrays by physically bending to a circular arc of radius equal to the test distance. With improvements in near-field measurements, intermediate-field techniques have become less important, with the exception of field monitoring of electronically steerable phased arrays.[17]

Large phased arrays with transmit/receive modules at each element are non-reciprocal, and therefore need separate testing in transmit and receive mode. In either case, active elements are always present as well as a combining network, and sometimes even special signal-processing stages between the aperture and the output/input port to what is now an antenna system. Testing during development calls on special techniques to isolate the strictly radiating structure from other circuit components of the system, but monitoring during operation also depends on special methods to identify and locate failures.[18] Later we will look at one example, using antenna noise measurements, to determine effective aperture and radiation efficiency in an array with apparently inseparable active components, but for the moment we will briefly review some methods of operational monitoring, based on intermediate-field techniques.

In-service testing of large arrays is essential, because the large number of elements means that, even if individual devices have very long mean times between failure, the probability of a single failure in many thousands of identical devices may become unacceptably high. Monitoring at line voltage level and through internally injected test signals can check individual active component failure, but an overall system test is also essential. Neither near- nor far-field testing is convenient: in the former case because the close proximity of probes would preclude continued use of the equipment during testing, and in the latter because a far-field source can rarely be placed at a suitable part of an operating environment; e.g. when the array is installed on a ship. Intermediate-field monitoring is possible if the array can be re-focused to a nearby source or receiver, preferably integral with the antenna structure. Phase-path lengths to the fixed near source or receiver are measured during commissioning, but remain fixed thereafter. A regular test cycle, implemented on a daily basis, checks the response of the active array against phase-shifter changes. In large arrays, a single phase-shift step of perhaps $22 \cdot 5°$ in only one element would not be detectable within the dynamic range of a monitoring system, but it may be possible to select sub-arrays with smaller numbers of elements to a level where single-phase-shifter steps are detectable. To focus the array on a point within the Fresnel zone, the insertion phase in each element must be set to give in-phase addition of the fields radiating to that point. In Fig. 12.32, the path difference S between the centre ray to the source or receiver and a ray from a position at x from the centre, is

$$S = \sqrt{R^2 + x^2} - R$$

If element separation in the array is d and $x = nd$, the insertion-phase retardation at the nth element is

$$\phi = \frac{2\pi}{\lambda}(R^2 + n^2d^2)^{1/2} \tag{12.70}$$

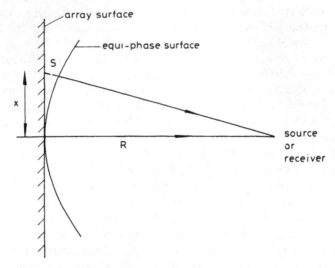

Fig. 12.32 *Phase in intermediate-field measurement*

It is not always necessary for the source or receiver to be displaced perpendicularly from the array. An edge pick-up or probe can provide the required information at the price of a more complicated phase programme across the elements. Monitoring times for arrays of 1000 elements are typically less than 1 s, which allows regular checks to be made between operational tasks.

12.9 Near-field ranges

The reactive near-field zone for an aperture antenna is usually taken from the inner Fresnel boundary, or R_F in eqn. 12.58, to the radiating surface. Solution of eqn. 12.3 for the electric-field vector becomes very complex at distances comparable to, or less than, a major linear dimension of the aperture. In particular, eqn. 12.7 for the phase cannot readily yield solutions for F in this form. The attractions of measuring near rather than far fields include the possiblity of creating indoor laboratory control of experiments designed to eliminate many of the site-related problems of far-field ranges. Since near fields contain all the information for calculating the far field, it should be possible, in principle, to obtain greater pattern accuracy, provided sufficient information is captured from the measurements.

There are three principal methods of making near-field measurements, based on rectangular, cylindrical and spherical co-ordinates, respectively. In rectangular, or planar, probing, a probe is moved to sample phase and amplitude in a plane parallel and close to the radiating aperture. Clearly not all the fields will be sampled unless the sampling area be extended to infinity, and since this is impracticable, overlap of sampled aperture area, illustrated in Fig. 12.33a, determines the percentage and level of the total measured field and defines one

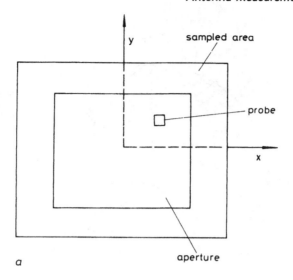

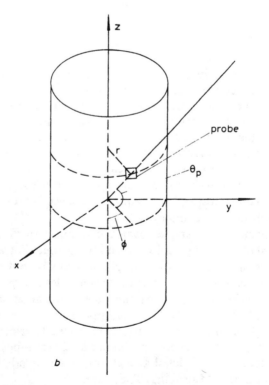

Fig. 12.33   *Near-field testing*
a Planar
b Cylindrical
c Spherical

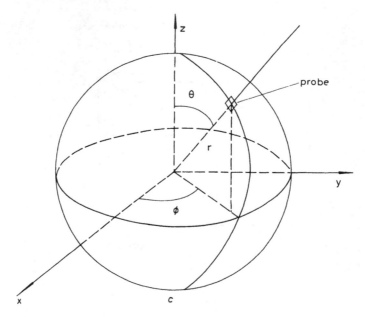

Fig. 12.33 *Continued*

accuracy criterion. Other criteria are the degree of precision required on probe positioning and the frequency and number of probing points. Measurement time, and therefore cost, increases with the number of samples, leading to a trade-off between cost and accuracy. Ideally, probes are point isotropic as a source or receiver, but, in practice, their patterns are shaped and also affected by a connecting feed line. Their responses are therefore characterised in amplitude and phase in a separate far-field measurement, usually in an anechoic chamber, and complex transformations are applied to the near-field results to correct for probe non-isotropy. The main consequence is an increase in measurement time and software complexity.[19]

Cylindrical probing takes samples over a cylindrical surface enclosing the test antenna, as shown in Fig. 12.33*b*. Positioning precision and number of samples have the same implications as in planar probing, but overlap criteria relate to the strengths of fields along the *z*-axis and lead to probing cylinders extending a considerable distance above and below the antenna. With spherical probing (Fig. 12.33*c*), there is the possibility of finding the total field near the antenna, though often the probed area is restricted to selected parts of the sphere, e.g. to include a main beam and higher-level side lobes.

We must now look in more detail at features, some already identified, that are common to all near-field probing. These are:

● Spatial frequency of probing positions
● Positional precision of probe settings

- Probed area
- Accuracy of amplitude and phase measurement
- Models for near/far field transformation.

12.9.1 Spatial frequency of probing positions

The minimum sampling condition for planar probing can be found by investigating field conditions close to an aperture in the reactive near-field region.[19,20] Following the references, we begin with the radiation half-space for an aperture as in Fig. 12.4 and Fourier-transform the field $E(x, y, z)$ with respect to x and y, to give

$$E(k_x, k_y, z) = \int\!\!\!\int_{-\infty}^{+\infty} E(x, y, z) \exp j(k_x x + k_y y)\, dx\, dy \qquad (12.71)$$

But $E(k_x, k_y, z)$ also satisfies the wave equation

$$\left[\frac{\partial^2}{\partial z^2} + k_0^2 - (k_x^2 + k_y^2)\right] E(k_x, k_y, z) = 0 \qquad (12.72)$$

which has a solution

$$E(k_x, k_y, z) = f(k_x, k_y) \exp -jk_z z \qquad (12.73)$$

The electric field can now be written as the inverse transform

$$E(x, y, z) = \frac{1}{4\pi^2} \int\!\!\!\int_{-\infty}^{+\infty} f(k_x, k_y) \exp -jk \cdot r\, dk_x\, dk_y \qquad (12.74)$$

with

$$k_0 = \left|k_x^2 + k_y^2 + k_z^2\right|^{1/2} \quad \text{and} \quad \left|k_x a_x + k_y a_y + k_z a_z\right| = |k|$$

That eqn. 12.74 shows $E(x, y, z)$, the field at any point in the radiation half-space of the aperture, as a superposition of plane waves can be seen by considering a model, illustrated in Fig. 12.34, of propagation in a waveguide formed from two infinite and parallel conducting planes separated by distance a. ED and FB define parts of the planes that lie parallel to the y, z plane. A wave propagating in the z-direction, with transverse electric field in the y-direction, has wavelength $\lambda_z = AG$ given by the distance between two wave positive peaks shown as solid lines labelled with a $+$ sign. To ensure correct boundary conditions, the z-directed wave can be made from a superposition of two other plane waves propagating at $\pm\theta$ from the z-direction, such that at the wall a positive peak in one wave is cancelled by an equal negative peak in the other. There are also two x-directed waves to form a standing wave between the plates with wavelength $\lambda_x = 2a = 2EF$ and propagation constants $k_x = \pi/a$. Clearly, both λ_z and λ_x are greater than the free-space wavelength $\lambda_0 = 2AP$, where AP is the perpendicular distance between adjacent positive and negative peaks in the oblique plane waves. We can find the relation between these wavelengths by

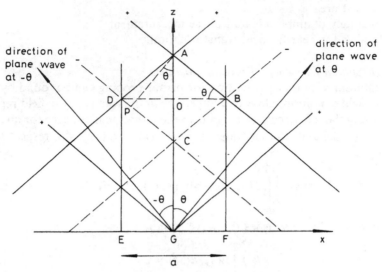

Fig. 12.34 *Showing addition of plane waves in the far field*

considering triangle AOB in which

$$\frac{AO}{OB} = \tan \theta$$

and triangle APC in which

$$\frac{AP}{AC} = \cos \theta \qquad (12.75)$$

to give on multiplying with $AC = 2AO$

$$\frac{AP}{2OB} = \sin \theta$$

But $AP = \lambda_0/2$ and $OB = a/2$; therefore

$$\frac{\lambda_0}{2a} = \sin \theta \qquad (12.76)$$

or

$$k_x = \frac{\pi}{a} = \frac{2\pi}{\lambda_x} = \frac{2\pi}{\lambda_0} \sin \theta \qquad (12.77)$$

If we add eqns. 12.75 and 12.76 after first squaring each and substituting $AP = \lambda_0/2$, $AC = \lambda_z/2$, we have

$$\frac{\lambda_0^2}{4a^2} + \frac{\lambda_0^2}{\lambda_z^2} = 1$$

or

$$k_0^2 = k_x^2 + k_z^2 \tag{12.78}$$

which is a similar form to that for k_0 in eqn. 12.74, except for the extra term k_y to signify a second transverse wave in the y-direction. A waveguide analogy could be continued by placing a second pair of conducting planes orthogonally to the first ones, to generate a second standing wave in the y-direction from the waves with propagation constant k_y, but it is simpler, and loses no validity, if we continue with our two-dimensional model.

If in eqn. 12.76, θ is zero, the two oblique waves coalesce to a single plane wave, perpendicular to the aperture a. Thus, since λ_0 is finite, $a \to \infty$. On the other hand, when $\theta = 90°$, $\lambda_0 = 2a$, and the z-directed wave is cut off, leaving only two waves propagating parallel to the aperture surface and reflecting perpendicularly at the walls. A further increase in λ_0, or decrease in a, causes an exponential decay of the surface-wave fields in the z-direction, so that the evanescent mode in a cut-off waveguide can be regarded as a surface wave at the launch port, with increasing surface-field concentration as the transverse wavelength, $\lambda_x = 2a$, decreases relative to the free-space wavelength. In general, cut-off modes, with reactive stored fields, have transverse wavelengths shorter than the corresponding free-space wavelength, whereas for z-propagating modes they are longer. At the launch position in a waveguide, matching seeks to compensate for evanescent modes generated by the structure of the launching probe. Transverse waves are generated with wavelengths in half-wave multiples of the transverse guide dimension, all with wavelengths less than that in free space, with the exception of the dominant mode for which the transverse dimension must allow λ_z to be greater than λ_0. When a parallel-plate waveguide is over-moded, z-propagating modes have transverse standing waves in multiples of the transverse wavelength, and each one fixes the direction of propagation of its corresponding pair of oblique waves according to eqn. 12.76 with $\lambda_x = \lambda_0/\sin\theta$. This can be made clear by noting that, for each wave pair, it is always possible to insert a set of parallel plates at the nulls of the transverse standing wave.

We can transfer these ideas to free-space antennas by considering that there is an apparent spectrum of waves on the aperture with wavelengths $\lambda_x = \lambda_0/\sin\theta$. Each wave is related to a particular angle θ in the far field along which a plane wave is propagating, and the summation of all these plane waves constitutes a single plane wave in the radiation half-space.[7] If the single plane wave propagates along the z-direction, the aperture spectrum is a set of standing waves to generate symmetrical propagating pairs at angles $\pm\theta$; but a single plane wave at angle α to the z-axis, being composed of two travelling waves, in the x- and z-directions, respectively, has both travelling and standing aperture waves. Aperture modes are simply the consequence of a particular way of inverting the Fourier transform; their physical reality is not mathematically necessary, but can sometimes be found in the detailed antenna structure. Their usefulness here is in establishing a spatial sampling frequency and a minimum distance between the probe and aperture.

Transverse waves, with wavelengths less than the free-space wavelength, suffer a cut off with exponentially decaying fields in the z-direction. They therefore carry no information about the far field, and it is not necessary to measure them in the near field. If the sampling plane is moved forward from the aperture by about three wavelengths, all cut-off fields become negligibly small, leaving only the propagating components in the sampled data. This can be shown from eqn. 12.74, in which each plane wave

$$f(k_x, k_y) \exp -jk \cdot r$$

propagating in direction r, has transverse resolved components with propagation constants k_x, k_y and a perpendicular component, with propagation constant k_z, such that

$$k_z = [k_0^2 - (k_x^2 + k_y^2)]^{1/2} = [k_0^2 - k_t^2]^{1/2} \tag{12.79}$$

where k_t is a transverse wave number. Cut off occurs at $k_0 = k_t$, to give for propagating modes

$$k_t \leqslant k_0, \qquad \lambda_t \geqslant \lambda_0$$

and for non-propagating modes

$$k_t \geqslant k_0, \qquad \lambda_t \leqslant \lambda_0$$

If we write $jk_z = -\alpha_z$ for the non-propagation attenuation constant,

$$\alpha_z = (k_t^2 - k_0^2)^{1/2} \tag{12.80}$$

If $z = N\lambda_0$

$$\alpha_z \approx 343 \cdot 5N \left(\frac{\lambda_0^2}{\lambda_t^2} - 1 \right)^{1/2} \tag{12.81}$$

At $N = 1$

$$\alpha_z \approx -60 \, \text{dB} \quad \text{for} \quad \frac{\lambda_t}{\lambda_0} \sim 0 \cdot 985$$

Thus at one wavelength from the aperture the attenuation is $-60 \, \text{dB}$ and the transverse wavelength is only $1 \cdot 5\%$ less than free-space wavelength. The smallest transverse wavelength that can be sampled, when non-propagating modes are eliminated, is the free-space wavelength, and from the sampling theorem, all essential information is captured, if samples are no greater than $\lambda_0/2$ apart. It is normal to sample at $3\lambda_0$ from an aperture and at $\lambda_0/3$ separations, to guarantee that all the energy is propagating and to ensure no loss of information.

12.9.2 Positional precision of probe settings

The far field of a radiating aperture, given by eqn. 12.74 with $f(k_x, k_y)$ restricted to z-propagating modes only, is an infinite summation of plane waves, to be determined from measurements at $\lambda_0/3$ intervals across the aperture. The measured accuracy of transverse modal wavelengths depends on precise positioning of the probe at each sampled point, and also determines the angular accuracy of the spectrum of plane waves through the relationship

$$\sin \theta_n = \frac{\lambda_0}{\lambda_x} \tag{12.82}$$

where n is the number of wavelengths λ_x in one dimension of the aperture. Angular error for a single mode can be found by differentiating eqn. 12.82 to give

$$\frac{d \sin \theta_n}{d\lambda_x} = -\frac{\lambda_0}{\lambda_x^2}$$

and on substituting for λ_x, we have

$$\Delta\theta_n \approx \frac{-\sin^2 \theta_n}{\cos \theta_n} \frac{\Delta\lambda_x}{\lambda_0} \tag{12.83}$$

Each modal error contributes to both angular and amplitude error at the observation point x, y, z, but, because of the statistical nature of the additions and the uncertainty of the number and amplitudes of the modes for a particular antenna, it is difficult to make estimates of the required accuracy in probe-positioning. Comparison of far- and near-field results on the same antennas indicates that, for ± 1 dB error on a -40 dB side lobe, the probe must be set to at least $0.01\lambda_0$ in all three dimensions. This can be relaxed to lower values when null depths and side-lobe specifications allow.

12.9.3 Probed area

Planar probing is appropriate only for narrow-beam antennas with field levels below -40 dB with respect to the main beam in the unprobed region. Similar arguments apply to cylindrical probing, except that the beam may now be fan shaped and perpendicular to the z-axis, since the only unprobed region is at the cylinder ends. In this case the test antenna rotates in azimuth and the probe moves along the z-axis. Probing area is optional in spherical systems, depending on the antenna pattern and the degree of accuracy required. There is also the advantage of having a fixed probe whilst the antenna rotates. A form of spatial spectral leakage occurs when only part of the total area is sampled, because of the effects of discontinuities on the near/far-field transformation.

12.9.4 Accuracy of amplitude and phase measurements

Each sample carries amplitude and phase errors due to the measuring equipment. These errors are likely to be small, perhaps ± 0.3 dB in amplitude and

$\pm 2°$ in phase. Measurement error is at best reduced proportionally to the square root of the number of samples, through an effective random addition during transformation. Thus equipment errors are negligible compared to probe positional ones.

12.9.5 Models for near/far-field transformation

We have already discussed the planar model as an infinite angular spectrum of plane waves. Determination of the far field is by a two-dimensional Fourier transform of the samples' amplitudes and phases.

Spherical near-field probing takes measurements over a minimum sphere, that encloses the antenna with a minimum clearance of $3\lambda_0$ from radiating parts.[21,23] On this sphere of radius r_0 in Fig. 12.35, there is an infinite number of spherical modes, but only those with transverse wavelengths greater than λ_0 are able to propagate in the radial direction. Field θ-dependency on the sphere follows the associated Legendre functions, and the θ variation is an integral number of cosine or sine repetitions. The electric field at (r_0, θ, ϕ) can be written as

$$E(r_0, \theta, \phi) = \sum_m \sum_n a_{mn} P_n^m Q_m \qquad (12.84)$$

where P_n^m is the associated Legendre function whose order n gives the number of integral variations in θ, and Q_m is a function in $\cos m\phi$ or $\sin m\phi$, where m gives the number of ϕ variations. r_0 dependency is a spherical Bessel function not shown explicitly, but necessary in transforming $E(r_0, \theta, \phi)$ to its equivalent at infinity.

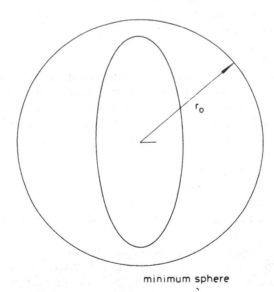

minimum sphere

Fig. 12.35 *Minimum sphere for near-field spherical scanning*

The frequency of the highest spherical modes is determined by the number of free-space wavelengths in any great circle of the minimum sphere. With each P_n is associated a group of m integers, where $m = -N, -(N-1) \cdots (N-1), N$, where N is the highest order of n. There are even and odd alternatives in each order n, giving a total of $2N(N+2)$ equations with known functions $P_n^m Q_m$, but unknown coefficients a_{mn}. The highest spatial frequency is associated with N variations in the θ-direction and $2(N+2)$ variations in the ϕ-direction. Propagation conditions restrict sampling separations to $\leqslant \lambda_0/2$; θ varies from 0 to π; and ϕ from 0 to 2π. In half a polar great circle

$$\frac{\pi r_0}{N} \leqslant \frac{\lambda_0}{2}$$

or

$$N \geqslant k_0 r_0 \tag{12.85}$$

and in the equatorial plane

$$\frac{2\pi r_0}{2(N+2)} \leqslant \frac{\lambda_0}{2}$$

which for large N gives

$$N \geqslant k_0 r_0 \tag{12.86}$$

$2N(N+2)$ samples taken on the minimum sphere at $\lambda_0/2$ intervals on polar great circles, separated by $\lambda_0/2$ on the equator, are therefore sufficient to specify $M = 2N(N+2)$ modes which propagate to the far field. The corresponding system of M equations with M unknown coefficients a_{mn} can be solved from measured field values at M properly spaced points, and the far field found by weighting the asymptotic spherical Bessel-function mode amplitudes at infinity with the experimentally derived coefficients before summing their values at a required angle.

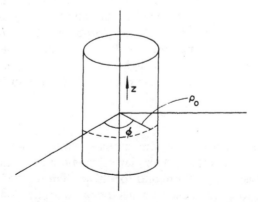

Fig. 12.36 *Minimum radius for near-field cylindrical scanning*

Cylindrical models of near-field probing consist of a linear aperture along the z-axis that determines θ variations and cylindrical modes for ϕ variations.[23] In Fig. 12.36 the minimum circle is in the x, y plane, where $\lambda_0/2$ spacing gives

$$\frac{2\pi r_0}{N} = \frac{\lambda_0}{2} \tag{12.87}$$

and sample separation in the z-direction is $\lambda_0/2$.

Near-field ranges may well become the most accurate technique available, as well as providing the most comprehensive information from a single scan in two polarisations. Once the data have been taken, it is a matter of software provision for the derivation of some or all of the following:[24]

● Gain comparison, pattern display
● Three-antenna polarisation measurement
● Probe-polarisation-corrected measurements
● Probe-pattern-corrected measurements
● Polarisation-based output presentation

12.10 Bore-sight alignment

The electrical bore-sight of an antenna is the direction in which some feature, e.g. the peak of the main beam or the null of a monopulse pattern, points with reference to a mechanical axis. Optical instruments, such as telescopes, mirrors, lasers and special indicators, are used to set up a mechanical axis, and electrical measurements are employed to determine the alignment of the chosen pattern feature.[25]

Errors in a bore-sight alignment are caused in parabolas by surface tolerances, feed misalignment, asymmetric phase distributions in primary or secondary patterns and other objects in close proximity. Radomes, particularly the supporting structure, are an important source of bore-sight error. In phased arrays errors are caused by phase-shifter quantum levels and errors, distribution-network asymmetries, element failure, beam distortion at wide angles and by irregularities in protecting surfaces. Bore-sight of a phased array usually refers to a beam at broadside bore-sight, but it is equally important to verify scanned bore-sight positions, because phase shifters may suffer different errors according to demanded scan.

12.10.1 Sum-pattern bore-sight
When all the elements of an array are in phase, a main beam is formed perpendicularly to the array aperture. A similar beam is found for any plane aperture with an equal phase distribution, and for this reason it is referred to as a sum pattern. A difference pattern occurs when the radiating aperture produces a null at the bore-sight angle; in a phased array by dividing the elements into two

anti-phase sets and in parabolas by illuminating from two anti-phase feeds. Accuracy of bore-sight determination depends on instrument dynamic range and positional angular error of the probe in a near field and of the antenna mount in a far-field system. The two main far-field methods for a sum pattern are by beam shift and conical scan, but we will begin with the near-field method and show how probe data accuracy affects the calculated bore-sight angle.

12.10.1.1 Near-field bore-sight

A calculation of bore-sight angle from near-field data depends on the accuracy of amplitude in the main beam, which, in turn, is dependent on errors in the scanning probe positions. For convenience and simplicity we consider a uniformly illuminated aperture, with far-field pattern.

$$A = A_0 \frac{\sin\{(\pi D/\lambda_0)\sin\theta\}}{(\pi D/\lambda_0)\sin\theta} \tag{12.88}$$

that, with a small angular change $\delta\theta$, has amplitude

$$A + \delta A = A_0 \frac{\sin\{(\pi D/\lambda_0)\sin(\theta + \delta\theta)\}}{(\pi D/\lambda_0)\sin(\theta + \delta\theta)} \tag{12.89}$$

which can be expanded and approximated to give

$$A + \delta A \approx$$

$$A_0 \frac{\sin\left(\frac{\pi D}{\lambda_0}\sin\theta\right)\cos\left(\frac{\pi D}{\lambda_0}\cos\theta\delta\theta\right) + \cos\frac{\pi D}{\lambda_0}\sin\theta\sin\left(\frac{\pi D}{\lambda_0}\cos\theta\delta\theta\right)}{\frac{\pi D}{\lambda_0}(\sin\theta\cos\delta\theta + \cos\theta\sin\delta\theta)} \tag{12.90}$$

At the first zero next to the main beam $A = 0$ and $(\pi D/\lambda_0)\sin\theta = \pi$, giving

$$\delta A \approx - A_0 \frac{\sin\{(\pi D/\lambda_0)(\cos\theta)\delta\theta\}}{(\pi D/\lambda_0)\sin\theta} \tag{12.91}$$

If the main beam is narrow, this can be written

$$\delta A \approx - A_0 \frac{\delta\theta}{\theta} \tag{12.92}$$

But $\theta \approx \theta_B$, since the first zero is one beam width, θ_B, from the peak of the main beam. Therefore

$$\frac{\delta A}{A_0} \approx - \frac{\delta\theta}{\theta_B} \tag{12.93}$$

It is not easy to find the position of the peak amplitude A_0 of a main beam because of the zero slope condition. It is usual to measure the average of the positions of the -3 dB points or of the first two zeros, where the greater slope ensures a higher accuracy. Eqn. 12.93 shows the fractional shift in a zero due to an amplitude error ratio $\delta A/A_0$, that might be caused by probe positional errors in a near-field measurement. The relation between amplitude changes and probe positional error in the spherical case follows from the functional dependence of spherical modes on the angular positions of the samples. Positional error in the probe leads to error in the solution of eqn. 12.84 for the M coefficients, and makes the modes uncertain in the angular frame. If there are M samples and $\delta\theta_s$ is the angular uncertainty of a single sample, the mean-square angular uncertainty of a single mode is $\delta\theta_s/\sqrt{M}$. Mode errors are normally distributed if their number is large enough, and, if equally weighted, add as $\sqrt{M}$ in the far field to give a final bore-sight accuracy equal to that of a single sample. Because mode amplitudes are related to antenna structure and do not make equal contributions to amplitude error, it is very difficult to make a generalised theoretical estimate of bore-sight accuracy in near-field measurements. Eqn. 12.93, with $\delta\theta = \delta\theta_s$, is therefore only an approximate guide and, in practice; estimates are the result of comparisons with conventional far-field measurements of the same antenna.

Eqn. 12.93 provides a useful, if optimistic, insight into the relation between minimum detectable pattern amplitude and probe-positional error. If the probe-transverse setting accuracy is Δx, a sample angular error on the minimum sphere is

$$\delta\theta_s = \frac{\Delta x}{r_0} \tag{12.94}$$

and eqn. 12.93 becomes

$$\frac{\delta A}{A_0} = \frac{\Delta x}{\theta_B r_0}$$

In a spherical scan the principal dimension of the antenna is $2r_0$ and the minimum beam width, as for uniform illumination, is therefore $\theta_B = \lambda_0/2r_0$, to give

$$\frac{\delta A}{A_0} = 2\frac{\Delta x}{\lambda_0} \tag{12.95}$$

Dynamic range due to probe positional error is therefore approximately equal to the transverse probe-positional accuracy as a ratio of the free-space wavelength. If $\Delta x = 0.01\lambda_0$, the amplitude ratio is

$$\frac{\delta A}{A_0} = 0.02$$

Expressed as a dynamic range this is 34 dB. Since the typical dynamic range of a probe amplifier is at least 60 dB, pattern accuracy is primarily dependent on probe positioning. Similar arguments can be developed for radial displacements of the probe, and result in equally stringent requirements. In practice near-field techniques do not reach levels of accuracy much below -40 dB from the main beam, unless extreme care is taken with probe positioning.

12.10.1.2 Far-field bore-sight—beam shift

A beam-shift method is usually adopted to find the effects of a protecting radome on beam alignment. The antenna under test (Fig. 12.37) is inside the radome and transmits to three probing antennas in the far field. The centre probe B tests for radome transmission loss and the outer pair, A and C, give the bore-sight shift. Amplitudes at A and C are between 1 and 3 dB below that at B, and on careful alignment should be equal for a symmetrical beam. A and C are connected to the test and reference channels of a two-channel phase and amplitude receiver, or to a network analyser, and the output amplitude is set to zero by means of a calibrated attenuator. Shifts in the beam caused by removing or installing the radome, or by adjustment of the mount, can then be detected by the relative levels of the signals received by A and C. To make an estimate of measurement accuracy, we assume the test antenna is a uniformly illuminated aperture with a $(\sin X)/X$ pattern in the far field. Thus

$$E = E_0 \frac{\sin x}{x}$$

where, for a linear aperture dimension a, the parametric pattern angle x equals $(\pi/\lambda_0)a \sin \theta$, with $\sin \theta \to 0$ in the main beam of a high-gain antenna. Accuracy depends on pattern slope at the -1 dB levels with respect to the main-beam peak, and can be found through a slope sensitivity defined as

$$\frac{S}{E_0} = \frac{1}{E_0} \frac{dE}{dx} = \frac{\cos x}{x} - \frac{\sin x}{x^2}$$

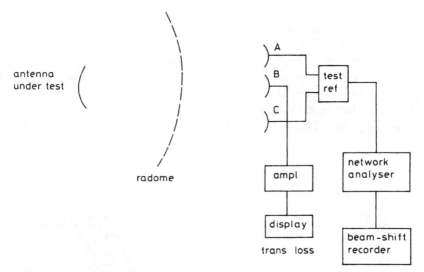

Fig. 12.37 *Test range for bore-sight measurement*

At the -1 dB points $x \approx \pm\pi/4$ to give

$$\frac{1}{E_0}\frac{\Delta E}{\Delta x} \approx \pm\frac{1}{2}$$

or an accuracy of

$$\Delta x \approx 2\frac{\Delta E}{E_0}$$

In practice, the minimum detectable change ΔE is set by instrumentation noise level and the maximum main-beam peak amplitude E_0 by amplifier saturation. Accuracy is therefore twice the dynamic range of the measurement system, provided this is not less than the accuracy of mechanically aligning the antennas. For a dynamic range of 40 dB,

$$\frac{\Delta E}{E_0} = 0{\cdot}01 = \frac{\Delta x}{2} = \frac{\pi}{2}\frac{\lambda_0}{a}\,\Delta\theta$$

or as a fraction of a beam width θ_B

$$\frac{\Delta\theta}{\theta_B} = \frac{0{\cdot}02}{\pi} \approx 0{\cdot}01$$

12.10.1.3 Far-field bore-sight—Conical scan

Conical-scan antennas rotate the main beam in a circle around the bore-sight, so that a displacement in electrical bore-sight is observed as a modulation of the received amplitude at the scan frequency. Rotation is by mechanical movement of a feed or a sub-reflector in paraboloids, or by phase-shifter adjustment in phased arrays. The principle is illustrated in Fig. 12.38, where, for mechanical and electrical coincidence, the amplitude remains constant at the crossover level if the beam is symmetrical, whilst any misalignment causes the appearance of a component at the scan frequency.

An estimate of measurement accuracy can be made using Fig. 12.39, in which θ_0 is the mechanical bore-sight line, θ_0' is the beam-pointing angle and ϕ is the angular position of the nose of the beam at instant t as it rotates about θ_0'. The beam is first displaced by $\Delta\theta$ from its bore-sight and then rotated around it at the scan frequency. This is equivalent to an orthogonal time modulation in the

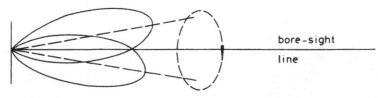

Fig. 12.38 *Conically scanned beam*

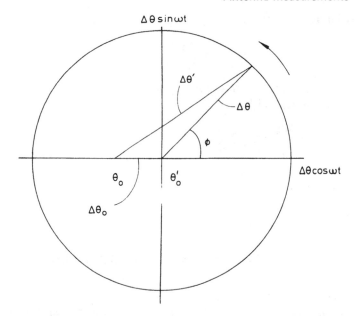

Fig. 12.39 *Bore-sight from conical scan*

horizontal and vertical directions. If the error between mechanical and electrical axes is $\Delta\theta_0 = \theta_0' - \theta$ and $\Delta\theta'$ is the angle between the beam and the mechanical bore-sight, application of the cosine rule in Fig. 12.39 gives

$$\Delta\theta' = \sqrt{(\Delta\theta_0)^2 + (\Delta\theta)^2 + 2\Delta\theta_0\,\Delta\theta\,\cos\omega t} \qquad (12.96)$$

where $\phi = \omega t$ and ω is the scan frequency.

For small errors in bore-sight position $\Delta\theta_0 \ll \Delta\theta$ and

$$\Delta\theta' \approx \Delta\theta\left(1 + \frac{\Delta\theta_0}{\Delta\theta}\cos\omega t\right) \qquad (12.97)$$

Now beam pattern can be approximated by

$$E = E_0 \cos\sqrt{3}\,\frac{\Delta\theta'}{\theta_B}$$

where θ_B is the -3 dB beam width. Therefore, on substituting for $\Delta\theta'$ from eqn. 12.97 and for $(\Delta\theta_0/\theta_B) \to 0$,

$$E \approx E_0\left[\cos\left(\sqrt{3}\,\frac{\Delta\theta}{\theta_B}\right) - \sin\left(\sqrt{3}\,\frac{\Delta\theta}{\theta_B}\right)\sqrt{3}\,\frac{\Delta\theta_0}{\theta_B}\cos\omega t\right]$$

The antenna response consists of a constant term

$$E(0) = E_0 \cos\left(\sqrt{3}\,\frac{\Delta\theta}{\theta_B}\right)$$

and an oscillating component

$$E(\omega) = E_0 \sin\left(\sqrt{3}\,\frac{\Delta\theta}{\theta_B}\right)\sqrt{3}\,\frac{\Delta\theta_0}{\theta_B}$$

When the scan is symmetrical about the bore-sight line $E(\omega) = 0$ and a bore-sight sensitivity is defined by

$$\frac{E(\omega)}{E(0)} = \sqrt{3}\,\frac{\Delta\theta_0}{\theta_B}\,\tan\sqrt{3}\,\frac{\Delta\theta}{\theta_B} \tag{12.98}$$

For a dynamic range of 40 dB, $E(\omega)/E(0) = 0{\cdot}01$; and for a quarter beam-width conical offset, $\Delta\theta/\theta_B = 0{\cdot}25$, to give

$$\frac{\Delta\theta_0}{\theta_B} \approx 0{\cdot}01$$

showing an accuracy similar to that in the beam-shift method.

12.10.2 Difference-pattern bore-sight

Monopulse antennas with split beams produce a null on bore sight, so that testing depends on null-seeking methods. Accuracy is again limited by error in the mechanical turning gear, and also the relation between the difference-pattern split sensitivity and dynamic range of the instrumentation. Consider, for example, a uniformly illuminated circular aperture with two feeds offset from the axis, as in Fig. 12.40, and connected to a magic-T to produce sum-and-difference outputs of the separate patterns in Fig. 12.41. Feed offset displaces the two beams from the axis by angle α, with displacement parameter $u_0 = 2\pi(a/\lambda_0)\sin\alpha$, where a is the radius of the paraboloid. At position x,

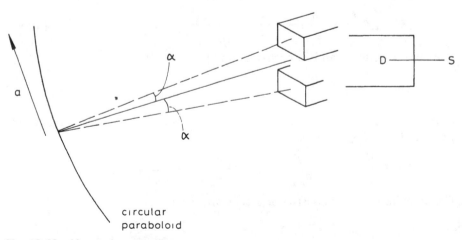

Fig. 12.40 *Monopulse antenna*

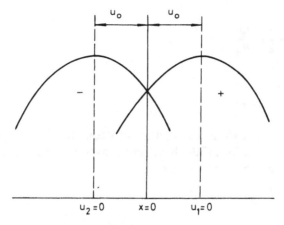

Fig. 12.41 *Monopulse bore-sight*

corresponding beam parameters are

$$u_1 = x - u_0, \qquad u_2 = x + u_0$$

with far-field amplitudes

$$E(u_1) = 2\frac{J_1(u_1)}{u_1}, \qquad E(u_2) = 2\frac{J_1(u_2)}{u_2} \qquad (12.99)$$

to give a difference output as

$$E(x) = 2\frac{J_1(x - u_0)}{(x - u_0)} - 2\frac{J_1(x + u_0)}{(x + u_0)}$$

where $J_1(u)$ is a first-order Bessel function. The slope of $E(x)$ at $x = 0$ is the split sensitivity. Therefore

$$\left[\frac{dE(x)}{dx}\right]_{x=0} = 4\frac{J_2(u_0)}{u_0}$$

For small-feed offset $u_0 \to 0$ and

$$\left[\frac{dE(x)}{dx}\right]_{x=0} \approx \frac{u_0}{2}$$

But $x = 2\pi(a/\lambda_0)\sin\theta \pm u_0$, which, for small θ, gives

$$\frac{dx}{d\theta} = \frac{2\pi a}{\lambda_0}$$

and

$$\frac{dE(x)}{d\theta} = \pi\frac{a u_0}{\lambda_0}$$

or

$$\Delta E(x) = \frac{\pi}{2} u_0 \frac{\Delta \theta}{\theta_B}$$

where

$$\theta_B \approx \frac{\lambda_0}{2a}$$

Since the peak amplitude of each pattern in eqn. 12.99 is $E(u_0) = 1$, the sum peak at $x = 0$ is $\sim 1 \cdot 4 E(u_0)$ if the beam cross-over is at the -3 dB points, with $u_0 \approx \pi/2$. $\Delta E(x)$ can be expressed as

$$\frac{\Delta E(x)}{E(u_0)} \approx \frac{0 \cdot 7}{4} \pi^2 \frac{\Delta \theta}{\theta_B} \qquad (12.100)$$

Again for a dynamic range of 40 dB

$$\frac{\Delta \theta}{\theta_B} \approx 0 \cdot 01$$

All the methods described depend on seeking a null at bore sight. Errors will occur if stray reflections are greater than the minimum detectable depth of the null. Thus, with a dynamic range of 40 dB, reflections at -40 dB to the peak of the main beam will, in worst-case addition, raise the null by 6 dB. Other sources of error include those in the following list:[25]

- Co-ordinate-axis misalignment
- Antenna mount errors
- Load stresses and environmental effects
- Phase curvature and amplitude taper
- Spurious interfering radiation

12.11 Antenna noise temperature

All objects appearing in an antenna pattern have a physical temperature above absolute zero, and therefore radiate noise power into the terminals. Noise temperatures of antennas became important during the early developments of ground stations, when satellite radiated powers were low and ground-receiving antennas were consequently very large. Small reductions in antenna size, resulting from reduced noise temperature, gave large cost savings. As satellite radiated powers have increased, emphasis on low noise has fallen, but remains of great significance in deep-space probes, high-capacity links and some radar applications. Antenna noise-temperature measurement follows procedures similar to those described in a previous Chapter, but the sources are usually more remote. We begin with a general description of the origins of antenna noise, and conclude with one example of its measurement that can also give information about the radiation efficiency.

12.11.1 Origins of antenna noise

Non-isotropic antennas do not have equal sensitivities in all directions and are characterised by their gain. In a rather special sense an antenna is an amplifier which is not fully specified until its noise level is known; and a good analogy with signal/noise ratio is the antenna gain/noise-temperature ratio. The chief sources of antenna noise are listed below:

Galactic noise: Non-black-body radiation from our own galaxy which is greater in the direction of the galactic poles. Its equivalent temperature follows a frequency law $f^{-2.5}$ and at 1 GHz it becomes negligible.

Tropospheric noise: Oxygen and water vapour molecules in the troposphere absorb and re-radiate energy as noise with maxima occurring at 22·2 and 60 GHz, the water-vapour and oxygen resonances, respectively. Tropospheric noise is negligible below 1 GHz and reaches 290°K near oxygen resonances. Rain increases noise temperature at frequencies greater than 3 GHz.

Celestial bodies: The sun and radio stars are discrete noise sources that are important when the main beam points directly at them. The sun's temperature is $>6000°K$ and the radio star *Cassiopeia A* is about two orders of magnitude lower at frequencies greater than 1 GHz.

Ground noise: The earth is a semi-black body with an emissivity of $(1 - \rho)$, where ρ is its reflectivity. Thus, for earth temperature T_E, its equivalent noise temperature is

$$T = T_E(1 - \rho) \tag{12.101}$$

Sea water, at glancing incidence, with $\rho = 1$, is not a source of direct noise, whereas an absorbing land mass approaches 290°K. An antenna therefore 'sees' a different noise temperature in different parts of its pattern. If the main beam points straight up into a cold sky, i.e. no celestial bodies in the beam, the total noise temperature is low because the only significant noise radiation is that received from the ground via the far out side lobes. When the beam points horizontally, the low close-in side lobes contribute a greater proportion of ground noise. Temperature is therefore a function of the antenna polar angles θ, ϕ; and the total received noise temperature is the integral of $T(\theta, \phi)$, the noise temperature at θ, ϕ weighted by the directivity $D(\theta, \phi)$ of the antenna at θ, ϕ. Thus, if T_a is the noise temperature at the antenna aperture,

$$T_a = \frac{\int_0^\pi \int_0^{2\pi} T(\theta, \phi) D(\theta, \phi) \sin \theta \, d\theta \, d\phi}{\int \int D(\theta, \phi) \sin \theta \, d\theta \, d\phi} \tag{12.102}$$

Rough estimates of T_a can be made from

$$T_a = \sum_{i=1}^n a_i T_i \tag{12.103}$$

where a_i is the fraction of the total integrated polar pattern containing a region at temperature T_i.

Typical figures above 1 GHz are 3—30°K contribution from the sky, for elevations between 90° and 5°, arriving via the main beam, and a 10°K ground contribution via the side and back lobes. In addition, the antenna's conductors and insulators have dissipative losses which produce noise. Again, at the aperture face, if L_a is the antenna loss, the effective noise temperature due to loss is $T(L_a - 1)$, giving a total antenna noise temperature at the aperture of

$$T_e = T_a + T(L_a - 1) \tag{12.104}$$

where T is the physical temperature of the antenna, usually close to $T_0 = 290°$K. An earth-station antenna might be 12 m in diameter with a typical gain of 57 dB. If its noise temperature is 10°K and the feed loss is 0·5 dB, $L_a = 1·12$ and $T_e = 45·4°$K, to give a gain/noise temperature ratio G/T of 40·4 dB.

The G/T ratio of an earth-station antenna and receiving subsystem can be measured by the Y-factor method from the proportional increase in noise power in the receiver when a radio star of known properties passes through the beam of the antenna. The three brightest radio sources are *Cassiopeia A*, *Taurus A* and *Cygnus A*, with radiated power densities such that antenna temperatures are in the range 50—200°K. Since earth terminals, when not pointed at radio sources, have noise temperatures in the range 50—80°K, depending on elevation, Y-factors are small and in the range 2—3 dB. Thus 0·1 dB of measurement uncertainty produces an error of 0·2 dB in the calculated G/T.

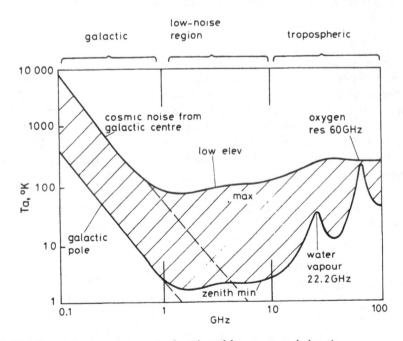

Fig. 12.42 *Sky-noise temperature as a function of frequency and elevation*

Figure 12.42 shows antenna noise temperature as a function of frequency. Maximum and minimum values are for horizontal and vertical main beams, respectively, but the latter can only be realised by reducing ground illumination to a negligible amount. Further information on the design of low-noise antennas can be found in the literature.[26-30]

12.11.2 Radiometric measurement of antenna radiation efficiency

Radiation efficiency is the fractional power loss between the input terminals and the aperture. In eqn. 12.26 it is a significant parameter for finding the directivity of an antenna whose gain is measured by substitution with a standard-gain horn or dipole. It can be defined as

$$\eta = \frac{P_r}{P_i} = \frac{\int\int P(\theta, \phi) \sin\theta \, d\theta \, d\phi}{P_i} \tag{12.105}$$

where P_r is the total power radiated from the aperture and P_i is the power input to the terminals. $P(\theta, \phi)$, the power per unit solid angle, can be measured at each θ, ϕ and numerically integrated, but measurement of $P(\theta, \phi)$ over all angles is both difficult and lengthy. These problems of complete solid-angle coverage and pattern integration can be simply overcome by means of extended noise sources.[31] If the temperature in direction θ, ϕ is $T(\theta, \phi)$, the antenna noise temperature is

$$T_a = \frac{\int\int T(\theta, \phi)D(\theta, \phi) \sin\theta \, d\theta \, d\phi}{\int\int D(\theta, \phi) \sin\theta \, d\theta \, d\phi} \tag{12.106}$$

But an extended source, such as the cold sky or a screen of radar absorbing material (RAM) that totally fills the pattern, radiates with the same temperature $T(\theta, \phi) = T_t$, from all directions towards the antenna, so that, in eqn. 12.106, $T_a = T_t$. At the output port this becomes ηT_t.

The extended source is particularly appropriate for finding the directivity of phased arrays, with separate element amplifiers feeding a combining network as shown in Fig. 12.43. It is possible to measure only an overall gain at the combining output port, where element amplifier gains and combining network losses cannot be separated from aperture directivity. The importance of separating directivity from gain is that it is a direct indication of the soundness of the original aperture design, and allows a better allocation of the contributions of the network and amplifiers to the final far-field pattern. If the antenna is therefore directed at an extended source which fills the whole pattern at a constant temperature, there is an automatic integration. Convenient extended sources are the cold sky, seen by directing the antenna vertically upwards, and an extended radar absorber, usually at room temperature. These become the cold and warm sources of a system noise-temperature measurement, which includes the antenna and a low-noise radiometer receiver, as shown in Fig. 12.44.

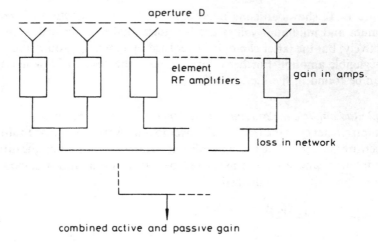

Fig. 12.43 *Active phased array*

At the aperture the effective noise temperature is

$$T_{ea} = T_A(1/\eta - 1) + T_t \qquad (12.107)$$

where T_A is the physical temperature of the test antenna, with loss η, equal to the radiation efficiency, and T_t is the temperature of the extended source. At the input to the low-noise amplifier, the temperature is

$$T_e = T_A(1 - \eta) + T_t\eta + T_n \qquad (12.108)$$

where T_n is the internal temperature of the radiometer given by its noise figure as

$$T_n = (F - 1)T_0 \qquad (12.109)$$

with $T_0 = 290°K$. Setting T_t equal to T_w and T_c for the warm and cold sources, the measured ratio of the temperatures is

$$\varepsilon = \frac{T_A(1 - \eta) + T_w\eta + T_n}{T_A(1 - \eta) + T_c\eta + T_n} \qquad (12.110)$$

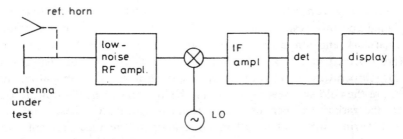

Fig. 12.44 *Radiometer for antenna noise temperature*

Temperatures T_A, T_c and T_w are quantities known to varying degrees of accuracy and T_n is given by the radiometer noise figure previously measured. Thus from the ratio ε, radiation efficiency can be calculated as

$$\eta = \frac{(1 - \varepsilon)(T_A + T_n)}{\varepsilon(T_c - T_A) + (T_A - T_w)} \tag{12.111}$$

Uncertainty can be reduced by making $T_w = T_A$ and repeating the measurement with a high-efficiency horn for which $\eta \approx 1$. Measured temperature ratio in this case is

$$\delta = \frac{T_n + T_A}{T_n + T_c} \tag{12.112}$$

None of the temperatures now needs to be known, because η can be written as an explicit function of ε and δ as

$$\eta = \frac{\delta(\varepsilon - 1)}{\varepsilon(\delta - 1)} \tag{12.113}$$

Temperature difference between the extended sources should be as high as possible, and, for greatest accuracy, the radiometer receiver should have a low noise figure. Accuracy has been checked by measuring the effect of an attenuator between the horn and the receiver on the measured temperature, and gave differences between attenuator settings and calculated values[31] of $0.2\,$dB. Sources of error are the lack of effectiveness and temperature uncertainty of extended sources, especially the absorber.

12.12 Polarisation

Maximum power transfer occurs between antennas that are polarisation matched. There are three important applications of polarisation measurements. First, polarisation discrimination increases channel capacity by frequency re-use on satellite links. Up and down links, offset in frequency, with overlapping sidebands, are orthogonally polarised to minimise crosstalk. Antenna axial ratios of 40 dB for linear and $0.15\,$dB for circular components are typical, with system discrimination ratios of 25—30 dB. Secondly, near-field probe characterisation requires precision polarisation measurements over wide angular ranges. Gain and frequency stability is of a high order and calls for frequency synthesisers and phase-lock techniques. Finally, in radar detection, rain cancellation depends on the proper use of circular polarisation in the illuminating and scattered radiation. We begin this Section by defining the co- and cross-polar directions in the radiation field of an aperture, continuing with a description of the Poincaré sphere as a method of defining polarisation efficiency, and ending with some practical aspects of polarisation measurements, including the use of circular polarisation in rain-reflection cancellation.

12.12.1 Co- and cross-polar patterns

It is usual to define the polarisation at a point in a field as the direction of the electric vector at that point. There are extra considerations for the general elliptic case, but, for the present, we will restrict the discussion to fields with a single linear polarisation perpendicular to the direction of propagation. For clarity, and because we have already developed some basic theory, we will take as our example the far-field polarisation of a rectangular radiating aperture: in this case a horn in the co-ordinate system of Fig. 12.45. In a typical far-field measurement, the test horn A would rotate about a vertical axis, whilst the distant probe remains fixed in a common horizontal plane. If the aperture field at the origin of the co-ordinate system is polarised in the y-direction, it can be written as E_y. In general, this radiates components E_θ and E_ϕ given by eqns. 12.15 and 12.16, with a linearly polarised far field in a direction determined by the spatial vector addition of E_θ and E_ϕ. In so far as this direction is in line with the aperture field E_y, the far field is called co-polar, whilst any perpendicular component becomes cross-polar. This deliberately vague statement reflects the fact that there is uncertain agreement about the most useful form of definition, though we shall choose Ludwig's third definition[32] because of its convenient relation to the normal facilities of an antenna far-field test range. By this definition, a co-polar measurement in a cut at angle ϕ is made by rotating the test antenna in Fig. 12.45 about the z-axis from A to position B, i.e. by $90 - \phi$ degrees in a clockwise direction along z. The probe antenna is aligned along the z-axis, looking directly at the origin, and then rolled about its bore sight until its polarisation direction corresponds to the E_y direction in the test antenna.

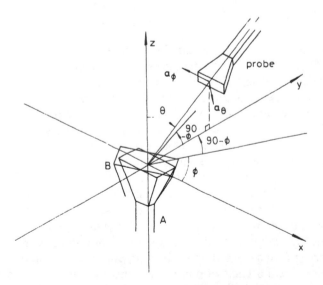

Fig. 12.45   *Test and source positioning for co- and cross-polar patterns*

There is an implicit assumption that the probe is an *H*-plane rectangular horn with electric vector parallel to the narrow dimension. Movement of the probe in the *yz*-plane, or, more practically, rotation of the test antenna about the *x*-axis, produces a co-polar pattern cut at angle ϕ. The relation to a far-field site is most easily seen by rotating the co-ordinate system so that the *yz*-plane is horizontal, and setting $\phi = 90°$ to align E_y in both test and probe antennas. A co-polar vertically polarised pattern results if the test-antenna mount axis is along *x*.

For each ϕ cut, the orientation of unit vectors a_θ and a_ϕ remain fixed, as shown in Fig. 12.46a, with respect to the probe aperture as it moves in the *yz*-plane. E_y at the probe is the measured amplitude of the co-polar far field $E_p(\theta, \phi)$, and in terms of E_θ and E_ϕ is

$$E_p(\theta, \phi) = E(\theta, \phi) . [\sin \phi a_\theta + \cos \phi a_\phi] = E(\theta, \phi) . a_{co} \qquad (12.114)$$

Cross-polar patterns are found by rotating the probe through $90°$ to give the orientation shown in Fig. 12.46b with measured amplitude

$$E_q(\theta, \phi) = E(\theta, \phi) . [\cos \phi a_\theta - \sin \phi a_\phi] = E(\theta, \phi) . a_{cross} \qquad (12.115)$$

In practice, test antennas may have cross-polar aperture fields E_x, and the effects of these can be seen by transforming the expression for E_p and E_q into the rectangular unit vectors of the test-antenna co-ordinate system. Thus

$$a_\theta = \cos \theta \cos \phi a_x + \cos \theta \sin \phi a_y - \sin \theta a_z$$

$$a_\phi = -\sin \phi a_x + \cos \phi a_y \qquad (12.116)$$

and eqns. 12.114 and 12.115 have

$$a_{co} = \sin \phi a_\theta + \cos \phi a_\phi$$

$$= -(1 - \cos \theta) \sin \phi \cos \phi a_x + [1 - \sin^2 \phi(1 - \cos \theta)]a_y - \sin \theta \sin \phi \, a_z$$

$$(12.117)$$

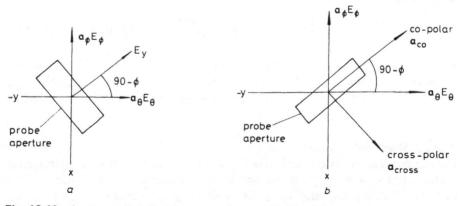

Fig. 12.46 *Aperture orientations for co- and cross-polar patterns*

$$a_{cross} = \cos \phi a_\theta - \sin \phi a_\phi$$

$$= [1 - \cos^2 \phi(1 - \cos \theta)]a_x - (1 - \cos \theta) \sin \phi \cos \phi a_y - \sin \theta \cos \phi a_z$$

$$(12.118)$$

For small θ the major contribution to a_{co} is the co-polar component a_y with increasing cross-polar content for larger θ. Similarly, a_{cross} consists predominantly of the cross-polar aperture component at small θ. Thus, for high-gain narrow-beam antennas, polarisation in the main beam, measured by this method, corresponds very accurately to the true polarisation in the aperture. At wider angles, usually in the side lobes, cross-polar terms become significant.

Radiated far fields, from eqns. 12.15 and 12.16, are dependent on the test-aperture dimensions, but the measured fields depend on the polarisation of the probe as receiver, as well as the radiated amplitudes. The co- and cross-polar fields in eqns. 12.114 and 12.115, in matrix form, are

$$\begin{bmatrix} E_p(\theta, \phi) \\ E_q(\theta, \phi) \end{bmatrix} = \begin{bmatrix} \sin \phi & \cos \phi \\ \cos \phi & -\sin \phi \end{bmatrix} \begin{bmatrix} E_\theta \\ E_\phi \end{bmatrix} \qquad (12.119)$$

and, since the probe would be oriented to observe either E_p or E_q, we can obtain complete expressions for an ideal far-field-range response by substituting for E_θ and E_ϕ from eqns. 12.15 and 12.16, to give[2]

$$\begin{bmatrix} E_p(\theta, \phi) \\ E_q(\theta, \phi) \end{bmatrix} = \frac{j \exp(-jk_0 r)}{\lambda r}$$

$$\times \cos^2 \frac{\theta}{2} \begin{bmatrix} \dfrac{1 - \tan^2 \theta}{2} \cos 2\phi & \dfrac{\tan^2 \theta}{2} \sin 2\phi \\ \dfrac{\tan^2 \theta}{2} \sin 2\phi & \dfrac{1 + \tan^2 \theta}{2} \cos 2\phi \end{bmatrix} \begin{bmatrix} f_y(\theta, \phi) \\ f_x(\theta, \phi) \end{bmatrix} \qquad (12.120)$$

For a high-gain antenna and angles close to the main beam $\tan^2 \theta/2 \to 0$, $\cos^2 \theta/2 \to 1$, and

$$E_p \approx j \frac{\exp(-jk_0 r)}{\lambda r} f_y(\theta, \phi), \qquad E_q \approx j \frac{\exp(-jk_0 r)}{\lambda r} f_x(\theta, \phi) \qquad (12.121)$$

f_x and f_y, recognised as Fourier transforms in eqns. 12.12 and 12.13, give the expected result that E_p depends on E_y and E_q on E_x; and both justifies the simple Fourier transformation of aperture fields and indicates its wide-angle limitations.

12.12.2 Wave polarisation

Co-/cross-polar ratios in transverse waves specify polarisation as orthogonal vector pairs in a plane perpendicular to the direction of propagation. Non-orthogonal ratios may also serve the same purpose, but we will restrict this discussion to linear, linear diagonal and circular orthogonal pairs; and we will also adopt the convention of horizontal and vertical vectors, E_H, E_V, in a horizontally propagating wave. Thus we may define the specifying methods as in Fig. 12.47, and write an equivalent co-/cross-polarisation ratio for each as

Linear

$$p_L = p_L \exp j\delta_L \quad \text{with} \quad p_L = \frac{E_V}{E_H} \tag{12.122}$$

Linear diagonal

$$p_D = p_D \exp j\delta_D \quad \text{with} \quad p_D = \frac{E_{135}}{E_{45}} \tag{12.123}$$

Circular

$$p_C = p_C \exp j\delta_C \quad \text{with} \quad p_C = \frac{E_R}{E_L} \tag{12.124}$$

The co-polar may be defined as either one of each pair, and we have chosen E_H, E_{45} in the first two cases and the left-hand rotating vector of amplitude E_L for the circularly polarised pair. In circular polarisation, the electric vectors are orthogonal in that they rotate in opposite senses with constant amplitude at the wave angular frequency ω. δ_L and δ_D are the relative phases between the linear vectors; and, in the case of circular polarisation, δ_c is the phase of the right-hand vector at the instant when the left-hand one is horizontal.

Any of these pairs can be used to completely describe a polarised field. For example, elliptical polarisation can be expressed in terms of E_V and E_H as

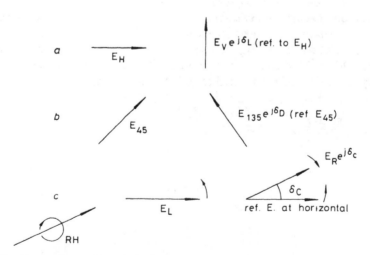

Fig. 12.47 *Linear, diagonal and circular polarisation*
a Linear
b Linear diagonal
c Circular

shown in Fig. 12.48. The lengths of E_x and E_y are time-varying sinusoids, as indicated by the time exponent with angular frequency ω, and the locus of the instantaneous amplitude of the resultant vector can be found from real parts of E_x and E_y. Thus

$$E_x = E_H \cos \omega t$$

$$E_y = E_V \cos (\omega t + \delta_L)$$

On expanding $\cos (\omega t + \delta_L)$ in the second equation and substituting for $\cos \omega t$ in the first, we have, after a little manipulation,

$$\frac{E_x^2}{E_H^2} + \frac{E_y^2}{E_V^2} - \frac{2E_x E_y}{E_H E_V} \cos \delta_L = \sin^2 \delta_L \qquad (12.125)$$

This is the equation of an ellipse with tilt angle determined by E_H, E_V and δ_L. For instance, if $\delta_L = \pi/2$,

$$\frac{E_x^2}{E_H^2} + \frac{E_y^2}{E_V^2} = 1$$

which is an ellipse with horizontal and vertical major and minor axes, respectively, if $E_H > E_V$. A special case occurs when $E_H = E_V$ and

$$E_x^2 + E_y^2 = E^2$$

The locus becomes circular, with the instantaneous vector rotating anti-clockwise around the origin. Similarly, right-hand, or clockwise, circular polarisation requires the phase to be $\delta_L = -\pi/2$.

As a second example we can show that elliptical polarisation is formed from the counter-rotating circular vectors in Fig. 12.49a. If $E_R = E_L$, the result is linear polarisation, as in Fig. 12.49b, tilted from the horizontal at an angle $\tau = \delta_c/2$. Otherwise the result is elliptical with the major axis, in Fig. 12.49c, tilted by the same angle. The axial ratio is found by dividing the minor into the major axis as

$$r = \frac{E_R + E_L}{E_R - E_L} \qquad (12.126)$$

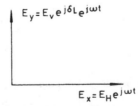

Fig. 12.48 *Showing polarisation as relative amplitude and phase of spatially orthogonal vectors*

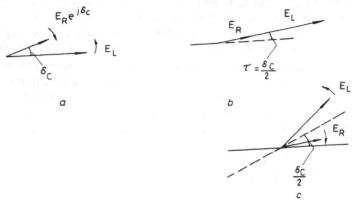

Fig. 12.49 *Tilt angle in elliptical polarisation*

A practical implementation of circular polarisation uses a nearly square wave-guide into which a diagonal vector E is launched. Its resolved horizontal and vertical components are equal in amplitude but have different phase velocities. If E_V is retarded by 90° relative to E_H in its progress along the waveguide in Fig. 12.50, a right-hand circularly polarised wave emerges. Errors in phase difference result in ellipticity with a 45° tilt angle, whilst amplitude errors cause ellipticity with zero or 90° tilt angle. The emerging circularly polarised wave might be radiated from an antenna attached to the polariser, with E defined as

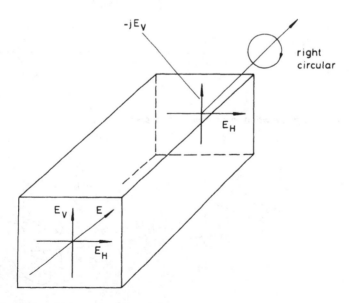

Fig. 12.50 *Rectangular-waveguide polariser*

the transmitted input vector. In reception, the wave propagates in reverse from the antenna to the input/output of the polariser, E becomes the received output vector, and total energy transfer from antenna to detector occurs only when the received wave is right-hand circularly polarised. This is illustrated in Fig. 12.51a, where a left-hand circular receive wave suffers a phase reversal in E_V on passing through the polariser, so that the received linear component E_r is perpendicular to the original diagonal direction of E, now labelled E_l in Fig. 12.51, and is therefore not detected. On receiving a right-hand circular wave in Fig. 12.51b, the depolarised linear output is along E_r. In the latter case the wave and receiving antenna are said to be polarisation matched.

It follows that antennas are polarisation matched when they each transmit the same hand of circular polarisation. For the general case of elliptically

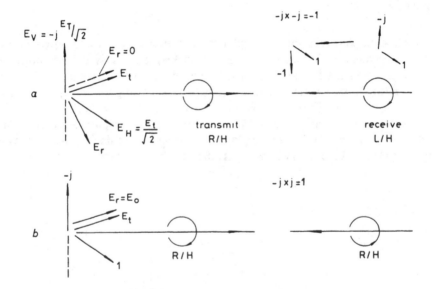

Fig. 12.51   *Circular polarisation in transmit and receive*

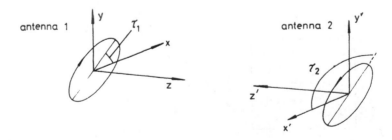

Fig. 12.52   *Polarisation matching*

polarised antennas, they are polarisation matched if each transmits an elliptical wave with the same axial ratio and sense of polarisation, and the tilt angles are related as

$$\tau_1 = 180° - \tau_2 \tag{12.127}$$

This is illustrated in Fig. 12.52, where the antennas transmit along z and z', respectively, and τ is measured from the x-axis in each case. In these circumstances, received linear vectors are aligned with input transmitted vectors and the antennas can mutually receive/transmit with no polarisation loss.

12.12.3 Poincaré sphere

In general, when antennas are not polarisation matched, it becomes necessary to find the polarisation loss factor p. Given the large number of possible polarisation pairs, this would be a difficult and time-consuming task if no generalised method were available, but by plotting all possible polarisations on a Poincaré sphere, it is possible to quickly find the loss from the tilt angles, axial ratios and polarisation senses of the receive and transmit antennas. All possible polarisations can be represented on the sphere with each point on its surface uniquely describing a particular polarisation,[1,3,13] The direction of propagation is outward, with left-hand rotation in the northern hemisphere and right hand in the south, as illustrated in Fig. 12.53. In this diagram, tilt angle τ is referenced to the direction of the θ unit vector in the antenna co-ordinate system, as shown in Fig. 12.54, and the following rules are used to locate the polarisations:

● Longitude equals twice the tilt angle, 2τ
● Latitude is twice the angle whose co-tangent is the negative of the axial ratio. $2 \cot^{-1}(-r)$.

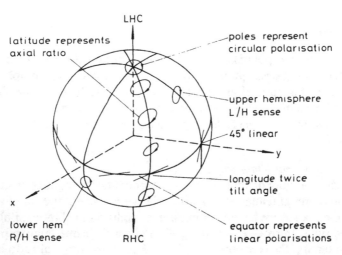

Fig. 12.53 *Poincaré sphere*

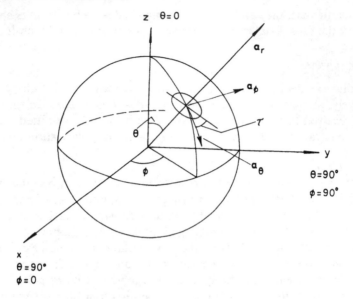

Fig. 12.54   *Tilt angle on Poincaré sphere*

Polarisation loss factor, defined as

$$p = \frac{\text{Power received by antenna}}{\text{Power received if it were polarisation matched}}$$

can be found from the angular separation 2ϕ in the great-circle plane containing the transmit and receive polarisations on the Poincaré sphere, in Fig. 12.55, with

$$p = \cos^2 \phi \qquad\qquad (12.128)$$

When $\phi = 0$ the two polarisations are coincident and the loss is zero.

A full theoretical description of the Poincaré sphere[1,13] is complex and not appropriate in this discussion, but it is essential to fully grasp the significance of polarisation matching and its relation to polarisation factor, if the rules are to be correctly applied. The following two examples are meant to illustrate this relationship.

12.12.3.1 Orthogonal phase shifter
In Fig. 12.56, spatially orthogonal field components are radiated from separate parts of the same antenna system with relative phase α_A and amplitudes $K_x E$ and $K_y E$. On the left we have transmission of right-hand elliptical polarisation, and on the right, in receive, assuming K_y is given the same relative phase shift before combining, the recovery of $K_x E$ and $K_y E$ in the original ratio is possible only if the receive wave is right-hand rotating and with the same axial ratio. In the general case, return amplitudes will not be in the ratio K_x/K_y nor will the

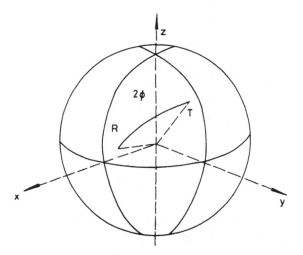

Fig. 12.55 *Polarisation factor on Poincaré sphere*

phase difference be α_A. This is illustrated in Fig. 12.57, where the incoming wave has components E_x and $E_y \exp(j\alpha_W)$, and the polariser has equal amplitude weighting, as in the square wave-guide polariser previously described, with phase shift α_A in the y channel. Assuming that only the 45° vector is able to enter the detecting circuits, the detected output is

$$E_{det} = \frac{1}{\sqrt{2}}[E_x + E_y \exp(j\alpha_W + \alpha_A)] \qquad (12.129)$$

In a square-law detector this becomes

$$|E_{det}|^2 = \frac{1}{2}[E_x^2 + E_y^2 + 2E_x E_y \cos(\alpha_W + \alpha_A)] \qquad (12.130)$$

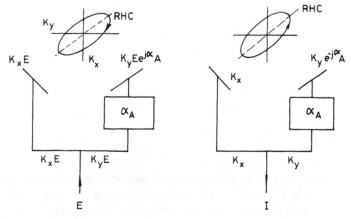

Fig. 12.56 *Spatially orthogonal radiators*

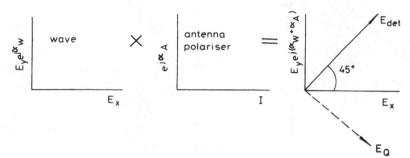

Fig. 12.57 *Receive waves from orthogonal radiators*

Depending on the relative phase of E_x and E_y, there will be a quadrature component E_Q in Fig. 12.57 given by

$$|E_Q|^2 = \frac{1}{2}[E_x^2 + E_y^2 - 2E_xE_y \cos(\alpha_W + \alpha_A)]$$ (12.131)

The total power in the wave is

$$\text{Total power} = |E_{det}|^2 + |E_Q|^2 = E_x^2 + E_y^2$$ (12.132)

It follows that if $\alpha_W = -\alpha_A$ and $E_x = E_y$

$$|E_{det}|^2 = \frac{1}{2}[E_x + E_y]^2 \quad \text{and} \quad |E_Q|^2 = \frac{1}{2}[E_x - E_y]^2 = 0$$

and all the power appears in the detector.

The polarisation loss factor, as defined in eqn. 12.128, is found from the ratio of the detected power, given by eqn. 12.130 for any value of $(\alpha_W + \alpha_A)$, to the total power in the wave, or

$$p = \cos^2 \phi = \frac{\frac{1}{2}[E_x^2 + E_y^2 + 2E_xE_y \cos(\alpha_W + \alpha_A)]}{E_x^2 + E_y^2}$$

$$= \frac{1}{2} + \frac{E_xE_y}{E_x^2 + E_y^2} \cos(\alpha_W + \alpha_A)$$ (12.133)

For $E_x = E_y$

$$p = \cos^2 \phi = \cos^2\left(\frac{\alpha_W + \alpha_A}{2}\right)$$ (12.134)

with ϕ defined on the Poincaré sphere in Fig. 12.55.

12.12.3.2 Radar return from rain
In satellite communications, ground station and satellite antennas are usually polarisation matched, but in radar, circular-polarisation returns from most targets are relatively insensitive to polarisation sense, whereas rain drops, as

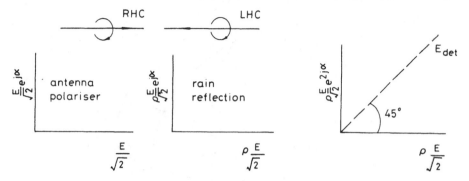

Fig. 12.58 *Rain cancellation in circular polarisation*

spheres, reflect with the opposite sense. This can be turned to advantage in reducing rain clutter with little effect on target returns. Thus, in Fig. 12.58, if $\alpha = \pi/2$, the polariser and antenna transmit right-hand circular polarisation. In reception, usually by the same antenna, rain reflection with coefficient ρ is left-hand polarised and appears after de-polarisation in the quadrature component, and is not detected. Thus, as before,

$$|E_{det}|^2 = \rho^2 \frac{E^2}{4} [2 + 2\cos 2\alpha] = \rho^2 E^2 \cos^2 \alpha \qquad (12.135)$$

and this is zero if $\alpha = \pi/2$.

In each of these examples, ϕ is the mean phase angle between the orthogonal polarisation vectors.

12.12.4 Polarisation measurements

With this background we are in a position to discuss some of the techniques for antenna-polarisation measurement. They can be divided into the following categories:

- Those that give all the information to completely characterise the polarisation by a unique point on the Poincaré sphere, but that rely on a standard for comparison.
- Those that completely characterise the polarisation by an absolute method.
- Those that give some of the ratios, but cannot uniquely specify a point on the Poincaré sphere.

In the following Sections we will give an example of each category.

12.12.4.1 Phase-amplitude method

Any of the orthogonal pairs defined in eqns. 12.122—12.124 is sufficient to locate a point on the Poincaré sphere. Amplitude and phase ratios for each pair have equivalent angles on the sphere in relation to the poles or equatorial plane. These are illustrated without proof for a point P in Fig. 12.59. The method

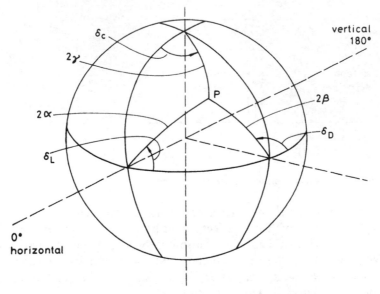

Fig. 12.59 *Angles on Poincaré sphere for orthogonal pairs*

$\hat{p}_c = p_c\, e^{j\delta_c}$ $\hat{p}_L = p_L e^{j\delta_L}$ $\hat{p}_D = p_D e^{j\delta_D}$

$p_c = \tan\gamma$ $p_L = \tan\alpha$ $p_D = \tan\beta$

relies on accurate measurement of such phase and amplitude pairs, and as an example we will take the case of the circularly polarised pair.

In Fig. 12.60 the test set consists of a dual circularly polarised receiving antenna connected to a dual-channel double-conversion phase-locked receiver. The test-antenna polarisation is therefore measured as the amplitudes E_R and E_L, and relative phase δ_c of two counter-rotating waves. These are related through eqn. 12.122 as

$$p_c = p_c \exp(j\delta_c) = \frac{E_R}{E_L}\exp(j\delta_c) \qquad (12.136)$$

with

$$\gamma = \tan^{-1} p_c$$

δ_c and γ can be used, as in Fig. 12.59, to uniquely plot the polarisation on the Poincaré sphere. The phase-amplitude receiving system is similar for linear polarisation pairs, except that the dual receiving antenna must now be linearly polarised. Even non-orthogonal pairs are possible, but they generally entail complicated software to separate out the orthogonal relationships. The dual polarised antenna is the standard in this experiment, and it must first have been calibrated by an absolute method.

antenna

under test

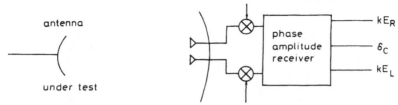

phase
amplitude
receiver

— kE_R

— δ_C

— kE_L

Fig. 12.60 *Phase-amplitude method*

12.12.4.2 Generalised three-antenna method

Absolute calibration is performed by means of a generalised three-antenna method. The three-antenna method for antenna-gain determination assumed each pair was polarisation matched, but it can be extended to a general case by including accurate phase measurement and repeating with each pair in two orthogonal alignments of the antennas.[8,34,35] The principle is illustrated in Fig. 12.61. Three pairs are selected and both phase and amplitude are measured in each case. These are then repeated with a 90° twist between each pair. Ratios of the orthogonal pairs form three simultaneous equations in the form

$$\frac{A_{12}(0)}{A_{12}(90)} = q_1, \qquad \frac{A_{23}(0)}{A_{23}(90)} = q_2 \qquad \frac{A_{31}(0)}{A_{31}(90)} = q_3 \qquad (12.137)$$

where the qs are the functions of p_{c1}, p_{c2} and p_{c3}. Further theoretical details can be found in the references.

12.12.4.3 Rotating-source method

It is often unnecessary to determine completely the polarisation of an antenna. In testing circularly polarised antennas a measurement of axial ratio is a sufficient indication of performance, and also allows the integrated cancellation ratio to be found in the volume of the main beam. In Fig. 12.62 a rotating linearly polarised dipole is the source of radiation at the test antenna. During

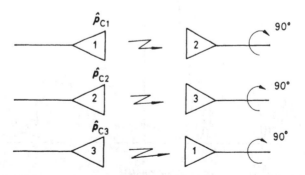

Fig. 12.61 *Generalised three-antenna method*

Fig. 12.62 *Axial ratio from rotating dipole*

continuous rotation of the source, a conventional azimuth-pattern cut is taken and has an appearance similar to that in Fig. 12.63 for a nominally circularly polarised test antenna. Increasing ellipticity in the test antenna is indicated by increasing ripples as the angle departs from the peak of the main beam. The width of the ripples equals the axial ratio in decibels, provided the rotation rate does not exceed the slew rate of the receiving and recording system, whilst remaining greater than the pattern scan rate of the test antenna.

Neither polarisation sense nor tilt angle is available from the pattern, but the latter can be found from the response at a fixed test-antenna angle. In Fig. 12.64, the dotted ellipse represents the test-antenna polarisation and the received amplitude follows the full line. Circular polarisation would give a circular response, and for other cases the tilt angle is as shown in the Figure.

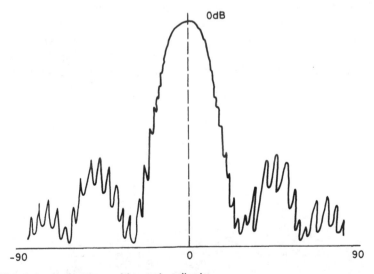

Fig. 12.63 *Aximuthal pattern with rotating dipole*

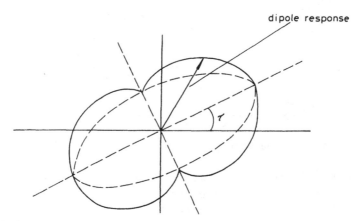

dipole response

Fig. 12.64 *Test-antenna polarisation*

12.12.5 Integrated cancellation ratio

Integrated cancellation ratio is a measured factor which quantifies rain cancellation in a nominally circularly polarised antenna. It is not difficult to achieve circular polarisation near the peak of the main beam of a radar antenna, but in the skirts and even more so in the side lobes, particularly in the 45° planes, cross-polar terms, in this case a small component of the opposite circularity, are more difficult to eliminate. Rain fills a wide angular range of an antenna pattern, whereas the target gives returns over a small angle within the main beam. Rain reflections, due to the cross-polar circular wave from the antenna as transmitter, return and are accepted into the receive channel. Since they are from an extended volume, their total effect is the result of integration over the significant parts of the pattern.

To simplify the discussion, we consider that lack of circularity is due only to phase error in the transmit and receive circular polariser. To see how the spinning-dipole method can be modified to find the integrated cancellation ratio (*ICR*), consider the response of the polariser to the field from the dipole when its instantaneous angle is ϕ to the horizontal. At the left of Fig. 12.65, the dipole field is E and the response of the polariser to this can be found from the method used in Section 12.12.3.1 as

$$|E_{det}|^2 = \frac{E^2}{2}[1 + \sin 2\phi \cos \alpha) \qquad (12.138)$$

Since we are allowing only phase errors in the polariser, its maximum and minimum responses occur at $\phi = \pm 45°$, whatever phase shift α is produced in the y-directed vector, and are

$$P_{max} = \frac{E^2}{2}[1 + \cos \alpha], \qquad P_{min} = \frac{E^2}{2}[1 - \cos \alpha] \qquad (12.139)$$

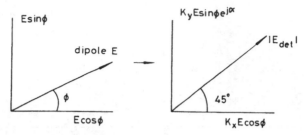

Fig. 12.65 *Response to dipole at angle ϕ*

Now eqn. 12.134 gives the polarisation loss factor for an incoming wave, with orthogonal phase α_W, entering a receiver with polariser phase α_A. For a radar antenna, which transmits and receives,

$$\alpha = \alpha_W = \alpha_A$$

and the polarisation loss factor is

$$p = \cos^2 \alpha$$

But from eqn. 12.139,

$$\cos^2 \alpha = \left(\frac{P_{max} - P_{min}}{P_{max} + P_{min}} \right)^2 \tag{12.140}$$

P_{max} and P_{min} are the peaks and troughs of the ripples at angle (θ, ϕ) on the pattern in Fig. 12.63, but since they are recorded in decibel ratio, they have first to be converted to power ratios to the peak of the main beam. Circular polarisation gives perfect cancellation when $P_{max} = P_{min}$, because α is then $\pi/2$, indicating a perfect polariser. Because received polarisation depends on the pattern as well as the polariser, the ratio in eqn. 12.140 is integrated over the pattern by taking azimuthal cuts for different elevation angles.[33] It is common practice to define integrated cancellation ratio as the integral of the reciprocal of eqn. 12.140, to give

$$ICR = \frac{\int_{\theta'} \int_\phi (P_{max} + P_{min})^2 \cos \theta' \, d\phi \, d\theta'}{\int_{\theta'} \int_\phi (P_{max} - P_{min})^2 \cos \theta' \, d\phi \, d\theta'} \tag{12.141}$$

where θ' is the elevation, $90 - \theta$, and ϕ is the azimuthal angle in Fig. 12.66. Since the patterns are usually analysed in numerical form, this expression becomes

$$ICR = \frac{\sum_{\theta'\phi} (P_{max} + P_{min})^2 \cos \theta'}{\sum_{\theta'\phi} (P_{max} - P_{min})^2 \cos \theta'} \tag{12.142}$$

*ICR*s of better than 20 dB are typically specified, but measurements over the whole pattern can lead to optimistic results, since accuracy diminishes rapidly

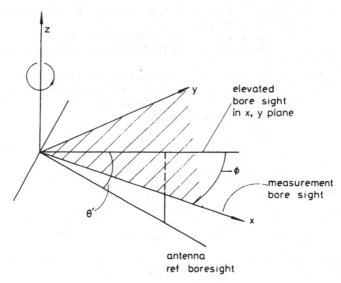

Fig. 12.66   *Co-ordinate system for integrated cancellation ratio*

away from the main beam. Site reflections can be a source of errors which are difficult to evaluate and correct. For this reason, measurements on smaller antennas are usually made in an anechoic chamber.

12.13 Conclusions

Many antenna evaluations involve some or all of the techniques described in previous Chapters, such as input-reflection, spectral-analysis, transmission-loss and noise-temperature methods, but in this Chapter we have concentrated on the special problems of radiating structures in which circuit boundaries extend throughout the radiating region, and where often there are effects, such as site reflections, external interference or propagation inhomogeneities, that are difficult to eliminate or even fully to take into account. Near-field methods allow better control of the environment, but are limited by probe-positional accuracies and the trade-off between measurement time for large probed areas and the truncation distortion as the area is reduced. The greatest difficulties are experienced in measuring low side-lobe levels at wide angles from the main beam, and in determining the gain of large antennas by substitution methods.

12.14 References

1 KUMMAR, W. H., and GILLESPIE, E. S.: 'Antenna measurements—1978', *Proc. IEEE*, 1978, **66**, (4)

2 RUDGE, A. W., MILNE, K., OLVER, A. D., and KNIGHT, P.: 'The Handbook of Antenna Design: Vol. 1' (Peter Peregrinus, 1982) Chap. 8

3 'IEEE standard test procedures for antennas'. IEEE Standard 149, 1979

4 BOWMAN, R. R.: 'Field strength above 1 GHz; Measurement procedures for standard antennas', *Proc. IEEE*, 1967, **55**, pp. 981–990

5 COLLIN, R. E., and ZUCKER, F. J.: 'Antenna theory; Pt. 1' (McGraw-Hill, 1969)

6 BALANIS, C. A.: 'Antenna theory: Analysis and design' (Harper & Row, 1982)

7 JOY, E. B., and PARIS, D. T.: 'Spatial sampling and filtering in near-field measurements', *IEEE Trans.*, 1972, **AP-20**, (3), pp. 253–261

8 NEWELL, A. C., BAIRD, R. C., and WACKER, P. F.: 'Accurate measurement of antenna gain and polarisation at reduced distances by an extrapolation technique', *IEEE Trans.*, 1973, **AP-21**, pp. 418–431

9 SILVER, S.: 'Microwave antenna theory and design', (Reprint) (Peter Peregrinus, 1984) p. 586

10 ARNOLD, P. W.: 'The "slant" antenna range', *IEEE Trans.*, 1966, **AP-14**, pp. 658–659

11 JOHNSON, R. C., ECKER, H. A., and HOLLIS, J. S.: 'Determination of far-field antenna patterns from near field measurements', *Proc. IEEE*, 1973, **61**, pp. 1668–1694

12 GALAGAN, S.: 'Understanding microwave absorbing materials and anechoic chambers', *Microwaves*, Pt. 1, Dec. 1969, pp. 38–41; Pt. 2, Jan. 1970, pp. 44–49; Pt. 3, April 1970, pp. 47–50; Pt. 4, May 1970, pp. 69–73

13 HOLLIS, J. S., LYON, T. J., and CLAYTON, Jnr., L. (Eds.): 'Microwave antenna measurements' (Scientific Atlanta Inc., 1970) pp. 7.1–7.9

14 VOKURKA, V. J.: 'Seeing double improves indoor range', *Microwave & RF*, Feb. 1985, pp. 71–76 and 94

15 EMERSON, W. H.: 'Electromagnetic wave absorbers and anechoic chambers through the years', *IEEE Trans.*, 1973, **AP-21**, pp. 484–490

16 GILLETTE, M. R., and WU, P. R.: 'RF Anechoic chamber design using ray tracing', IEEE/AP-S Symposium Digest, June 1977, pp. 246–252

17 SANDER, W.: 'Monitoring and calibration of active phased arrays'. IEEE International Radar Conference, 1985, pp. 45–51

18 BRYANT, G. H., and STOWE, H.: 'The automatic alignment and measurement of beam pointing angle in a phased array'. ICAP85, IEE Conference Publication 248, April 1985, pp. 202–212

19 PARIS, D. T., LEACH, Jnr., W. M., and JOY, E. B.: 'Basic theory of probe compensated near-field measurements', *IEEE Trans.*, 1978, **AP-26**, pp. 373–379

20 COLLIN, R. E., and ZUCKER, F. J.: 'Antenna theory: Part 1' (McGraw-Hill, 1969)

21 LUDWIG, A. C.: 'Near-field/far-field transformation using spherical wave expansions', *IEEE Trans.*, 1971, **AP-19**, pp. 214–220

22 BOOKER, H. G., and CLEMMOW, P. C.: 'The concept of an angular spectrum of plane waves and its relation to that of polar diagram and aperture distribution', *Proc. IEE*, 1950, **97**, Part III, pp. 11–17

23 LEACH, Jnr., W. M., and PARIS, D. T.: 'Probe compensated near-field measurements on a cylinder', *IEEE Trans.*, 1973, **AP-21**, pp. 435–445

24 HESS, D. W., *et al.*: 'Spherical near field antenna measurements improve through expanded software features', *MSN & CT*, March 1985, pp. 64–83

25 BRYANT, G. H.: 'Measurement of mean square pointing error in a phased array'. IEE Colloquium Digest 1985/9, Jan. 1985, pp. 9/1–9/4

26 HANSEN, R. C.: 'Low noise antennas', *Microwave J.*, June 1959, p. 19

27 BLAKE, L. U.: 'Low noise receiving antennas', *Microwaves*, March 1966, p. 19.

28 REED, H. H.: 'Communication satellite ground station antennas', *Microwave J.*, June 1967, p. 63

29 'Design and construction of large steerable aerials'. IEE Conference Publication 21, June 1966

30 HOWELL, T. F.: 'Test set for measurement of G/T', p. 169; and DOREY, J. D.: 'FM method of determining the gain/temperature ratio of an earth station antenna and receiving sub-system', p. 175 *in* 'Earth station technology'. IEE Conference Publication 72, Oct. 1970

31 ASHKENAZY, J., LAWRIE, E., and TREVIS, D.: 'Radiometric measurement of antenna efficiency', *Electron. Lett.*, 1985, **21**, pp. 111–112
32 LUDWIG, A. C.: 'The definition of cross polarisation', *IEEE Trans.*, 1973, **AP-21**, pp. 116–119
33 JASIK, H., and JOHNSON, R. C.: 'Antenna engineering handbook' (McGraw-Hill, 1984) pp. 23–28
34 NEWELL, A. C.: 'Improved polarisation measurements using a modified three antenna technique'. IEEE AP-S International Symposium, 1975, pp. 337–385
35 JOY, E. B., and PARIS, D. T.: 'A practical method of measuring the complex polarisation ratio of arbitrary antennas', *IEEE Trans.*, 1973, **AP-21**, pp. 432–435

12.15 Examples

1   A receiving rectangular phased array has 900 elements, each separated by $\lambda/2$. Each element is connected to an amplifier with a gain of 15 dB and having 1 dB compression when the output is at -10 dBm. A transmitting antenna, with effective aperture of 1 m^2, is at a distance of 500 m. Both antennas are oriented for maximum signal transmission. What is the transmitter power at the 1 dB compression point?

2   Calculate the lowest frequency for which the antennas in example 1 can operate in the far field.

3   A receiving antenna is mounted on an elevation over azimuth mount, and the line of site to the transmitter is elevated by 10°. A pattern cut at $\phi = 45°$ is required for θ in the range $\pm 45°$. Find expressions for the elevation and azimuth angles, and give their values at $\theta = \pm 45°$.

4   The element receivers in example 1 have a noise figure of 3 dB and a bandwidth of 300 MHz. The antenna temperature at its aperture is 100°K and the combining network loss is 1 dB. What is the dynamic range of the array?

5   By dividing the array about a vertical centre line and comparing the two halves in anti-phase, the antenna becomes a one-dimensional monopulse system with a null on a horizontal bore sight perpendicular to its face. What is the tracking accuracy in milliradians?

6   A horn antenna has a circular aperture of diameter 10λ. It is probed in a planar scanner and also over the whole sphere of a spherical scanner. The data transformation gives far-field patterns which are identical in a central region including the main beam and first two side lobes, provided the planar scanning area extends at least 3λ beyond the aperture edge. Calculate the number of samples in the two cases. Comment on your results.

7   Calculate the ellipticity loss factor for a radar rain cancellation when the orthogonal polariser has arbitrary amplitudes and phases. Calculate the ratio

$$\left(\frac{P_{max} - P_{min}}{P_{max} + P_{min}}\right)^2$$

for the response to a spinning dipole, and compare this with the polarisation efficiency. Comment on your result.

8   The far field of an aperture antenna with field E_y is investigated with a linearly polarised probe in the $\phi = 45°$ cut. Explain how the co- and cross-polar components vary as θ changes from 0 to 90°.

9   A phase-amplitude antenna measurement gives a right-/left-hand circular amplitude ratio of 0·2 and a phase $\delta_c = 0·9$ rad. What is the longitude and latitude of the equivalent point on the Poincaré sphere?

Performance characteristics of a spectrum analyser

In recognition of the dependence of modern measurements on swept frequencies, this book begins by examining the bandwidth limitations imposed on spectrum analysers when they are operated with swept sources. It is not possible at that stage to give a complete account of the spectrum analyser, because its great versatility means it can serve many of the different functions of other specialised instruments.[1] We have seen, in Chapter 9, how it can be used as a noise meter; but it is also adaptable as a power meter, network analyser, frequency meter or simply as a detector. In Chapter 1, further consideration of the applications and limitations of spectrum analysers would have required many of the concepts still to come in the later Chapters, and therefore not yet available to all readers. This Appendix seeks to fill partially that omission by completing the discussion of Chapter 1 at a point where it may be assumed the reader is conversant with the relevant background information. A second purpose of this Appendix is to look at the meaning and interpretation of specifications in as general a sense as possible, but also in the full recognition that specifications are specific. The wide application and versatility of the spectrum analyser allows us to be both general and specific to a particular instrument.

In a mature subject, like spectrum analysis, specifications follow the guidelines of national and international standards. This should avoid confusion over the terms and meanings of experimental quantities based on published test procedures. In the United States the IEEE publishes national standards: e.g. IEE Standard 62 IRE 7.S2, 1962 on 'Noise definitions and measurement methods'; IEEE Standard 149, 1979 on 'Test procedures for antennas'; IEEE Standard 474, 1973 on 'Test methods for fixed and variable attenuators, D.C. to 40 GHz'; and, of most relevance in this instance, IEEE Standard 478, 1979 on 'Spectrum analysers'. In order to give substance to our discussions we will examine the specifications of one spectrum analyser, namely the Tektronix 492. The instrument is available in a programmable form through a GPIB interface to an external computer. It has a frequency range from 50 kHz to 21 GHz in a coaxial output, or from 18 GHz to 220 GHz with external waveguide mixers. In Table A1.1, some important features are extracted from an abbreviated specification, given at the end of this Appendix, and taken from a Tektronix publication.[2]

Table A1.1   *Performance factors of a spectrum analyser*

Frequency accuracy	$\pm 0.2\%$ or 5 MHz (max) plus 20% of span per division
Long-term drift	3 kHz per 10 min after 30 min
Noise sidebands	-75 dBc at $30 \times$ resolution offset from centre frequency
Residual FM	1 kHz peak to peak for 2 ms duration
Input sensitivity	-115 dBm in 1 kHz bandwidth
Third-order intermodulation	70 dB down from two full-screen signals
Resolution bandwidth	-6 dB points 1 MHz to 1 kHz
Resolution shape factor	60/6 dB—7.5 : 1 or less

We will now take each of these performance factors in order to show what they mean and to give some examples of their applications.

A1.1 Frequency accuracy and long-term drift

Frequency accuracy can be affected by the long-term drift of the swept local oscillator, which is given as 3 kHz per 10 min. Thus at 50 kHz with a low sweep rate (see fuller specification) the accuracy is 2 per cent of 50 kHz or 100 Hz, corresponding to a mean drift time of 20 s. Accuracy is reduced when scan rates are comparable to the mean drift rate. Thus if the minimum resolution bandwidth R is 100 Hz (option 03), and the scan is 50 kHz, eqn. 1.13 gives the scan time $T = 5$ s. The corresponding drift is 25 Hz.

A1.2 Noise sidebands

These depend on short-term local-oscillator phase noise. They degrade the low-frequency response and cause carrier-sideband pedestal break-out, thus setting a limit to the resolution of two closely spaced carriers. They are specified in terms of the resolution offset and a close-to-carrier dynamic range. For example, if the resolution bandwidth is 1 kHz, the phase noise level might be

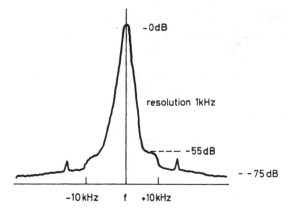

Fig. A1.1 *Phase-noise sidebands*

75 dB down on the carrier peak at 30 KHz and would then be specified as −75 dBc at 30 times the resolution offset from the carrier frequency. Phase-noise sidebands appear as two shoulders in Fig. A1.1 at about −55 dBc at 10 kHz from the carrier, and then decline to −75 dBc at 30 kHz. The two pips may be due to spurious interference from control circuits for signal conditioning of the source. If the resolution had been 100 Hz, the specification, with similar ratios, would have been −75 dBc at 3 kHz.

Similar noise sidebands, that occur with respect to zero frequency as a carrier, degrade low-frequency performance due to feed-through in the local oscillator. At low frequency, 50 kHz is at the centre of the screen with a span of 10 kHz per division, or in the phase-locked version 2·5 kHz is at the centre with 500 Hz per division. The span therefore extends nominally to zero frequency, but with an input sensitivity decreasing with lower frequency as shown in Fig. A1.2, where, for the phase locked 100 Hz resolution bandwidth, the sensitivity has deteriorated to −70 dBm at about 800 Hz from a

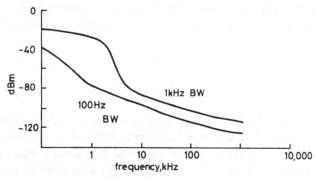

Fig. A1.2 *Low-frequency sensitivity*

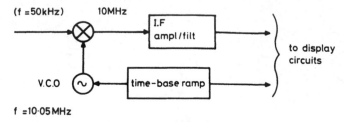

Fig. A1.3 *Up-conversion with swept local oscillator*

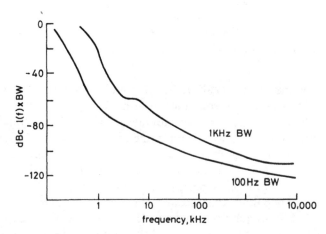

Fig. A1.4 *Local-oscillator-induced carrier-phase-noise sidebands*

high-frequency average of $-120\,\text{dBm}$. Similarly, at 1 kHz resolution, the sensitivity is $-90\,\text{dBm}$ at 50 kHz against an average of $-110\,\text{dBm}$. Low-frequency noise is up-converted in the mixer in Fig. A1.3 and appears in the IF at 10 MHz in addition to the VCO phase sidebands on 10·05 MHz. Away from zero frequency, i.e. for input carriers f_s at frequencies greater than 50 kHz, the only significant noise is local-oscillator-induced carrier phase-noise sidebands, shown in Fig. A1.4 as a two-sided power-density spectral distribution, $\mathscr{L}(f)$.

A1.3 Residual FM

The average phase fluctuation at each offset frequency contributes to a short-term frequency fluctuation of the carrier, referred to as residual FM. But the averaging time sets a lower limit on the offset frequency contributing to that average. Thus, when residual FM is specified as 1 kHz peak-to-peak for 2 ms, changes occurring in intervals longer than 2 ms are not observed; or phase sidebands up to 500 Hz from the carrier do not contribute to the average. This is one way of making the distinction between long- and short-term noise. The spectral density of fractional frequency fluctuations $S_y(f)$ is related to the mean-square frequency fluctuation at offset f and bandwidth B through the carrier frequency f_0, from eqn. 10.24 as[3]

$$\overline{\delta f^2} = B f_0^2 S_y(f)$$

and to the spectral density of phase fluctuation as

$$\overline{\delta f^2} = B f^2 S_{\Delta\phi}(f)$$

But

$$S_{\Delta\phi}(f) = 2\mathscr{L}(f)$$

Therefore

$$\overline{\delta f^2} = 2 B f^2 \mathscr{L}(f) \tag{A1.1}$$

$2 f^2 \mathscr{L}(f)$ should be integrated over all offset frequencies in the sidebands to find the average square carrier-frequency fluctuation. From Fig. A1.4, $\mathscr{L}(f)B = 10^{-7}(-70 \text{ dB})$ at 15 kHz and for a bandwidth of 1 kHz. Therefore

$$2 f^2 \mathscr{L}(f) = 2 \times (1 \cdot 5)^2 \times 10^8 \times 10^{-7} \times 10^{-3}$$

$$\approx 5 \times 10^{-2}$$

Assuming $2 f^2 \mathscr{L}(f)$ is constant with f from the carrier to the noise floor at 1 MHz, the total mean-square frequency fluctuation is

$$\sqrt{\overline{\Delta f^2}} \approx \sqrt{5 \times 10^{-2} \times 10^6} = 224 \text{ Hz}$$

In view of the very coarse approximation in integrating the sideband noise, this figure is close enough to the residual FM of 1 kHz, quoted in the specification.

A1.4 Input sensitivity

This is the minimum detectable signal or the equivalent input noise level. Considering the spectrum analyser as a receiver, we can write for its noise figure

$$F = F_M + L_c(F_{IF} - 1) \tag{A1.2}$$

where f_M is the noise figure, L_c is the conversion loss of the input mixer, and F_{IF} is the noise figure of the IF amplifier. Eqn. A1.2 is for cascaded two-ports with first stage 'gain' $G_c = 1/L_c$. It applies to random noise only; and any spurious deterministic noise such as harmonics or some forms of residual FM, e.g. mains-induced FM, must be considered separately. In this sense, this noise figure is an ideal that is not always achieved at all parts of an operating frequency band.

Mixer noise is usually expressed in terms of a noise-temperature ratio defined as[4]

$$t_r = \frac{\text{Actual available noise power}}{\text{Available noise power in an equivalent pure resistance}}$$

$$= F_M \frac{kTB_N G_c}{kTB_N} = F_M G_c = \frac{F_M}{L_c} \tag{A1.3}$$

where B_N is the noise bandwidth. On substituting for F_M in eqn. A1.2, we have

$$F = L_c(t_r + F_{IF} - 1)$$

With typical front ends consisting of a tracking-filter selector (6 dB loss), a matching circuit (5 dB), a very broad-band mixer (fundamental conversion loss of 10 dB and noise-temperature conversion loss between 1 and 2), sometimes a second conversion loss and then an IF amplifier with a 6 dB noise figure, noise figures in spectrum analysers are high at about 30 dB. In terms of noise temperature, we have

$$F = 1 + \frac{T_e}{T_0}$$

to give in bandwidth B_N, a noise power

$$N = kT_e B_N = (F - 1)kT_0 B_N$$

and an input noise spectral density of

$$N_i \approx F k T_0$$

or, for $F = 30$ dB,

$$N_i = -144 \text{ dBm per Hz}$$

The specification is -115 dBm in a 1 kHz bandwidth away from the carrier, which, at -145 dBm per Hz, agrees sufficiently well with our estimate. Phase noise, close to the carrier, reduces input sensitivity, and further reduction can occur owing to harmonic distortion, mostly second harmonic at -30 dBm, at the first mixer. The latter is normally at -60 dBc, but with an input YIG pre-selector it becomes -100 dBc. Pre-selection actually reduces sensitivity by 5 dB, but, because of harmonic suppression, the dynamic range is increased.

To see how the dynamic range and sensitivity are related we consider the effects of non-linearities on large signals. As input signal level rises, output becomes distorted, primarily owing to the dependence of mixer conductance on amplitude. The -1 dB compression point is a useful measure of the upper distortion level. It is illustrated in Fig. A1.5, where the output power S_0 of an amplifier increases linearly with input power S_i up to the distortion level, defined as the input power at which the output level is -1 dB from its linear projection. Input dynamic range is therefore the ratio of the -1 dB compression level S_M to the input sensitivity N_i. A word of caution is necessary here

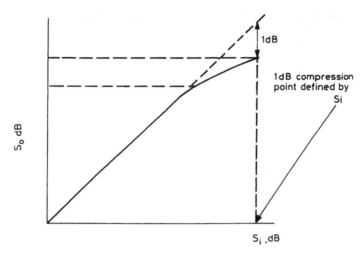

Fig. A1.5 *Dynamic range from input sensitivity to 1 dB compression point*

because there are two definitions of dynamic range. One is the ratio of the largest to the smallest signal that can be displayed on the screen, and the other is the linear range of the receiver, with respect to the -1 dB compression point. The 80 dB figure for the 492 is substantially identical for both. It is common to have logarithmic as well as linear displays; thus a linear envelope detector can give the latter, with a quoted accuracy of $\pm 5\%$ of full scale, and IF gain shaping produces the former, to an accuracy of ± 1 dB at the 10 dB level and ± 2 dB at 80 dB.

A1.5 Intermodulation distortion

Intermodulation distortion, a major limitation to dynamic range, occurs in all mixers, but at low enough drive powers its effects are below the system noise level. Mixing products of input signals and their harmonics consist of sum and difference frequencies. For instance, two frequencies f_1 and f_2 have intermodulation (IM) products at $f_1 + f_2, f_2 - f_1, 2f_2 - f_1, 2f_1 - f_2$ and higher-order terms. Closest to the input signals are the third-order terms $2f_2 - f_1$ and $2f_1 - f_2$. For two closely spaced input frequencies, such IM products are difficult to filter, and therefore have a significant effect on the dynamic range, that directly affects the minimum separation or frequency resolution of a spectrum analyser.

Third-order components originate from the cubic term in the mixer-output polynomial defined in eqn. 1.32, from which

$$v_3 = cv^3 \tag{A1.4}$$

where v_3 is the output voltage, and the input voltage, $v = v_1 + v_2 + v_0$, is the addition of two signals, v_1 and v_2, plus the local oscillator voltage v_0. The corresponding frequencies are $f_1, f_2,$ and f_0, respectively. On substituting for v in eqn. A1.4 and expanding, we have

$$\frac{v_3}{c} = v_1^3 + 3(v_0 + v_2)v_1^2 + 3(v_0 + v_1)v_2^2 + 3(v_1 + v_2)v_0^2 + 6v_1v_2v_0 + v_2^3 + v_0^3$$

$$(A1.5)$$

The sum and difference terms of interest are

$$\frac{v_\pm}{c} = 3v_2v_1^2 + 3v_1v_2^2 \tag{A1.6}$$

If we let $v_1 = V_1 \cos \omega_1 t$ and $v_2 = V_2 \cos \omega_2 t$, eqn. A1.6 is

$$\frac{2v_\pm}{c} = V_2 V_1^2 \cos \omega_2 t + V_2^2 V_1 \cos \omega_1 t + \frac{V_2 V_1^2}{2} [\cos (2\omega_1 + \omega_2)t$$

$$+ \cos (2\omega_1 - \omega_2)t] + \frac{V_2^2 V_1}{2} [\cos (2\omega_2 + \omega_1)t + \cos (2\omega_2 - \omega_1)t]$$

from which we select the third order-difference terms

$$v_- = \frac{3c}{4} V_2 V_1^2 \cos (2\omega_1 - \omega_2)t + \frac{3c}{4} V_2^2 V_1 \cos (2\omega_2 - \omega_1)t \tag{A1.7}$$

There is a standard method for determining third-order distortion, known as the two-tone test, in which the distortion ratio, or dynamic range, is found for specified separations of two equal-amplitude signals closely spaced in frequency. The distortion-free dynamic range in this test is expressed as the ratio squared of the amplitudes of third-order mixer products to the amplitude of one of the input tones. During an IM test of this type $V_1 = V_2 = V$ and

$$\text{IM distortion ratio} = \left| \frac{\text{Amplitude of third-order difference product}}{\text{Amplitude of one of two equal inputs}} \right|^2$$

$$= \left| \frac{3c}{4} \right|^2 V^4 \tag{A1.8}$$

Third-order IM power ratio therefore increases as the square of the input signal power, and we shall see that in general, for any order n, the corresponding distortion components increase as the $(n-1)$th power of the signal power. The coefficient c and its equivalents for any other order are generally very small, but as input power increases, non-linear distortions increase at a faster rate than the signals, and eventually cause the distortion saturation illustrated in Fig. A1.6. Dynamic range D_M is the linear range in decibels from the system noise level N_i to the saturation limit S_i at the -1 dB compression point. It has the same range on the output S_0 or input S_i, but is shifted on the output scale by the gain G of the receiver. Similarly, the ratio N_0/N_i is also the receiver gain.

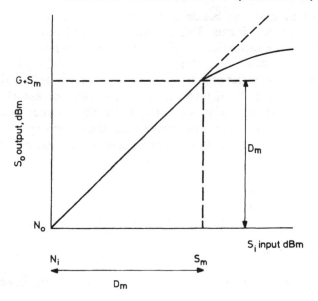

Fig. A1.6 *Third-order distortion saturation*

First-order 'distortion' is the gain ratio of the output power P_0 to the input power P_i, defined as

$$g = P_0/P_i$$

or in decibels

$$S_0 = G + S_i \qquad (A1.9)$$

Dynamic range, apparently infinite in first order, in practice, has lower and upper bounds set by the system noise power and the onset of saturation distortion in the receiver. If P_n is the effective input distortion noise of order n, the ratio of this to the signal power is

$$\frac{P_n}{P_i} = g_n P_i^{n-1}$$

where g_n is the nth-order gain ratio. Thus, whilst the ratio increases to the $(n-1)$th power, the distortion noise increases as the nth power according to

$$P_n = g_n P_i^n$$

or in decibels

$$N_n = G_n + nS \qquad (A1.10)$$

For third-order intermodulation products, $n = 3$; and eqn. A1.10 shows that noise caused by third-order distortion increases at three times the rate of increase of input signal power. Therefore, even though, for low input signals, IM noise may be insignificant beside system noise, it will always exceed it for sufficiently high input signals. This is shown for third order by the line $S_M I$ in Fig. A1.7. If the distortion noise, in dBm, had increased with the same slope as the signal, along $S_M A$, the dynamic range would have remained unchanged, whereas, by increasing the slope threefold, third-order IM reduces the dynamic range to zero at the intercept point I for an increase of $D_M/2$ in the input signal. Inspection of Fig. A1.7 shows that the distance to the intercept point on the output axis equals the input signal at the intercept, which, in turn, is the sum of the maximum distortionless input signal and half the dynamic range. Thus

$$I = S_I = S_M + \frac{D_M}{2}$$

It is easy to show that in general, for order n, the intercept is

$$I = \frac{D_M}{n-1} + S_M \tag{A1.11}$$

and that this can be extended by substituting $S_M = N_i + D_M$ in eqn. A1.11 to give

$$D_M = \frac{(n-1)(I - N_i)}{n}$$

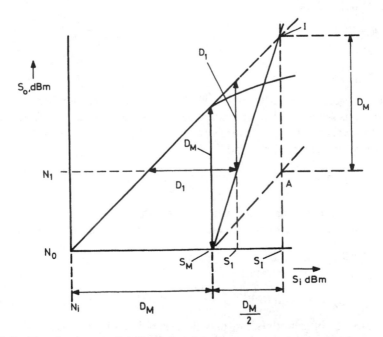

Fig. A1.7 *Showing increase in third-order distortion*

and

$$S_M = \frac{(n-1)I + N_i}{n} \qquad (A1.12)$$

These two equations give the dynamic range and maximum input signal from the input sensitivity N_i and the intercept I. N_i can be obtained directly from the specification, but I has generally to be derived from quoted dynamic range for a specified input level. To see how this is done we find the equation of the line $S_M I$ in Fig. A1.7 by considering the point at S_1, N_1, for which

$$I = \frac{D_1}{n-1} + S_1 \qquad (A1.13)$$

with $n = 3$ for third order, and substitute the third-order IM specification figures of 70 dBc from a -30 dBm input signal, to give[2]

$$I = \frac{70}{2} + (-30)$$

$$= +5 \text{ dBm}$$

But the best minimum signal for maximum dynamic range is found by substituting, for I and the quoted N_i of -110 dBm, in the second of eqns. A1.12 to give

$$S_M = \frac{2 \times 5 - 110}{3}$$

$$= -33 \cdot 3 \text{ dBm}$$

The optimum dynamic range, from the first of eqns. A1.12, follows as

$$D_M = \frac{2 \times 115}{3}$$

$$= 76 \cdot 7 \text{ dBc}$$

The input sensitivity has been taken for a pre-selected input (see option 01 in the specification abstract), and similar calculations would be necessary for other options.

A1.6 Resolution bandwidth

This is stated as a -6 dB bandwidth, because, if two equal-amplitude signals cross over at their -6 dB points, as in Fig. A1.8, their peaks are distinguishable with a $\geqslant 3$ dB notch between them. Resolution bandwidth and scan range are

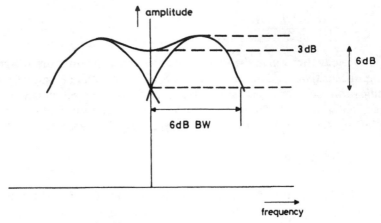

Fig. A1.8 *Resolution bandwidth*

related through the minimum resolution bandwidth, $\sqrt{S/T}$. With a frequency span per division of 10 kHz to 200 MHz and a sweep time of 20 μs to 10 s, S is from 100 kHz to 2 GHz and T, from 200 μs to 100 s. Two limiting cases are of interest; namely, the maximum scan obtainable at the smallest-resolution bandwidth and the minimum time for maximum scan at the largest-resolution bandwidth. Quoted resolution bandwidth is from 1 MHz to 1 kHz. In the first case we take the slowest scan with $T = 100$ s with a bandwidth of 1 kHz to give $S/100 = 10^6$, or a maximum scan of 100 MHz. In the second case $S = 2 \times 10^9$ and the bandwidth is 1 MHz, or $10^6 = \sqrt{2 \times 10^9/T}$, giving $T = 2$ ms. It is therefore possible to extend information about performance factors by matching specification figures to known theoretical limitations.

A1.7 Resolution shape factor

The shape factor is defined as the ratio between the -60 dB and the resolution bandwidths. Shape factor is most important when signals of unequal amplitudes are to be resolved. Steep-sided frequency responses have smaller shape factors, giving greater resolution of unequal amplitude signals. In Fig. A1.9, if Δf_{-60} and Δf_{-6} are the -60 and -6 dB bandwidths, respectively, the shape factor is $\Delta f_{-60}/\Delta f_{-6}$.

There are other bandwidths such as noise or impulse bandwidth, so that resolution specification is not a simple matter. For instance, residual peak-to-peak FM should be less than the theoretical resolution bandwidth, and we have seen that phase noise and intermodulation also affect close-in small signals.

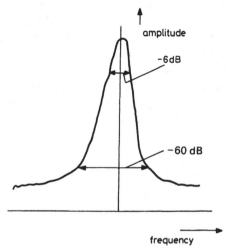

Fig. A1.9 *Resolution shape factor*

A1.8 Summary

This appendix has drawn together some ideas from several chapters to analyse the performance of one instrument from its specification. The analysis was not exhaustive since its purpose was to demonstrate an attitude that seeks out inconsistencies and extends knowledge of performance beyond what is written in the specification. Further information on spectrum analysers can be found in References 5 and 6.

A1.9 References

1 LAVERGHETTA, T. S.: 'Handbook of microwave testing' (Artech, 1981)
2 'Tektronix 490 Series Spectrum Analysers'. Abbreviated specification in Publication 26W-4817-1, 1983
3 HOWE, D. A.: 'Frequency domain stability measurements: A tutorial introduction'. NBS Technical Note 679, National Bureau of Standards, Boulder, Colorado, March 1976
4 SKOLNIK, S. I.: 'Introduction to radar systems' (McGraw-Hill, 1962)
5 ENGELSON, M.: 'Modern spectrum analyser theory and applications' (Artech House, 1984)
6 PECCOLO, T.: 'The spectrum analyser comes of age'. *Microwave J.*, April 1983, pp. 125–133

A1.10 Examples

1 Calculate the optimum input signal and dynamic range for 2nd harmonic distortion using the 492 specification. Assume 1 kHz resolution and no pre-selection.

2   Explain why spectrum analysers have poor noise figures. The noise figure of a spectrum analyser is 37 dB and a lower-noise RF amplifier is placed at the input to improve overall noise figure. Explain why the dynamic range is reduced by the gain of the amplifier, but increased by the noise-figure improvement.

If a noise-figure improvement of 15 dB is required, specify the gain and noise figure of the input amplifier if the dynamic-range reduction must be $\leqslant 3$ dB.

3   Estimate the residual FM for a phase-lock option using the graph in Fig. A1.4 of $\mathscr{L}(f)$ against offset frequency.

Extract from Tektronix type 492/492P spectrum analyser specification*

A2.1 492/492P characteristics

The following characteristics and features apply to the 492/492P spectrum analyser after a 30 min warm-up period, unless otherwise noted.

A2.1.1 Frequency related
Centre-frequency range: 50 kHz to 21 GHz standard; amplitude specified coverage to 220 GHz with optional Tektronix waveguide mixers

Frequency accuracy: ±0·2% or 5 MHz, whichever is greater, +20% of span/div

Frequency-readout resolution: Within 1 MHz 492P TUNE Command accuracy (±7% or ±150 kHz)N, whichever is greater

Frequency span per division: 10 kHz to 200 MHz plus zero and full-band max span, down to 500 Hz with option 03 in 1–2–5 sequence

Frequency-span accuracy: ±5% of span/div, measured over center eight divisions

Resolution bandwidth (−6 dB) points: 1 MHz to 1 kHz (100 Hz for option 03) in decade steps, plus an auto position. Resolution is within 20% of selected bandwidth

Resolution shape factor (60/6 dB): 7·5 : 1 or less

Residual FM: 1 kHz peak-to-peak for 2 ms time duration: improves to (50 Hz) for 20 ms with phase-lock option 03

Long-term drift (at constant temperature and fixed centre frequency): 3 kHz/10 mins after 1 h warm-up with option 03 for fundamental mixing

Noise sidebands: At least -75 dBc at $30\times$ resolution offset from the centre frequency (-70 dBc for 100 Hz resolution bandwidth)

A2.1.2 Spurious responses

Residual (no input signal referenced to mixer input): -100 dBm or less

Harmonic distortion (CW signal, minimum distortion mode): At least -60 dBc for full-screen signal in the minimum distortion mode to 21 GHz. At least -100 dBc for pre-selected option 01. 1·7 to 21 GHz

Third-order intermodulation distortion (minimum distortion mode): At least 70 dB down from two full-screen signals within any frequency span. At least 100 dB down for two signals spaced more than 100 MHz apart from 1·7 to 21 GHz for pre-selected option 01

LO emissions (referenced to input mixer): -10 dBm maximum; -70 dBm maximum for option 01

A2.1.3 Amplitude related

Reference-level range: Full screen, top of graticule -123 dBm to $+40$ dBm ($+40$ dBm includes maximum safe input of $+30$ dBm and 10 dB gain of IF gain reduction) for 10 dB/div. and 2 dB/div. log modes. 20 nV/div. to 2 V/div. (1 W maximum safe input) in the linear mode

Reference-level steps: 10 dB, 1 dB and 0·25 dB for relative level (Δ) measurements in log mode. 1–2–5 sequence and 1 dB equivalent increments in linear mode

Reference level accuracy: Accuracy is a function of changes in RF attenuation, resolution bandwidth, display mode and reference level (see amplitude accuracies of these functions). The RF attenuator steps 10 dB for reference-level changes above -30 dBm (-20 dBm when minimum noise is active) unless minimum RF attenuation is greater than normal. The IF gain increases 10 dB for each 10 dB reference-level change below -30 dBm (-20 dBm when minimum noise is active)

Display dynamic range: 80 dB at 10 dB/div., 16 dB at 2 dB/div. and 8 div. in linear mode

Display-amplitude accuracy: ± 1 dB/10 dB to maximum of ± 2 dB/80 dB; ± 0.4 dB/2 dB to maximum of ± 1 dB/16 dB; $\pm 5\%$ of full screen in linear mode

Resolution-bandwidth gain variation: ± 0.5 dB.

A2.1.4 Input-signal characteristics

RF input: Type N female connector

Input impedance: 50 Ω

Maximum VSWR* with ≥ 10 dB attenuation

Frequency range	Typical	Specified maximum
DC to 2·5 GHz	1·2 : 1	1·3 : 1
2·5 GHz to 6·0 GHz	1·5 : 1	1·7 : 1
6·0 GHz to 18 GHz	1·9 : 1	2·3 : 1
18 GHz to 21 GHz	2·7 : 1	3·5 : 1

*At type N female connector to internal mixer, with 10 dB attenuation.

Input level (optimum level for linear operation): -30 dBm referenced to input mixer. Full screen not exceeded and minimum distortion control settings

1 dB compression point: -10 dBm except -28 dBm at 1·7—2 GHz for option 01 only

Maximum safe input level (RF attenuation at 0 dB): $+13$ dBm without option 01. $+30$ dBm (1W) with option 01

Maximum input level (with 20 dB or more RF attenuation): $+30$ dBm (1 W) continuous, 75 W peak for 1 μs or less pulse width and 0·001 maximum duty factor (attenuation limit). DC must never be applied to RF input

A2.1.5 Sensitivity and frequency response

Frequency range	Mixing number (n)	Average noise level for 1 kHz resolution		Frequency response with 10 dB attenuation	
		No pre-selection	Pre-selected option 01	No pre-selection	Pre-selected option 01
50 kHz—1·8 GHz*	1	-115 dBm	-110 dBm		$\pm1·5$ dB
50 kHz—4·2 GHz*	1	-115 dBm	-110 dBm	$\pm2·5$ dB	
1·7—5·5 GHz	1	-115 dBm	-110 dBm	$\pm1·5$ dB	$\pm2·5$ dB
3·0—7·1 GHz	1	-115 dBm	-110 dBm	$\pm1·5$ dB	$\pm2·5$ dB
5·4—18 GHz	3	-100 dBm	-95 dBm (12 GHz) -90 dBm (18 GHz)	$\pm2·5$ dB	$\pm3·5$ dB
15—21 GHz	3	95 dBm	-85 dBm	$\pm3·5$ dB	$\pm5·0$ dB
100 MHz— 18 GHz***				$\pm3·5$ dB	$\pm4·5$ dB

With Tektronix optional high-performance waveguide mixers

Frequency range	Mixing number (n)	No pre-selection	No pre-selection
18—26 GHz	6	−100 dBm	±3·0 dB
26—40 GHz	10	−95 dBm	±3·0 dB
40—60 GHz	10	−95 dBm	±3·0 dB
60—90 GHz	15	−95 dBm at 60 GHz†	±3·0 dB**†
		−85 dBm at 90 GHz†	±3·0 dB**†
90—140 GHz	23	−85 dBm at 90 GHz†	±3·0 dB**†
		−75 dBm at 140 GHz†	±3·0 dB**†
140—220 GHz	37	−65 dBm at 220 GHz†	±3·0 dB**†

*   Low frequency and performance does not include effects due to 0 Hz feedthrough

**   Over any 5 GHz bandwidth

***Includes frequency-band switching error of 1 dB maximum

†   Typical

A2.1.6 Output characteristics

Calibrator (cal. out): −20 dBm ±0·3 dB at 100 MHz ±1·7 kHz

1st and 2nd local oscillation (LO): Provides access to the output of the respective local oscillators (1st LO +7·5 dBm minimum to a maximum of +15 dBm, 2nd LO −16 dBm minimum to a maximum of +15 dBm). These ports must be terminated in 50 Ω at all times

Vertical out: Provides 0·5 V ±5% of signal per division of video above and below the centre line

Horizontal out: Provides 0·5 V either side of centre. Full range −2·5 V to 2·5 V ±10%

Pen lift: TTL, 5 V nominal to lift pen

IF out: Output of the 10 MHz IF. Level is approximately −16 dBm for a full-screen signal at −30 dBm input reference level. Nominal impedance 50 Ω

492P only: IEEE Standard 488–1978 Port (GPIB)—In accordance with IEEE 488 Standard

A2.1.7 General characteristics

Sweep time: 20 µs to 5 s/div. (10 s/div. auto) in 1–2–5 sequence

CRT readout: Displays: reference level, centre frequency, frequency range, vertical display mode, frequency span/division, resolution bandwidth, RF attenuation

CRT: 8 × 10 cm, P31 phosphor

Power: 210 W maximum with all options, at 115 V and 60 Hz

Input voltage: 90—132 V AC or 180—250 V AC, 48—440 Hz

Configuration: Portable, 20 kg (44 lb) (all options), 17·5 × 32·7 × 49·9 cm (6·9 × 12·9 × 19·7 in) without handle or cover

A2.1.8 *Environmental characteristics*
Per MIL–T–28800C type III, class 3 style C

Temperature: −15°C to +55°C operating; −62°C to +75°C non-operating storage

Humidity: 95% operating; 120 h per MIL–Std. 810 non-operating

Rain resistance: Drip-proof at 16 litres h⁻¹ ft⁻²

Altitude: 15 000 ft operating; 40 000 ft non-operating

Vibration: 15—55 Hz at 0·025 in excursion

Shock: 30 g of half-sine, 11 ms duration

Drop: 12 in.

Electromagnetic compatibility: 490 series spectrum analysers meet the requirements of MIL–Std.–461B, operating from 48 Hz to 440 Hz power sources, with the exceptions shown below

Conducted emissions: CE01—15 dB relaxation for first 10 harmonics of power-line frequency. CE03 (narrow band)—full limits. CE03 (broad band)—15 dB relaxation from 15 kHz to 50 kHz

Conducted susceptibility: CS01—full limits. CS02—full limits. CS06—full limits

Radiated emissions: RE01—10 dB relaxation for first 10 harmonics of power-line frequency, and exceptioned from 30 kHz to 36 kHz. RE02—full limits

Radiated susceptibility: RS01—full limits. RS02–1—full limits. RS02–2—to 5 A only. RS03—up to 1 GHz only

A2.2 492/492P specification options

A2.2.1 *Option 01: Internal pre-selection*
With this option, internally generated image and harmonic-mixing spurious responses are effectively eliminated. This results in a display that is much easier to interpret. In the frequency range of 50 kHz to 1·7 GHz, a low-pass filter is used to limit spurious responses. In the range of 1·7—21 GHz, a tracking YIG pre-selector is used. Internal calibrated pre-selection reduces the requirement to examine each signal to verify authenticity.

Measurement capability is enhanced with option 01 by an increase in dynamic range from 80 dB in the basic analyser to 100 dB (for signals separated by 100 MHz). This is because the automatic tracking pre-selector rejects signals outside its bandwidth by 70 dB or more.

Option 01 also includes a limiter to provide +30 dBm input protection to the first mixer up to 1·7 GHz. Above 1·7 GHz the input mixer is protected by the pre-selector.

A2.2.2 Option 03: Frequency-stabilisation/100 Hz-resolution

With this option, phase-locked local-oscillator stabilisation provides exceptional display stability and low-noise sidebands, and results in less frequency drift and less residual FM. Thus the 492/492P user can observe and measure characteristics of lower modulation frequencies. As part of option 03, improved resolution (100 Hz) and narrow span of 500 Hz/div. provide increased measurement capability for close-in-sidebands analysis. The spectral purity of clean oscillators may thus be measured directly at microwave frequencies. The 492/492P retains its one-knob centre-frequency control with constant tuning rate (CTR), even with phase lock.

Option 03 is recommended when the 492/492P will be used at spans less than 50 kHz per division, and is required for spans of less than 10 kHz per division. Phase lock occurs automatically, and is a function of the setting of the span/division control. For convenience in operating the analyser in fixed-tuned receiver (zero-span) mode, phase lock may be de-activated by a front-panel control.

Symmetrical discrete Fourier transforms

In discrete methods the significant section of an infinite function is turned into a periodic function, on order to convert a Fourier transform into a Fourier series. The repeated section is sampled at a number of discrete points, and may, of course, be sampled through any number of complete periods. Calculation time is minimised by sampling in the frequency domain at multiples of the lowest frequency, given by the reciprocal of the time period, but it is essential to preserve the cyclic nature of the approximating function. Time waveforms of finite length may be assumed to repeat in an infinite train with spectra recoverable, with no aliasing, from regular samples at time separations equal to at least the reciprocal of twice the highest frequency in the waveform. If the first sample is at zero time, there is always an even number of samples in a period, if one sample lies at the centre of the period. Thus the right-end sample in Fig. A3.1 is the beginning of the next sequence, not the end of the current one. The value of even sampling is that many waveforms have maximum energy at their centres

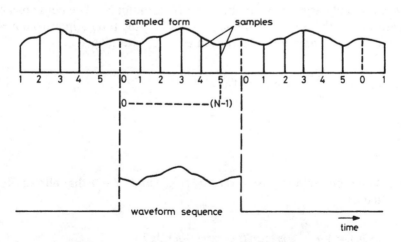

Fig. A3.1 *Sampled periodic function for section of infinite function*

that could otherwise be under-weighted. A disadvantage would appear to be the apparent asymmetry in the sampling, because of the removal of the right-end sample from each sequence, but we shall see that this asymmetry is avoided by the special nature of the discrete Fourier transform; and that the chief effect is a windowing of the sampled function, that serves to reduce the resolution and the ripples due to spectral leakage.

We begin with a function $f(t)$ that can be sampled at NT points separated by intervals T. The finite Fourier series gives the spectrum $F(\omega_k)$ as

$$F(\omega_k) = \sum_{n=0}^{N-1} f(t)\delta(t - nT) \exp(-j\omega_k nT) \qquad \text{with } N \text{ even}$$

where ω_k is the frequency of the kth spectral line in the DFT.[1] $F(\omega_k)$ is the spectrum for waveforms sampled from 0 to $N - 1$ in which the right-end sample is removed as in Fig. A3.1. The sampling function is

$$S(t) = \sum_{n=0}^{N-1} \delta(t - nT) \tag{A3.1}$$

with a spectrum

$$S(\omega_k) = \sum_{n=0}^{N-1} \exp(-j\omega_k nT) \tag{A3.2}$$

and it windows the waveform sequence in the time domain with equal weighting on each sample. The waveform, which in reality stretches from $n = 0$ to N, has even symmetry about the $N/2$th sample, though this may not be the case for all waveforms. A similar sequence of equal-amplitude sampling impulses is always of even symmetry about the centre impulse, whilst the sequence defined by eqn. A3.1 is normally asymmetrical about the $N/2$th impulse. The consequences of this are seen in the frequency domain by shifting the time origin from $n = 0$ to $n = N/2$ in the summation for $S(\omega_k)$ in eqn. A3.2 to give

$$S(\omega_k) = \sum_{n=-M}^{M-1} \exp(j\omega_k mT) \tag{A3.3}$$

with

$$M = \frac{N}{2}$$

There is an even symmetry between $-(M - 1)$ and $M - 1$ that allows $S(\omega_k)$ to be written as

$$S(\omega_k) = 1 + \sum_{1}^{M} \cos \omega_k mT + \exp(-j\omega_k MT) \tag{A3.4}$$

The right-hand term, $\exp(-j\omega_k MT)$, is due to the sample originally defined as at $n = 0$, but now at $m = -M$, whereas the remaining samples from 1 to M, including $M = 0$, give real components. Deletion of the Nth sample therefore causes odd symmetry through the imaginary part of the exponential, since

$$\exp(-j\omega_k MT) = \cos \omega_k MT - j \sin \omega_k MT \qquad \text{(A3.5)}$$

Transformation of the first two terms of the right-hand side of eqn. A3.4 gives the set of time samples between $n = 1$ and $N - 1$ (inclusive), as indicated in Fig. A3.2a with the values shown in the upper line of eqn. A3.6; whilst transformation of the two terms in eqn. A3.5 produces the end samples at $n = 0$ and N, or $m = -M$ or M, as in the lower line[2] in eqn. A3.6 and in Fig. A3.2b.

$$f(t) = \qquad\qquad \delta(t - T) + \delta(t - 2T) \ldots \delta[t - (N - 1)T]$$

$$N = \qquad 0 \qquad\qquad\qquad 1 \qquad 2 \ldots \qquad\qquad N - 1 \qquad N$$

$$f(t) = \tfrac{1}{2}\delta(t + MT) + \tfrac{1}{2}\delta(t + MT) \qquad\qquad \tfrac{1}{2}\delta(t - MT) - \tfrac{1}{2}\delta(t - MT)$$

$$\text{(A3.6)}$$

The two end samples add to one at $n = 0$ and to zero at $n = N$, but a special feature of the fast Fourier transform is that the only frequencies calculated are

$$\omega_k = \frac{2\pi k}{NT}, \qquad k = 0, 1, 2, \ldots, N - 1$$

from which

$$\omega_k MT = 2\pi \frac{Mk}{N} = \pi k$$

Thus sampling is both even and symmetrical, because the frequency-domain

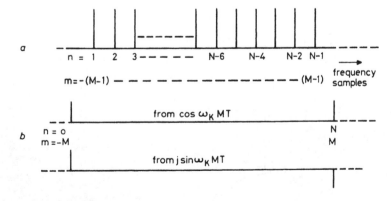

Fig. A3.2   *Symmetrical windowing by sample removal*

sine term in eqn. A3.4 is always zero, leaving just two half-amplitude samples at $n = 0$ and $n = N$.

This kind of windowing—a consequence of the sampling processes adopted in the fast Fourier transform—reduces spectral leakage and degrades resolution, but has little effect on the general shape of the spectrum. It has particular reference to antenna design, where the far field can be calculated through Fourier transform of samples taken in the aperture field or vice versa, because patterns are often symmetrical with peak amplitudes at the centres.

A3.1 References

1 LYNN, P. A.: 'An introduction to the analysis and processing of signals' (Macmillan, 1982) pp. 59–64
2 HARRIS, F. J.: 'On the use of windows for harmonic analysis with the discrete Fourier transform', *Proc. IEEE*, 1978, **66**, pp. 55–83

Notes and solutions to examples

Chapter 1: Swept-frequency principles

The use of Fresnel integrals for the swept spectrum is a deliberate attempt to generalise the description and to bring out the analogy with a radiating aperture suffering a quadrature phase error. Other forms of nonlinear sweeps lead to higher terms appearing in the phase polynomial and again these have analogies in aperture illumination functions.

A lot of fundamental theory is assumed in this Chapter, all of which is covered in the References. Students should study these, particularly References 4, 6 and 7, for more information on the Fourier transform of a Gaussian and on convolution integrals. The important sections in Reference 6 are in Appendix A. There is an excellent account of spectrum analyser fundamentals in Reference 8.

Solutions to examples

Question 1
If T is the period of the sinusoid, the gate time is

$$\tau = T/2$$

The spectrum of a sinusoid consists of two δ-functions at frequencies

$$f = \pm f_0 = \pm 1/T$$

If the gate is a rectangular time function, its spectrum is $a(\sin X)/X$ with first zeros at $\pm 2/T$. The time and frequency functions for a gated half sinusoid are shown in the first three Figures overleaf.

The time functions are multiplied and the frequency functions convoluted as shown in the last Figure overleaf. The convolution shows a peak at zero frequency but no discrimination in the region of $\delta(f \pm f_0)$. When the gate time is

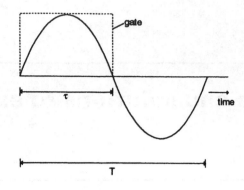

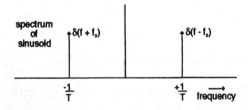

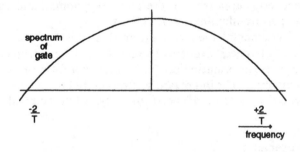

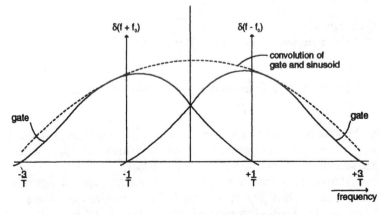

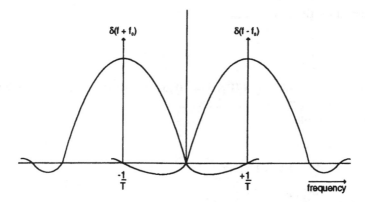

extended to $\tau = T$, there are two peaks at $f = \pm 1/T$ thus allowing f_0 to be determined. This is illustrated in the diagram above.

Question 2
The pulse length $t_0 = 2\,\mu\text{s}$

$$\frac{2}{t_0} = 1\,\text{MHz}$$

To observe the first sidelobes the scan is

$$S = 2\,\text{MHz}$$

If T is the scan period

$$\frac{S}{T} = 10^{10}\,\text{Hz}\,\text{s}^{-1}$$

$$T = \frac{2 \times 10^6}{10^{10}} = 0{\cdot}2\,\text{ms}$$

The pulse repetition frequency for ten Fourier lines in the main lobe is

$$\text{PRF} = \frac{2/t_0}{10} = \frac{10^6}{10} = 100\,\text{kHz}$$

Sweep time between spectral lines

$$= 0{\cdot}2 \times \frac{\text{PRF}}{S}$$

$$= 0{\cdot}2 \times \frac{100 \times 10^3}{2 \times 10^6}$$

$$= 10\,\mu\text{s}$$

Minimum resolution bandwidth

$$R = \sqrt{\frac{S}{T}}$$

$$= 100\,\text{kHz}$$

The observation time in the filter is

$$\Delta t \approx 1/R$$

$$\approx 10 \, \mu s$$

The time between pulses $= 1/\text{PRF} = 10 \, \mu s$. There is therefore one observed spectral line per pulse when $t_0 = 2 \, \mu s$.

When t_0 is doubled

$$t_0 = 4 \, \mu s$$

$$S = 2 \times 0 \cdot 5 = 1 \, \text{MHz}$$

$$T = 0 \cdot 1 \, \text{ms}$$

$$\text{PRF} = 50 \, \text{kHz}$$

The sweep time between spectral lines $= 5 \, \mu s$. The observation time,

$$\Delta t = 10 \, \mu s$$

and

$$\frac{1}{\text{PRF}} = 20 \, \mu s$$

There are two Fourier lines per observation window. These are integrated by the filter to give a single response. There are four Fourier lines between the pulses which are not observed.

The question is intended to produce considerable discussion about the choice of sweep rate, pulse width, etc. An extension might include, say, doubling the pulse repetition frequency. Frequency/time diagrams, similar to those at the end of the chapter, are a good method of illustrating the effects graphically.

Question 3
Let

$$f - f_1 = \Delta f$$

to give

$$v_1 = \sqrt{\frac{2T}{S}} \, \Delta f$$

$$v_2 = \sqrt{\frac{2T}{S}} \, (\Delta f - B)$$

where

$$B = \sqrt{\frac{S}{2T}} \, , \text{ the filter bandwidth}$$

therefore

$$v_2 = v_1 - 1$$

A range of v_1 from $-1\cdot4$ to $2\cdot4$ contains the main lobe response of the filter. Values of $C(v_1)$, $C(v_2)$ and $S(v_1)$, $S(v_2)$ can be found in the Fresnel integral tables or calculated directly for substitution in

$$|S(f)|^2 = V_0^2 \frac{T}{S}\{[C(v_1) - C(v_2)]^2 + [S(v_1) - S(v_2)]^2\}$$

with proper attention to the signs of the integrals.

Normalised $|S(f)|^2$ is plotted in the Figure below. The $-6\,\text{dB}$ bandwidth, expressed in Δv_1, is

$$\Delta v_1 = 2\cdot42$$

The peak response is at

$$v_1 = 0\cdot5$$

But

$$\Delta f = \sqrt{\frac{S}{2T}}v_1$$

Therefore the frequency at the peak is

$$f = f_1 + \frac{1}{2}\sqrt{\frac{S}{2T}}$$

and the bandwidth of the filter is from $v_1 = 0$ to $v_1 = 1$. The $-6\,\text{dB}$ bandwidth has increased from the equivalent minimum resolution bandwidth in the ratio of $2\cdot42$ to 1.

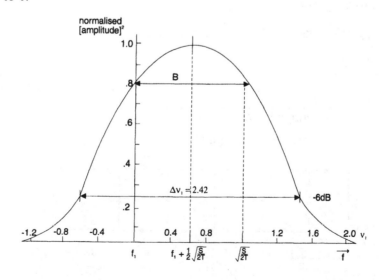

Chapter 2: Error models

An attempt has been made to use flowgraphs rather than matrices wherever possible. This is because physical significance is better maintained in flowgraph analysis, though detailed design with computer support is more likely to use matrix algebra. In explanation of all except the most complicated networks, flowgraphs might be preferred. An example of this approach is in Chapter 7 on six-port network analysers, where a complex subject is in need of some simplification.

Solutions to examples

Question 1
The missing terms are in the Δ_k, where loops 1 and 2 form a second-order loop. They may not be significant if they are products of small numbers.

Question 2
With no DUT the flowgraph is

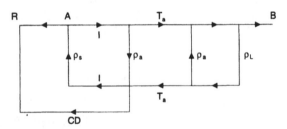

The nontouching loop rule gives

$$\frac{R}{A} = \frac{C(1 - \rho_a\rho_L) + CD\rho_a(1 - \rho_a\rho_L) + \rho_L T_a^2 CD}{\Delta}$$

$$\frac{B}{A} = \frac{T_a}{\Delta}$$

$$\frac{B}{R} = \frac{T_a}{C(1 - \rho_a\rho_L)(1 + D\rho_a) + \rho_L T_a CD} \tag{1}$$

With the DUT in place the flowgraph is

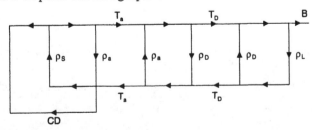

The ratios are now

$$\left(\frac{R}{A}\right)_D = \frac{\begin{array}{c}C(1 - \rho_a\rho_L T_D^2 - \rho_D\rho_L - \rho_a\rho_D + \rho_a\rho_D^2\rho_L)(1 + D\rho_a)\\ +CD(\rho_a T_a^2 + \rho_L T_a^2 T_D^2)\end{array}}{\Delta_D}$$

$$\left(\frac{B}{A}\right)_D = \frac{T_a T_D}{\Delta_D}$$

which gives

$$\left(\frac{B}{R}\right)_D = \frac{T_a T_D}{\begin{array}{c}C(1 - \rho_a\rho_L T_D^2 - \rho_D\rho_L - \rho_a\rho_D + \rho_a\rho_D^2\rho_L)(1 + D\rho_a)\\ +CD(\rho_a T_a^2 + \rho_L T_a^2 T_D^2)\end{array}} \qquad (2)$$

The measured transmission coefficient is the ratio of eqn. 2 to eqn. 1

$$T_M = \frac{B_D}{B} = T_D \frac{(1 - \rho_a\rho_L)(1 + D\rho_a) + \rho_L T_a D}{(1 - \rho_D\rho_L - \rho_a\rho_D - \rho_a\rho_L T_D^2 + \rho_a\rho_D^2\rho_L) + DT_a^2(\rho_a + \rho_L T_D^2)}$$

Note the T_D^2 in the denominator. Thus the error ratio depends on T_D as well as the internal reflection coefficients.

Question 3
The error ratio is

$$\frac{T_M}{T_D} = \frac{(1 + \rho_a\rho_L)(1 + D\rho_a) + \rho_L T_a D}{[(1 - \rho_D\rho_L)(1 - \rho_a\rho_D) - \rho_a\rho_L T_D^2] + DT_a^2(\rho_a + \rho_L T_D^2)}$$

Take ρ_L, ρ_a, ρ_S as 0·1, D as 0·02 and T_D as 20 dB for an attenuator and, say, 0·2 dB for a filter in its passband. Calculate

$$\left|\frac{T_M}{T_D}\right|_{MAX/MIN} = \frac{(1 \pm |\rho_a\rho_L|)(1 \pm |D\rho_a|) \pm |\rho_L T_a D|}{(1 \mp |\rho_D\rho_L|)(1 \mp |\rho_a\rho_D|) \mp |\rho_a\rho_L T_D^2| \mp |DT_a^2|(|\rho_a| \mp |\rho_L T_D^2|)}$$

For a maximum and minimum value.

Express in dB to give an uncertainty range for $T_D = 20$ dB and 0·2 dB.

Question 4
With no reference channel

$$\Delta\rho_{MAX} = \frac{D_2}{T} + |\rho_L|\left(|\rho_0| + \frac{|T^2\rho_S|}{2} + \left|\frac{D_2}{T}\right|\right) + |\rho_L|^2\left(|\rho_0| + \left|\frac{T\rho_S}{\tau}\right|\right)$$

$$D_2 = 0·018 \ (-35 \text{ dB})$$

$$\rho_0 = \rho_S = 0·1$$

$$T = 1, \quad \tau = 1$$

$$\Delta\rho_{MAX} = 0·018 + 0·218|\rho_L| + 0·2|\rho_L|^2$$

With a reference channel the effective input source reflection coefficient at the test port is

$$|\rho_0| + |TD_1| = 0\cdot 1 + |D_1| = 0\cdot 2$$

$$\text{or } |D_1| = -20\,\text{dB}.$$

Note that this is normally too low in a good reflectometer.

Question 5
The levelled ratio is

$$\frac{\delta a_L}{a_L} \approx \frac{20}{KCa_L \ln 10} \left[\frac{\delta a_i}{a_i} + \rho_S \frac{A'Tb}{a_i} \right] - \frac{Db}{a_L}$$

For a matched load, $b = 0$ and

$$K = \frac{20}{Ca_L \ln 10} \frac{\delta a_i/a_i}{\delta a_L/a_L}$$

Take the variations as increases to give

$$\frac{a_i + \delta a_i}{a_i} = 1\cdot 26 \ (+2\cdot 0\,\text{dB})$$

$$\frac{\delta a_i}{a_i} = 0\cdot 26$$

and

$$\frac{\delta a_L}{a_L} = 0\cdot 06 \ (+0\cdot 5\,\text{dB})$$

which gives

$$K = 376 \text{ per mW}.$$

A $-30\,\text{dB}$ directivity gives an effective source reflection coefficient of $0\cdot 03$.

Question 6
The transmission uncertainty is

$$\frac{1 \pm |\rho_S\rho_D T_F|}{(1 \mp |\rho_S S_{11}|)(1 \mp |\rho_D S_{22} T_F|) \mp |\rho_S\rho_D S_{21} S_{12} T|}$$

With $\rho_S = \rho_D = 0\cdot 1$, $T_F = 0\cdot 97$, $S_{21} = S_{12} = 0\cdot 1$ and $S_{11} = S_{22} = 0\cdot 16$ the uncertainty limits are $\pm 0\cdot 36\,\text{dB}$.

Chapter 3: The Smith chart

The ideas of generalised wave functions and their representation on the Smith chart enable circuit representations to be translated between lumped component and wave function forms. The reference material is an essential minimum.

Solutions to examples

Question 1

A short placed at the load position would have zero resistance at the 180°
position on the Smith chart. The measured minimum is at $\lambda/8$ from this towards
the generator with a standing wave ratio of 7. By moving along the SWR = 7
circle towards the load by $\lambda/8$, the point P on the Smith chart gives the load
impedance as

$$\frac{Z}{Z_0} = 0\cdot3 - j0\cdot95$$

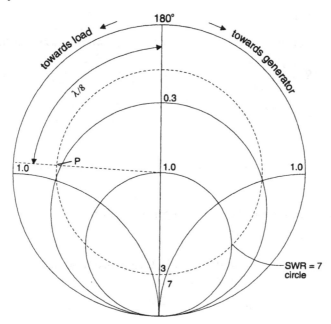

The reflection coefficient is $|\rho_L| = (S - 1)/(S + 1) = 0\cdot75$.
Return loss $= -20\log|\rho_L| = 2\cdot5\,\text{dB}$.
Mismatch loss $= -10\log(1 - |\rho_L|^2) = 3\cdot5\,\text{dB}$.

Question 2

The following points should be covered:

(i) The match to both source and load has to be considered.
(ii) Conjugate matching between source and load gives maximum power
transfer.
(iii) A Z_0 match means no reflected wave from the load.
(iv) Multiple reflections cause variations in received power as frequency
changes.

(v) Reflection from a load may cause source pulling.

(vi) Match for maximum power transfer over a frequency range has to be considered.

The reflection coefficient is

$$\rho_L = \frac{Z_L - Z_0}{Z_L + Z_0}$$

$$= \frac{1 + j0\cdot2 - 1}{1 + j0\cdot2 + 1}$$

$$|\rho_L| \approx 0\cdot1$$

Fraction of power absorbed by load $= 0\cdot99$. The standing wave ratio is

$$S = \frac{1 + |\rho_L|}{1 - |\rho_L|}$$

$$20 \log S = 1\cdot74\,\text{dB}.$$

Question 3

The load reflection ρ_L cancels the connector reflection ρ_c when

$$|\rho_L| = |\rho_c|$$

and they are in antiphase. The total reflection coefficient is

$$\rho = |\rho_c|e^{j\phi_c} + |\rho_L|e^{j(\phi_L + \phi_z)}$$

ϕ_c and ϕ_L are the connector and load reflection phases and ϕ_z is the effect of the airline. Taking the modulus with $|\rho_L| = |\rho_c|$

$$|\rho|^2 = 2|\rho_c|^2[1 + \cos(\phi_c - \phi_L - \phi_z)]$$

The maximum occurs when $(\phi_c - \phi_L - \phi_z) = 0$

$$|\rho|_{MAX}^2 = 4|\rho_c|^2$$

For $|\rho|_{MAX}^2 \equiv -20\,\text{dB}$,

$$|\rho_c|^2 = \frac{0\cdot01}{4} \quad \text{or} \quad |\rho_c| \equiv -26\,\text{dB}$$

The change in frequency

$$\frac{\delta f}{f} = -\frac{\delta\lambda}{\lambda} \quad \text{with} \quad \delta f = 0\cdot075$$

$$\frac{\delta\lambda}{\lambda} = -\frac{0\cdot075}{3\cdot0} = -0\cdot025$$

Total path length is 10λ and $10(\delta\lambda/\lambda) = -0.25$

$$\delta\phi_z = 2\pi 10\frac{\delta\lambda}{\lambda} = -\frac{\pi}{2}$$

Zero reflection occurs when $\phi_c - \phi_L - \phi_z = \pi$ and this changes to $(\pi - 0.5\pi)$ giving

$$|\rho|^2 = 2|\rho_c|^2[1 - 0]$$

$$= \frac{0.01}{2}$$

$$|\rho| \equiv -23\,\text{dB}$$

Chapter 4: Signal generation

An understanding of the design principles of a synthesised sweep generator is the essential component of this Chapter, but it is important to point out that not all measurements require that kind of high-quality instrument. The brief review of sources is intended to provide the background for a proper choice depending on the application. It is not a comprehensive description and the experimenter or test engineer would have to consult manufacturers' literature in detail before coming to a decision about a particular kind of oscillator. Reference 3 provides a starting point for further information on any of the source devices and Reference 6 gives some useful comparisons.

The description of indirect synthesis and oscillator phase noise may cause difficulty at this stage. There is a full explanation given in Chapter 10.

Solutions to examples

Question 1
Students may give any solution provided it meets the specification, and is technically and economically sound. The calculation given here is based on the example in the Chapter.

The poorest resolution occurs when the highest harmonic number is used at the highest frequency. Thus

$$m = \frac{\text{lower}}{\text{integer value}}\left[\frac{18\,000}{440}\right]$$

$$= 40$$

The required value of N_1 is

$$N_1 = \text{int}\left[\tfrac{1}{2}(440 - 396)\right]$$

$$= 22$$

The resolution is 2 MHz times the harmonic number or 44 MHz. A harmonic line is within ± 22 MHz of any frequency. Since capture can occur within ± 35 MHz, this is satisfactory. There are 23 levels in N_1 and the levels in N_2 remain unchanged at 1211.

Question 2
The question is deliberately open-ended. A minimum set of starting questions is:

(a) Is it a CW source?
(b) Are pulses required?
(c) Is a swept range necessary?
(d) Is phase stability important?

Answers should be given in a form which includes all these possibilities, so that clearly there is no single source. The exercise is a big one. Reference 6 gives some comparisons, but manufacturer's literature will have to be consulted.

Question 3
Fundamental oscillators have

(a) lower harmonic and subharmonic output levels, typically -40 dBc
(b) lower residual FM with wider sweeps, typically < 20 kHz peak for sweep widths of up to 50 MHz at frequency ranges up to 18 GHz
(c) greater synthesised accuracy
(d) no complicated YIG filter tracking circuits.

Harmonic multiplier and tracking filter oscillators have only one YIG oscillator to control.

The disadvantages of fundamental oscillators are the greater complexity in the phase lock circuits and the need for three separate YIG oscillators. The disadvantages of filtered multiplier oscillators are

(a) high harmonic and subharmonic output levels, typically -25 dBc
(b) synthesised accuracy which decreases proportionally with harmonic number
(c) higher residual FM, typically 100 kHz peak for sweep widths of up to 50 MHz at frequency ranges up to 18 GHz.

Chapter 5: Vector analysers

Practical experience with a vector analyser is an essential background to this Chapter. Measurements made over a frequency range, with and without error correction using a semi- or fully automatic analyser, will give a good understanding of the effects of uncalibrated test set reflection and transmission coefficients.

Solutions to examples

Question 1
The measured reflection coefficient as a function of the error terms is

$$S_{11M} = E_D + \frac{S_{11} E_R}{1 - E_S S_{11}}$$

The directivity error is zero in the reflectometer, therefore

$$E_D = 0$$

If the load at the test port is a short circuit

$$S_{11} = -1$$

The measured reflection coefficient is

$$S_{11M} = \frac{-E_R}{1 + E_S} = \frac{-1}{1 - L_A^2 L_D^2 (k - \rho_S)}$$

For an open circuit

$$S_{11M} = \frac{E_R}{1 - E_S} = \frac{1}{1 + L_A^2 L_D^2 (k - \rho_S)}$$

Therefore

$$E_R = 1$$

and

$$E_S = -L_A^2 L_D^2 (k - \rho_S)$$

Question 2
It is very important that students are able to make sensible approximations in arriving at performance estimates. The manipulations in this question are very cumbersome if second- and higher-order terms are retained, but since the errors ϵ_S and ϵ_0 are ≤ 0.1, their products can safely be ignored. The solution given here is only one of several ways to arrive at a sensible estimate.

Let $L_A^2 L_D^2 (k - \rho_S) = L e^{j\delta_L}$ and substitute the imperfect short- and open-circuit standards into the expressions for E_R and E_S to give

$$\frac{-E_R (1 + \epsilon_S) e^{j\delta_S}}{1 + E_S (1 + \epsilon_S) e^{j\delta_S}} = \frac{-1}{1 - L e^{j\delta_L}}$$

$$\frac{E_R (1 + \epsilon_0) e^{j\delta_0}}{1 - E_S (1 + \epsilon_0) e^{j\delta_0}} = \frac{1}{1 + L e^{j\delta_L}}$$

Divide one into the other to give

$$\frac{[1 - E_S(1 + \epsilon_0)e^{j\delta_0}](1 + \epsilon_S)(1 + \epsilon_0)^{-1}e^{j(\delta_S - \delta_0)}}{1 + E_S(1 + \epsilon_S)e^{j\delta_s}} = -\frac{1 + Le^{j\delta_L}}{1 - Le^{j\delta_L}}$$

This can be simplified to first order

$$e^{j(\delta_S - \delta_0)}[1 + \epsilon_S - \epsilon_0 - |E_S|e^{j\delta_E}(e^{j\delta_0} + e^{j\delta_s})] = -(1 + 2Le^{j\delta_L})$$

and the modulus is

$$|E_S| = \frac{2L\cos\delta_L - \epsilon_S + \epsilon_0}{2\cos^2(\delta_S - \delta_0)}$$

E_R can be found from the first equation

$$\frac{E_R(1 + \epsilon_S)e^{j\delta_s}}{1 + E_S(1 + \epsilon_S)e^{j\delta_s}} = \frac{1}{1 - Le^{j\delta_L}}$$

To first order this gives

$$|E_R|^2(1 + 2\epsilon_S) = 1 + 2|E_S|\cos(\delta_S + \delta_E) + 2L\cos\delta_L$$

and after further use of the binomial theorem

$$|E_R| \approx 1 - \epsilon_S + |E_S|\cos(\delta_S + \delta_E) + L\cos\delta_L$$

The worst case measured values are given by the maximum and minimum values of

$$S_{11M} = \frac{S_{11}|E_R|}{1 - |E_S|S_{11}}$$

$$= \frac{\dfrac{(1 - \epsilon_S \pm L)\cos^2(\delta_S - \delta_0)}{\pm 2L - \epsilon_S + \epsilon_0} \pm 0\cdot 5}{\dfrac{2\cos^2(\delta_S - \delta_0)}{\pm 2L - \epsilon_S + \epsilon_0} - 0\cdot 5}$$

When there are no errors in the standards (E_S) reduces to L and the original form is found. If $(\delta_S - \delta_0) = \pi/2$ the measured range is

$$S_{11M} = \pm 1$$

Substitution of other combinations of the $\pm$ signs demonstrates how damaging the use of inferior standards is. Measurements can become meaningless.

Question 3
The measured transmission coefficient is

$$S_{21M} \approx E_X + \frac{S_{21}E_T}{1 - E_S S_{11} - E_L S_{22}}$$

The uncertainty is expressed by taking maximum and minimum additions of the error terms and their products with the DUT S-parameters.

$$|S_{21M}|_{MAX} = |E_X| + \frac{|S_{21}||E_T|}{1 + |E_S||S_{11}| \pm |E_L||S_{22}|}$$

$$|S_{21M}|_{MIN} = \pm|E_X| \mp \frac{|S_{21}||E_T|}{1 \pm |E_S||S_{11}| \pm |E_L||S_{22}|}$$

The signs are chosen according to the relative magnitudes of the cross-talk error and the term involving products with the S-parameters.

Chapter 6: Scalar analysers 1

Phase information provides easier access to error correction in vector analysis. In consequence scalar analysis is usually more complicated, and there is a greater diversity of methods depending on how much extra information, above a simple amplitude determination, is actually required. There is a fine distinction between scalar and vector analysis and some overlaps, for instance, when phase is recovered from the dimensional features of the test set, as in standing wave indicators and multiports.

Solutions to examples

Question 1
The swept carrier, at frequency f_c, has a harmonic at twice its frequency. If P_c is the carrier and P_1 the harmonic power level, the three cases can be distinguished as follows:

(a) Neither carrier nor harmonic is in the notch

$$10 \log (P_c + P_1) = 0 \, \text{dB}$$

(b) The carrier is in the notch

$$10 \log \left(\frac{0 \cdot 01 P_c + P_1}{P_c + P_1} \right) = -6 \cdot 8 \, \text{dB}$$

(c) The harmonic is in the notch

$$10 \log \left(\frac{P_c + 0 \cdot 01 P_1}{P_c + P_1} \right) = -1 \cdot 0 \, \text{dB}$$

These are illustrated on the next page.
Taking the ratio of (b) and (c) gives

$$10 \log \left(\frac{0 \cdot 01 P_c + P_1}{P_c + 0 \cdot 01 P_1} \right) = -5 \cdot 8 \, \text{dB}$$

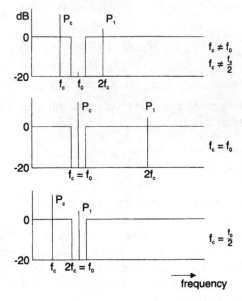

or

$$\frac{0.01P_c + P_1}{P_c + 0.01P_1} = 0.26$$

which gives a harmonic to carrier ratio of $-6\,\mathrm{dB}$.

Question 2
Since the directivity-coupled component adds in phase and amplitude to the true reflected component the measured reflectivity is

$$|\rho_M| = |D + \rho|$$

In most case conditions

$$|\rho_M| = |D| \pm |\rho|$$
$$= 0.03 \pm |\rho|$$

or

$$\frac{|\rho_M| \pm |\rho|}{|\rho|} = \frac{0.03}{|\rho|}$$

For > 10 per cent accuracy $|\rho| > 0.3$.

Question 3
Points to consider include

(a) Internal levelling depends on the quality of the components.
(b) There will be residual amplitude variations with frequency.
(c) The source reflection is reduced.

(*d*) External levelling can remove some of the effects of amplitude variation with frequency, by careful matching of couplers and detectors.

(*e*) Isolators or pads can improve source reflection effects but amplitude variations must be either calibrated or reduced by use of matched couplers.

Ratioing a coupled forward wave with a coupled reflected wave is similar to levelling. In the limit both reduce the effective source reflection to the directivity of the sampling coupler.

Question 4

In taking the average of $(S_{11M})^2$ the slow moving ripple due to $(\phi_c + \phi_D)$ is not averaged over the usual frequency sweep. It therefore gives an error of $\pm 2DR_L^2 R_c$. The fast moving terms due to $(\phi_L + \phi_c)$ and $(\phi_L - \phi_D)$ become zero in the average, but their amplitudes represent the ripple terms. The average is therefore

$$\left|\frac{S_{11M}}{E_R}\right|^2_{AV} = \frac{D^2 + D^2 R_L^2 R_c^2 + R_L^2 \pm 2DR_L^2 R_c}{1 - 2R_L R_c \cos(\phi_L + \phi_c) + R_L^2 R_c^2}$$

$$= (D^2 + D^2 R_L^2 R_c^2 + R_L^2 \pm 2DR_L^2 R_c)(1 - R_L^2 R_c^2)$$

and to sixth order

$$\left|\frac{S_{11M}}{E_R}\right|^2_{AV} = D^2 + R_L^2 - R_L^4 R_c^2 \pm 2DR_L^2 R_c$$

The reciprocal square root is

$$\left|\frac{S_{11M}}{E_R}\right|^{-1}_{AV} = D^{-1}\left[1 - \frac{1}{2}\frac{R_L^2}{D^2} + \frac{1}{2}\frac{R_L^4 R_c}{D^2} \pm \frac{R_L^2 R_c}{D}\right]$$

The ripple amplitude is

$$\left|\frac{S_{11M}}{E_R}\right|^2_{RIPPLE} = (2R_L D \pm 2D^2 R_L R_c)(1 + 2R_L R_c)$$

The second term $2D^2 R_L R_c$ is a smaller amplitude ripple appearing as an error on the main ripple due to the airline.

To fifth order

$$\left|\frac{S_{11M}}{E_R}\right|^2_{RIPPLE} = 2R_L D \pm 2D^2 R_L R_c + 4R_L^2 R_c D$$

$$= 2R_L D(1 \pm DR_c + 2R_L R_c)$$

The product

$$\frac{1}{2}\left|\frac{S_{11M}}{E_R}\right|^2_{RIPPLE}\left|\frac{S_{11M}}{E_R}\right|^{-1}_{AV} = R_L\left[1 - \frac{1}{2}\frac{R_L^2}{D^2} + \frac{1}{2}\frac{R_L^4 R_c}{D^2} \pm \frac{R_L R_c}{D}\right.$$

$$\left. \pm DR_c + 2R_L R_c \mp \tfrac{1}{2}R_L^2 R_c \cdots \right]$$

$$= R_L\left[1 - \frac{1}{2}\frac{R_L^2}{D^2} + \text{terms in } R_c\right]$$

The connector reflection must be less than the DUT reflection to give

$$|\rho_{LM}| \approx R_L\left[1 - \frac{1}{2}\frac{R_L^2}{D^2}\right]$$

Question 5

The relative reflected amplitude at $-50\,\text{dB} = 0.00316$ and the coupled amplitude at $-40\,\text{dB} = 0.01$. The received amplitude is 0.01 ± 0.00316, which is either $-43.3\,\text{dB}$ or $-37.6\,\text{dB}$ or an error range of $5.7\,\text{dB}$.

The sum of direct and site reflected waves can be written as

$$V_S = 1 + \rho e^{-j(2\pi/\lambda)z}$$

or

$$|V_S| = 2\cos\frac{\pi}{\lambda}z$$

This has zeros at

$$\frac{\pi}{\lambda}z = \frac{n+1}{2}\pi$$

or

$$\frac{f_z}{c} = \frac{n+1}{2}$$

For a frequency change Δf

$$\Delta n = 2\frac{\Delta f}{c}z$$

But $z = 40\,\text{m}$. Therefore

$$\Delta n = \tfrac{8}{3}\cdot 10^{-7}\Delta f$$

If the frequency is swept over the element bandwidth, $\Delta f = 300\,\text{MHz}$ and $\Delta n = 80$.

About 80 ripples occur in the sweep. The average level gives the true coupled level between elements. The conditions which make this possible are

(*a*) The element separations must be much less than the distance to the nearest site reflection.
(*b*) The element response should be relatively constant through the swept band.

Chapter 7: Six ports

The selective nature of the material in this Chapter reflects the fact that six-port analysis is as yet an unsettled technique with no universally accepted procedures.

Solutions to examples

Question 1
The sources of error are

(*a*)   less than 'perfect' standards
(*b*)   connector reflections
(*c*)   long lengths between ports giving rapid phase changes
(*d*)   instability in frequency
(*e*)   source pulling variations when different loads are placed at the test port
(*f*)   amplitude errors in the detectors
(*g*)   instrumentation errors
(*h*)   errors due to poor approximations in a theoretical error model.

Question 2
In the flowgraph the source wave is a and the indirect wave at the DUT is b. The sliding reflection coefficient is ρ_R and the DUT reflection ρ_D. b_0 and b_1 are the waves out of ports P_0 and P_1, respectively. The nontouching loop rule can be

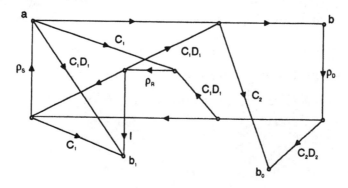

used to solve for b_0 and b_1 as a ratio to a.

$$\frac{b_1}{a} = \frac{[C_1 + C_1 D_1 (C_1 + 1)\rho_R]\rho_D + [C_1 D_1 + C_1(C_1 + 1)\rho_R]}{\Delta}$$

$$\frac{b_0}{a} = \frac{C_2 D_2 (1 + C_1^2 D_1 \rho_R)\rho_D + C_2(1 + C_1^2 D_1 \rho_R)}{\Delta}$$

which gives

$$\frac{b_1}{b_0} = \frac{[C_1 + C_1 D_1 (C_1 + 1)\rho_R]\rho_D + [C_1 D_1 + C_1(C_1 + 1)\rho_R]}{C_2 D_2 (1 + C_1^2 D_1 \rho_R)\rho_D + C_2(1 + C_1^2 D_1 \rho_R)}$$

$$= \frac{C_1 + C_1 D_1 (C_1 + 1)\rho_R}{C_2 D_2 (1 + C_1^2 D_1 \rho_R)} \cdot \frac{\rho_D - q_i}{\rho_D - q_0}$$

The power ratio becomes

$$\left|\frac{b_1}{b_0}\right|^2 = \left|\frac{A_0}{A_i}\right|^2 \left|\frac{\rho_D - q_i}{\rho_D - q_0}\right|^2$$

where

$$q_i = -\frac{C_1 D_1 + C_1(C_1 + 1)\rho_R}{C_1 + C_1 D_1 (C_1 + 1)\rho_R}$$

$$q_0 = -D_2$$

The state i depends on the setting of the sliding reflector with coefficient ρ_R.

Since q_0 is independent of ρ_R it is invariant with the state i, thus giving a multistate reflectometer which can be solved in the same way as for a six-port analyser. The advantage is that calibration software developed for the six-port can be applied directly to the multistate reflectometer.

Question 3

The S-parameter matrix describes the output wave at a port as a linear combination of the inputs to all ports. If we use the wave functions as defined in Fig. 7.2 and add a_S and b_S for the source port, the set of linear equations is as follows

$$b = S_{11}a + S_{12}a_0 + S_{13}a_1 + S_{14}a_2 + S_{15}a_3 + S_{16}a_S$$

$$b_0 = S_{21}a + S_{22}a_0 + S_{23}a_1 + S_{24}a_2 + S_{25}a_3 + S_{26}a_S$$

$$b_1 = S_{31}a + S_{32}a_0 + S_{33}a_1 + S_{34}a_2 + S_{35}a_3 + S_{36}a_S$$

$$b_2 = \ldots$$

$$b_3 = \ldots$$

$$b_S = \ldots$$

Writing

$$a_0 = \frac{b_0}{\rho_0} \quad a_1 = \frac{b_1}{\rho_1} \quad a_2 = \frac{b_2}{\rho_2} \quad a_3 = \frac{b_3}{\rho_3} \quad a_S = \frac{b_S}{\rho_S}$$

the equations become

$$b = S_{11}a + \frac{S_{12}}{\rho_0}b_0 + \frac{S_{13}}{\rho_1}b_1 + \frac{S_{14}}{\rho_2}b_2 + \frac{S_{15}}{\rho_3}b_3 + \frac{S_{16}}{\rho_S}b_S$$

$$0 = S_{21}a + \left(\frac{S_{22}}{\rho_0} - 1\right)b_0 + \frac{S_{23}}{\rho_1}b_1 + \frac{S_{24}}{\rho_2}b_2 + \frac{S_{25}}{\rho_3}b_3 + \frac{S_{26}}{\rho_S}b_S$$

$$0 = S_{31}a + \frac{S_{32}}{\rho_0}b_0 + \left(\frac{S_{33}}{\rho_1} - 1\right)b_1 + \frac{S_{34}}{\rho_2}b_2 + \frac{S_{35}}{\rho_3}b_3 + \frac{S_{36}}{\rho_S}b_S$$

$$0 = \ldots$$

$$0 = \ldots$$

$$0 = \ldots$$

This gives six equations with five unknowns. It is, therefore, possible to solve for b_0 to b_3 in terms of a linear combination of a and b.

Question 4

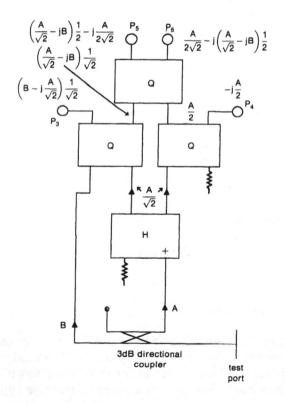

The connection of the 3 dB coupler and the output waves at ports P_3 to P_6 are shown in the Figure on the previous page. Calculation of the power level ratios proceeds as follows

$$P_5 = \left|\left(\frac{A}{\sqrt{2}} - jB\right)\frac{1}{2} - j\frac{A}{\sqrt{2}}\right|^2$$

$$= |A|^2\left|\frac{1}{2\sqrt{2}} - j\frac{B}{2A} - j\frac{1}{2\sqrt{2}}\right|^2$$

$$P_6 = \left|\frac{A}{2\sqrt{2}} - j\left(\frac{A}{\sqrt{2}} - jB\right)\frac{1}{2}\right|^2$$

$$= |A|^2\left|\frac{1}{2\sqrt{2}} - j\frac{1}{2\sqrt{2}} - \frac{1}{2}\frac{B}{A}\right|^2$$

But $B/A = \rho_L = \rho_R + j\rho_I$, giving

$$P_5 = |A|^2\left|\frac{1}{2\sqrt{2}} + \frac{\rho_I}{2} - j\left(\frac{\rho_R}{2} + \frac{1}{2\sqrt{2}}\right)\right|^2$$

$$P_6 = |A|^2\left|\frac{1}{2\sqrt{2}} - \frac{\rho_R}{2} - j\left(\frac{\rho_I}{2} + \frac{1}{2\sqrt{2}}\right)\right|^2$$

On expanding and subtracting

$$P_5 - P_6 = \frac{|A|^2}{\sqrt{2}}\rho_R$$

But

$$P_4 = \frac{|A|^2}{4}$$

Therefore

$$\frac{P_5 - P_6}{P_4} = 2\sqrt{2}\,\mathrm{Re}\,(\rho_L)$$

A similar procedure gives

$$\frac{P_5 + P_6 - P_3 - P_4}{2P_4} = 2\sqrt{2}\,\mathrm{Im}\,(\rho_L)$$

Chapter 8: Power measurement

Conjugate matching, Z_0 matching and their significance in estimating uncertainty show the dangers of an uncritical acceptance of measured results. Power measurement requires a feel for the power ranges of the different methods and the ability to make sensible choices from a range of available options.

Solutions to examples

Question 1
The power measured at the head is

$$P_L = P_{z_0}\eta\frac{(1 - |\rho_L|^2)}{|1 - \rho_L\rho_S|^2}$$

The required quantity is P_{z_0}

$$|\rho_L| = \frac{S - 1}{S + 1} = \frac{0\cdot2}{2\cdot2} \approx 0\cdot1$$

$$|\rho_S| = 0\cdot1$$

The range of P_{z_0} in the measurement is expressed by the uncertainty limits as

$$P_{z_0} = P_L\frac{(1 \pm |\rho_L||\rho_S|)^2}{\eta(1 - |\rho_L|^2)}$$

$$= \frac{8\cdot7(1 \pm 0\cdot01)}{0\cdot9(1 - 0\cdot01)}$$

$$= 9\cdot67\,\text{mW or } 9\cdot86\,\text{mW}$$

$$= 9\cdot76 \pm 0\cdot10\,\text{mW}$$

Question 2
This question is intended to test the student's ability to make reasonable assumptions and sensible approximations. Open-ended questions do not have 'correct' answers; it is the approach that matters. One possible answer is as follows.

A long transmission line, particularly in the open air, will suffer large phase changes due to temperature changes. Since the source is temperature controlled we can ignore that and assume that only ρ_L is affected by the temperature changes.

If $P'_L = 5\cdot1\,\text{mW}$ and $P_L = 4\cdot9\,\text{mW}$ and if the corresponding power head reflection coefficients are ρ'_L and ρ_L, the uncertainty relationship is

$$\frac{P'_L}{P_L} = \frac{1 - |\rho'_L|^2}{(1 \pm |\rho'_L||\rho_S|)^2} \cdot \frac{(1 \pm |\rho_L||\rho_S|)^2}{1 - |\rho_L|^2}$$

$$= \frac{1 - |\rho'_L|^2}{1 - |\rho_L|^2} \cdot \frac{(1 \pm 0\cdot1|\rho_L|)^2}{(1 \pm 0\cdot1|\rho'_L|)^2}$$

Assuming $|\rho_L|$ is small and that only phase changes are involved

$$|\rho'_L| = |\rho_L|$$

and

$$\frac{P'_L}{P_L} = \frac{1 \pm 0.2|\rho_L|}{1 \pm 0.2|\rho'_L|}$$

or

$$\frac{P'_L}{P_L} = 1 \pm 0.2|\rho_L| \pm 0.2|\rho'_L|$$

but

$$(P'_L/P_L) = 1.04.$$

Therefore

$$0.2(|\rho_L| \pm |\rho'_L|) = 0.04$$

Since $|\rho_L| = |\rho'_L|$, the uncertainty limits are from 0 to $2|\rho_L|$ or

$$2|\rho_L| = 0.2$$

$$|\rho_L| = 0.1$$

For ρ_L to change sign requires a line length change of $\lambda/2$ which might seem unlikely in most cases. Should the true length change be less than λ, then $|\rho_L|$ must be greater than 0.1. Then again, perhaps $|\rho_L|$ is changing with temperature after all, because power heads with $|\rho_L|$ much greater than 0.1 might also be unlikely.

If $|\rho'_L|$ changes then the other term

$$\frac{1 - |\rho'_L|^2}{1 - |\rho_L|}$$

must also be considered.

Question 3
The factors include source pulling, uncertainty, temperature gradients, calibration error and instrument error. Reference 3 gives a good discussion of most of these.

Question 4

$$\text{Peak power} = \frac{P_{MEAS} \times \text{coupling factor}}{\text{PRF} \times \text{pulse length}}$$

$$= \frac{1.2 \times 10^{-3} \times 10^2}{2 \times 10^3 \times 10^{-6}}$$

$$= 600 \text{ W}$$

Question 5
If P is the RF power an extra $0.1P$ is added by the DC power balance. Therefore

$$P + 0.1P = \frac{MSt}{60}$$

where M is in kg per minute, $S = 4 \cdot 2 \times 10^3$ joule per kg per °C and t is the temperature rise.

$$1 \cdot 01 P = \frac{0 \cdot 25 \times 4 \cdot 2 \times 10^3 \times 40}{60}$$

$$P = 700 \text{ W}$$

Question 6
See Reference 7 for further help.

Chapter 9: Noise

Solutions to examples

Question 1

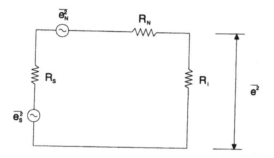

$$\overline{e^2} = \frac{4kT(R_N + R_S)\Delta f R_i^2}{(R_i + R_N + R_S)^2}$$

$\overline{e^2}$ is the amplifier output noise if multiplied by the gain. If

$$R_N = \frac{R_i}{2}$$

$$\overline{e^2} = 16kT\Delta f \frac{(R_N + R_S)R_N^2}{(3R_N + R_S)^2}$$

$$= 16kT\Delta f R_N \frac{(1 + R_S/R_N)}{(3 + R_S/R_N)^2}$$

In the Chapter R_i is implicitly assumed to be infinite. The equivalent plot of $\overline{e^2}$ against R_S/R_N for that case has an intercept at $R_S/R_N = -1$ and is $4kT\Delta f R_N$ at $R_S/R_N = 0$. When R_i is finite as in the expression for $\overline{e^2}$ given above, the plot is no longer a straight line, though its intercept is at $R_S/R_N = -1$.

There is an unreality about this exercise. It bears little relation to the methods of measuring noise. The assumption of a constant R_N as R_S varies is possible only because R_i is very large in voltage amplifiers. This is the essential point that the question should illustrate. When R_i is comparable to R_S, the measured results would give a different R_N for each R_S.

The noise in an equivalent current amplifier with a very low input impedance can be represented by a conductance G_N in parallel with the input terminals. It would remain constant for different source admittances. The general case is examined later in the Chapter.

Question 2
The effective input temperature $= 300 - 20 = 280\,\text{K}$

$$T_{ei} = T_1 + \frac{T_2}{G_1}$$

$$280 = T_1 + \frac{3000}{31\cdot6}$$

$$T_1 = 185\,\text{K}$$

The first stage noise figure is

$$F = 1 + \frac{185}{290}$$

$$= 2\cdot1\,\text{dB}$$

Question 3
The source reflection coefficient

$$|\rho_S| = \frac{0\cdot2}{2\cdot2}$$

$$|\rho_S|^2 = 0\cdot008$$

Available noise temperature $= T_a = 290\,\text{K}$

Effective noise temperature $= T_a(1 - |\rho_S|^2)$
$$= 290 \times 0\cdot992$$
$$= 287\cdot7\,\text{K}$$

For a 3 dB noise figure the effective input noise temperature of the amplifier is 290 K. Therefore, total noise temperature

$$= 290 + 287\cdot7$$
$$= 557\cdot7\,\text{K}$$

A totally reflecting short circuit would give zero noise temperature from the source, leaving just 290 K from the amplifier.

Question 4

$$F_0 = 2 \cdot 24 \quad (3 \cdot 5 \, \text{dB})$$
$$F_1 = 2 \cdot 82 \quad (4 \cdot 5 \, \text{dB})$$
$$F_2 = 4 \cdot 47 \quad (6 \cdot 5 \, \text{dB})$$

The radius

$$r_{F_1} = \sqrt{\frac{N}{N+1}\left(1 - \frac{|\rho_0|^2}{N+1}\right)}$$

The minimum noise figure occurs at $|\rho_0| = 0 \cdot 4$ and at

$$F_1 = 4 \cdot 5 \, \text{dB} \quad r_{F_1} = 0 \cdot 1$$

Therefore

$$0 \cdot 01 = \frac{N}{N+1}\left(1 - \frac{|\rho_0|^2}{N+1}\right)$$

which gives

$$N^2 + 0 \cdot 83 N - 0 \cdot 01 = 0$$

with

$$N = -0 \cdot 84 \quad \text{or} \quad +0 \cdot 012$$

N is positive and gives r_N from

$$N = \frac{F - F_0}{4 r_N}|1 + \rho_0|^2$$

$$r_N = \frac{(2 \cdot 82 - 2 \cdot 24)}{4 \times 0 \cdot 012}|1 + 0 \cdot 4 e^{j170}|$$

$$= \frac{0 \cdot 58}{0 \cdot 048}|1 + 0 \cdot 4 \cos 170 + j \sin 170|^2$$

$$= \frac{0 \cdot 58 \times 0 \cdot 37}{0 \cdot 048}$$

$$= 4 \cdot 47$$

For $F_2 = 6 \cdot 5 \, \text{dB}$

$$N = \frac{4 \cdot 47 - 2 \cdot 24}{4 \times 4 \cdot 47} \cdot 0 \cdot 37$$

$$= 0 \cdot 046$$

which gives for the radius

$$r_{F_2} = \sqrt{\frac{0 \cdot 046}{1 \cdot 046}\left(1 - \frac{0 \cdot 16}{1 \cdot 046}\right)}$$

$$= 0 \cdot 19$$

Question 5

The maximum gain is at S_{11}^* on the Smith chart.

$$S_{11}^* = 0 \cdot 45 \cos 142 + j0 \cdot 45 \sin 142$$

$$= -0 \cdot 35 + j0 \cdot 28$$

The noise circles lie on a line from the origin to the point corresponding to ρ_0

$$\rho_0 = 0 \cdot 4 \cos 170 + j0 \cdot 4 \sin 170$$

$$= -0 \cdot 39 + j0 \cdot 07$$

For $F_2 = 6 \cdot 5 \, \text{dB}$, $N = 0 \cdot 046$

$$d_{F_2} = \frac{|\rho_0|}{1 + N}$$

$$= \frac{0 \cdot 4}{1 \cdot 046}$$

$$= 0 \cdot 38$$

and $r_{F_2} = 0 \cdot 19$. Increasing r_{F_2} to an estimated $0 \cdot 22$ will cause the new noise circle to intersect S_{11}^*, the maximum gain point. The new N can be estimated from

$$r_{F_3} = \sqrt{\frac{N}{N+1}\left(1 - \frac{0 \cdot 16}{1 \cdot 046}\right)}$$

where the old N gives $1 \cdot 046$ in the second term to give

$$0 \cdot 22 = \sqrt{\frac{N}{N+1}} \times 0 \cdot 84$$

or

$$N = 0 \cdot 061$$

Substituting back for N in the $|\rho_0|^2/(N+1)$ term gives

$$r_{F_3} = \sqrt{\frac{0 \cdot 061}{1 \cdot 061}\left(1 - \frac{0 \cdot 16}{1 \cdot 061}\right)}$$

$$= 0 \cdot 221$$

which requires no further iteration.

The new d_{F_2} is

$$d_{F_3} = \frac{0 \cdot 4}{1 \cdot 061}$$

$$= 0 \cdot 38$$

The noise figure is found from

$$N = \frac{F - F_0}{4r_N}|1 + \rho_0|^2$$

$$0 \cdot 061 = \frac{F - 2 \cdot 24}{4 \times 4 \cdot 47} \cdot 0 \cdot 37$$

$$F = 5 \cdot 18 \quad (7 \cdot 15 \, \text{dB})$$

The Figure shows a polar plot of these circles.

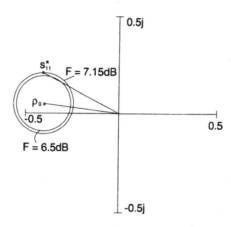

Question 6

$$F_0 = 2 \cdot 24 \quad (3 \cdot 5 \, \text{dB})$$

$$\rho_0 = 0 \cdot 4 \cos 164 + j0 \cdot 4 \sin 164$$

$$= -0 \cdot 38 + j0 \cdot 11$$

$$y_0 = \frac{1 - \rho_0}{1 + \rho_0} = 2 \cdot 15 - j0 \cdot 56$$

When

$$F = 3 \quad (4 \cdot 8 \, \text{dB})$$

$$\rho_S = -0 \cdot 29 + j0 \cdot 08$$

$$y_S = 1 \cdot 78 - j0 \cdot 31$$

$$\frac{F - F_0}{r_N} = \frac{|y_S - y_0|^2}{g_S}$$

$$\frac{3 - 2 \cdot 24}{r_N} = \frac{|1 \cdot 78 - j0 \cdot 31 - 2 \cdot 15 + j0 \cdot 56|^2}{1 \cdot 78}$$

$$\frac{0 \cdot 76}{r_N} = \frac{|-0 \cdot 37 + j0 \cdot 25|^2}{1 \cdot 78}$$

$$r_N = 6 \cdot 78$$

The lumped constant equation is

$$\frac{F - 2\cdot24}{6\cdot78} = \frac{|y_S - 2\cdot15 + j0\cdot56|^2}{g_S}$$

Question 7

The input reflection coefficient $|\rho_i| = 0\cdot33$. For the noise source $|\rho_{ON}| = 0\cdot048$ and $|\rho_{OFF}| = 0\cdot024$. The uncertainty

$$= \frac{1 \pm |\rho_{ON}| \, |\rho_i|}{1 \pm |\rho_{OFF}| \, |\rho_i|}$$

$$\approx 1 \pm |\rho_{ON}| \, |\rho_i| \pm |\rho_{OFF}| \, |\rho_i|$$

$$\approx 1 \pm |\rho_i|(|\rho_{ON}| \pm |\rho_{OFF}|)$$

$$\approx 1 \pm 0\cdot33(0\cdot024 + 0\cdot048)$$

$$\approx 1 \pm 0\cdot024$$

$$\approx \pm 0\cdot2\,\text{dB}$$

For the hot and cold temperatures the uncertainties are

$$T_H \rightarrow 1 \pm 0\cdot048 \times 0\cdot33 \rightarrow 0\cdot06\,\text{dB}$$

$$T_C \rightarrow 1 \pm 0\cdot024 \times 0\cdot33 \rightarrow 0\cdot07\,\text{dB}$$

Question 8

$$F = \frac{GKT_0 B_N + GKT_N B_N}{GKT_0 B_N}$$

But

$$GKT_0 B_N = 10^6 \times 1\cdot374 \times 10^{-23} \times 290 \times 10^6$$

$$= 3\cdot985 \times 10^{-9}$$

and

$$GKT_0 B_N + GKT_N B_N = 10^{-3}$$

$$F = \frac{10^{-3}}{3\cdot985 \times 10^{-9}}$$

$$= 0\cdot251 \times 10^6 \quad (54\,\text{dB})$$

Question 9

$$10 \log \frac{T_H - T_0}{T_0} = 15\cdot5$$

Therefore

$$T_H = 10\,580 \text{ K}$$

$$Y = \frac{2}{0\cdot1}$$

$$= 20$$

$$F = \frac{T_H - YT_c + YT_0 - T_0}{T_0(Y-1)}$$

If the cold temperature is assumed to be 290 K

$$F = \frac{T_H - T_0}{T_0(Y-1)}$$

$$= 1\cdot9 \quad (2\cdot7\,\text{dB})$$

Chapter 10: Frequency stability and measurement

The contents of this Chapter cannot be avoided in design laboratories where highly stable sources are used and measured.

Solutions to examples

Question 1

$$S_{\Delta\phi}(45\,\text{Hz}) = \left[\frac{100\,\text{nV Hz}^{-1/2}}{1\,\text{V/rad}}\right]^2$$

$$= 10^{-14}\,\text{rad}^2\,\text{Hz}^{-1} \quad (-140\,\text{dB})$$

Question 2
The reference signal is offset to a new frequency ω_1

$$v_R = V_R \sin(\omega_1 t + \pi/2)$$

This is mixed with the test oscillator to give the second-order product term

$$V(t)v_R = V_0 V_R \sin \omega_0 t \sin(\omega_1 t + \pi/2)$$

$$= \frac{bV_0 V_R}{2}\{\cos[(\omega_0 - \omega_1)t - \pi/2] - \cos[(\omega_0 + \omega_1)t + \pi/2]\}$$

The beat frequency is selected to give

$$V_b = b\frac{V_0 V_R}{2}\sin(\omega_0 - \omega_1)t$$

The amplitude of the beat frequency is

$$K_d = b\frac{V_0 V_R}{2}$$

The output of the phase detector at the zero carrier sideband frequency, ω, is

$$V_d = \phi b\frac{V_0 V_R}{2}$$

$$= K_d \phi$$

Question 3
At 18 GHz

$$0 \cdot 1 = 60 f_{VCO} - 18$$

$$f_{VCO} = \frac{18 \cdot 1}{60} = 301 \cdot 7\,\text{MHz}$$

At 4 GHz

$$N\,301 \cdot 7 = 4 \cdot 1$$

$$N = 13 \cdot 6$$

$$\text{take } N = 13$$

The separation of the harmonics is $301 \cdot 7\,\text{MHz}$ and the tuning range is $1/N$ of the harmonic separation for lowest N.

$$\text{Tuning range} = \pm\frac{1}{2} \cdot \frac{301 \cdot 7}{13} = \pm 11 \cdot 6\,\text{MHz}$$

The minimum time for 1 Hz accuracy is 1 second. At 18 GHz, $N = 60$. Therefore, the measuring time is 1 minute.

Chapter 11: Time-domain reflectometry

Fourier transform theory can be found in Reference 11 and there are further references in Chapter 1.

Solutions to examples

Question 1
Low pass

$$\int_0^B \cos 2\pi f t\, df = \frac{\sin 2\pi f t}{2\pi t}\bigg|_0^B$$

$$= B \cdot \frac{\sin 2\pi B t}{2\pi B t}$$

At first zero

$$2\pi Bt = \pm\pi$$

$$t = \pm\frac{1}{2B}$$

This is also approximately equal to the 3 dB bandwidth, since time is positive and negative from $t = 0$.

Band pass

$$\int_{B_1}^{B_2} \cos 2\pi ft\, df = \frac{\sin 2\pi ft}{2\pi t}\bigg|_{B_1}^{B_2}$$

$$= \frac{B}{2\pi Bt}[\sin 2\pi B_2 t - \sin 2\pi B_1 t]$$

$$= B \cdot \frac{2}{2\pi Bt}\cos 2\pi\left(\frac{B_2 + B_1}{2}\right)t \sin 2\pi\left(\frac{B_2 - B_1}{2}\right)t$$

$$= B\cos\pi(B_2 + B_1)t\frac{\sin\pi Bt}{\pi Bt}$$

where

$$B = B_2 - B_1$$

At first zero

$$\pi Bt = \pm\pi$$

$$t = \pm\frac{1}{B}$$

By analogy with array theory $\cos\pi(B_2 + B_1)t$ is the array factor and represents the interference of two 'elements' at $\pm((B_2 + B_1)/2)$. $\sin\pi Bt/\pi Bt$ is the 'element' pattern. The interference causes maxima and zeros in the time spectrum.

Question 2
The frequency scan is band-pass with frequency separation $\Delta f = 10\,\text{MHz}$ for 401 points. Site range is determined by the aliasing time $1/\Delta f$. If z is the site range

$$2z = v_p\frac{1}{\Delta f}$$

$$z = \frac{3 \times 10^8}{2 \times 10^7}$$

$$= 15\,\text{m}$$

The resolution for bandpass is

$$\Delta z = \frac{1\cdot2}{2S}v_p$$

$$= \frac{0\cdot6 \times 3 \times 10^8}{4 \times 10^9}$$

$$= 0\cdot045\,\text{m}$$

Time gating eliminates parts of the time response where unwanted reflections are greater. The Fourier transform gives a frequency response more typical of the DUT without interfering reflections.

Question 3

At the discontinuity

$$\rho = \frac{50 - 70}{50 + 70}$$

$$= \frac{-20}{120}$$

$$= 0\cdot17$$

If z is the length of the 70 ohm line

$$z = \frac{v_p \times \Delta t}{2\sqrt{\epsilon}}$$

$$= \frac{3 \times 10^{10} \times 2\cdot5 \times 10^{-9}}{2\sqrt{2\cdot54}}$$

$$= 23\cdot5\,\text{cm}$$

There will be capacitive fringing at the discontinuity.

Chapter 12: Antenna measurements

The practical aspects of antenna measurements are emphasised in this Chapter. Range alignment of the transmitter and receiver, the minimum separation for far field conditions, pattern cuts, gain measurements, etc. should be familiar topics to students who work seriously with the review material at the beginning of the Chapter.

The level of polarisation measurement depends on the application. For instance, in radar detection the partial methods are sufficient to determine integrated cancellation ratio, whereas in satellite communication a much more stringent measurement may be necessary.

Solutions to examples

Question 1
Maximum input is $-25\,$dBm per element

$$900 \text{ elements} = 29{\cdot}5\,\text{dB}$$

Total received power $= -25 + 29{\cdot}5 = 4{\cdot}5\,$dBm

The receive aperture is

$$\frac{30 \times 30\lambda^2}{4} = 225\lambda^2$$

Power at receiver

$$P_R = P_T \frac{G_T}{4\pi r^2} 225\lambda^2$$

where $P_T = $ transmitted power and $G_T = $ transmitter gain. But

$$G_T = \frac{4\pi}{\lambda^2} A_T$$

where A_T is the effective aperture of the transmitter

$$\frac{P_R}{P_T} = A_T \frac{225}{(500)^2}$$

$$10 \log \frac{P_R}{P_T} = 10 \log A_T - 30{\cdot}5\,\text{dB}$$

But A_T is one square metre, therefore $P_T = 4{\cdot}5 + 30{\cdot}5$, so transmitter power $= 35\,$dBm.

Question 2
The Fresnel boundary is at

$$\frac{2(D_T + D_R)^2}{\lambda}$$

Therefore

$$\frac{2(15\lambda + 1)^2}{\lambda} \geq 500$$

$$(15\lambda + 1)^2 = 250$$

$$225\lambda^2 - 220\lambda + 1 = 0$$

$$\lambda = \frac{220 \pm 217 \cdot 9}{2 \times 225}$$

$$= 0 \cdot 97 \, \text{m} \quad (\text{approx. 1 metre})$$

The lowest frequency is therefore approximately 300 MHz.

Question 3

$$\cos \phi = \frac{\cos \alpha \sin \eta}{\sin \theta}$$

For azimuth angle

$$\sin \eta = \frac{\cos \phi \sin \theta}{\cos \alpha}$$

$$\cos \theta = \cos \alpha \cos \eta \cos (\xi - \alpha)$$

For elevation

$$\cos (\xi - \alpha) = \frac{\cos \theta}{\cos \alpha \cos \eta}$$

at

$$\phi = 45°, \quad \theta = \pm 45°$$

$$\sin \eta = \pm \frac{1}{2 \cos \alpha}$$

$$= \pm \frac{1}{2 \cos 10}$$

Azimuth

$$\eta = \pm 30 \cdot 5°$$

$$\cos (\xi - \alpha) = \frac{1}{\sqrt{2} \cos 10 \cos 30 \cdot 5}$$

Elevation

$$\xi - \alpha = 33 \cdot 6°$$

Question 4
$F = 2$ giving an internal noise temperature of 290 K for the receiver.

$$T_{ea} = T_a + T_A \left(\frac{1}{\eta} - 1 \right)$$

where $T_a = 100 \, \text{K}$ for the antenna, $T_A = 290 \, \text{K}$ and $\eta = 0 \cdot 79$ for the attenuation in the combining network. Therefore

$$T_{ea} = 100 + 290 \times 0 \cdot 26$$

$$= 175 \cdot 4 \, \text{K}$$

The total input noise temperature at the receiver is

$$T_N = \eta T_{ea} + T_0$$

$$= 0{\cdot}79 \times 175{\cdot}4 + 290$$

$$= 428{\cdot}6 \text{ K}$$

Boltzmann's constant $K = -198{\cdot}6 \text{ dBm/K/Hz}$

$$428{\cdot}6 \text{ K} \equiv 26{\cdot}3 \text{ dB}$$

$$300 \text{ MHz} \equiv 84{\cdot}8 \text{ dB}$$

Therefore

$$kT_N B = -87{\cdot}5 \text{ dBm}$$

The 1 dB compression point is at an input of -25 dBm.

$$\text{Dynamic range} = 87{\cdot}5 - 25$$

$$= 62{\cdot}5 \text{ dB}$$

This assumes that each amplifier makes an equal contribution to the combining network.

Question 5

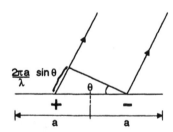

$$E = a \frac{E_0 \sin \frac{\pi a}{\lambda} \sin \theta}{\frac{\pi a}{\lambda} \sin \theta} e^{j(\pi a/\lambda) \sin \theta} \left(e^{j(\pi a/\lambda) \sin \theta} - e^{-j(\pi a/\lambda) \sin \theta} \right)$$

$$= a \frac{E_0 \sin \frac{\pi a}{\lambda} \sin \theta}{\frac{\pi a}{\lambda} \sin \theta} e^{j(\pi a/\lambda) \sin \theta} 2j \sin \frac{\pi a}{\lambda} \sin \theta$$

For small θ

$$E \approx j 2 a^2 E_0 \frac{\pi}{\lambda} \sin \theta$$

$$\frac{dE}{d\theta} \approx 2j a^2 E_0 \frac{\pi}{\lambda}$$

$$\left| \frac{\Delta E}{a E_0} \right| = 2\pi \frac{a}{\lambda} \Delta \theta$$

But $a = 15(\lambda/2)$ so

$$\left|\frac{\Delta E}{aE_0}\right| = 15\pi\Delta\theta$$

But if the dynamic range is

$$\left|\frac{\Delta E}{aE_0}\right|^2 \equiv -62 \cdot 5 \, \text{dB}$$

then

$$\left|\frac{\Delta E}{aE_0}\right| = 7 \cdot 5 \times 10^{-4}$$

Therefore

$$\Delta\theta = \frac{7 \cdot 5 \times 10^{-4}}{15\pi}$$

$$= 0 \cdot 016 \, \text{m radian}$$

Question 6

Planar scanned area $= 16\lambda \times 16\lambda = 256\lambda^2$

Number of samples $= 256 \times 4$ at $\lambda/2$ spacings

$$= 1024$$

Minimum sphere in spherical scanning

$$\rho_0 \approx 6\lambda$$

$$N = \frac{2\pi\rho_0}{\lambda} = 12\pi$$

Total number of samples $= 2N^2$

$$= 2842$$

It is probably unnecessary to scan the whole sphere since only the main beam and first sidelobes are of interest.

Question 7

$$E_{DET} = \frac{1}{\sqrt{2}}[K_x E \cos\phi + K_y E \sin\phi e^{j\alpha}]$$

$$|E_{DET}|^2 = \frac{E^2}{2}[K_x^2 \cos^2\phi + K_y^2 \sin^2\phi + K_x K_y \sin 2\phi \cos\alpha]$$

$$P_{MAX} \approx \frac{E^2}{2}\left[\frac{K_x^2 + K_y^2}{2} + K_x K_y \cos\alpha\right]$$

if $k_x \sim k_y$

$$P_{MIN} \approx \frac{E^2}{2}\left[\frac{K_x^2 + K_y^2}{2} - K_x K_y \cos\alpha\right]$$

$$P_{MAX} + P_{MIN} = \frac{E^2}{2}[K_x^2 + K_y^2]$$

$$P_{MAX} - P_{MIN} = E^2 K_x K_y \cos\alpha$$

$$\frac{P_{MAX} - P_{MIN}}{P_{MAX} + P_{MIN}} = \frac{2K_x K_y}{K_x^2 + K_y^2}\cos\alpha$$

But the polarisation efficiency is

$$p = \frac{1}{2} + \frac{K_x K_y}{K_x^2 + K_y^2}\cos\alpha$$

The integrated cancellation ratio is therefore equal to the polarisation efficiency only when $K_x = K_y$ and phase error is the cause of polarisation loss.

Question 8

At $\phi = 45°$ and $\theta = 0$ the co-polar measured fields are

$$R = E_y$$

and the cross-polar measured fields are

$$C = E_x$$

The measured values correspond with the aperture fields.

At $\phi = 45°$ and $\theta = 90°$

$$R = -\frac{E_x}{2} + \frac{E_y}{2}$$

$$C = \frac{E_x}{2} - \frac{E_y}{2}$$

The measured values respond equally to the two aperture polarisations, such that if $E_x = E_y$, $R = C = 0$.

The full expressions for i_{ref} and i_{CROSS} give the response to the intermediate angles.

Question 9

$$r = -\frac{1\cdot2}{0\cdot8} = -1\cdot5$$

$$\tau = \frac{0\cdot9}{2} = 0\cdot45\,\text{rad}$$

$$= 25\cdot8\,\text{degrees}$$

$$\text{Longitude} = 2\tau = 51{\cdot}7 \text{ degrees}$$

$$\text{Latitude} = 2\cot^{-1}(-r)$$

$$= 2\cot^{-1}(1{\cdot}5)$$

$$= 67{\cdot}5°$$

Appendix 1: Performance characteristics of a spectrum analyser

Spectrum analysers can be used instead of some of the specialised instruments previously described. They can be modified to be like other instruments, such as frequency meters or scalar analysers by externally adding phase-locked references, extra amplification, etc.

Solutions to examples

Question 1
The intercept point equation is

$$I = \frac{D}{n-1} + S_i$$

For second-harmonic distortion the specification gives $D = 60\,\text{dB}$ and the input to the mixer as $S_i = -30\,\text{dBm}$. Therefore

$$I = \frac{60}{2-1} + (-30) = +30\,\text{dBm}$$

The sensitivity noise level $N_i = -115\,\text{dBm}$ and the optimum input signal is

$$S_i = S_M = \frac{I + N_i}{2}$$

$$= \frac{30 - 115}{2}$$

$$S_M = -42{\cdot}5\,\text{dBm}$$

The maximum dynamic range is

$$D_M = \frac{(n-1)(I - N_i)}{n}$$

$$= \frac{30 + 115}{2}$$

$$= 72{\cdot}5\,\text{dBc}$$

Question 2

A typical front end might have a tracking filter selector (6 dB loss), a matching circuit (5 dB), a very broadband mixer (fundamental conversion loss of 10 dB and noise temperature conversion loss between 1 and 2), sometimes a second conversion loss and then an IF amplifier with about a 6 dB noise figure (Reference 5).

Gain increase raises the output signal closer to the distortion level, thus reducing dynamic range. Noise figure improvement lowers the system noise level (improves receiver sensitivity) thereby increasing the dynamic range.

Amplifier gain $= G$. Loss of dynamic range due to amplifier gain $= G$. Improvement in dynamic range due to noise figure reduction $= 15\,\mathrm{dB}$.

Overall reduction in dynamic range $= G - 15 = 3\,\mathrm{dB}$. Therefore $G = 18\,\mathrm{dB}$.

The original receiver noise figure is $F_2 = 37\,\mathrm{dB}$. The improved noise figure is

$$F = F_2 - 15$$

$$= 22\,\mathrm{dB}$$

But if F_1 is the noise figure of the added amplifier, the new noise figure is

$$F = F_1 + \frac{F_2 - 1}{G}$$

therefore

$$F_1 = F - \frac{F_2 - 1}{G}$$

$$= 158 \cdot 5 - \frac{5010 \cdot 9}{63 \cdot 1}$$

$$= 79 \cdot 1 \quad (19\,\mathrm{dB})$$

Therefore $G = 18\,\mathrm{dB}$ and $F_1 = 19\,\mathrm{dB}$.

Question 3

f_m is the phase modulation frequency. The amplitude of the side-band in phase modulation for small deviations is $J_1(\Delta\phi)$ which with $\Delta\phi \rightarrow 0$ gives

$$A_{SB} \approx \frac{1}{2}\Delta\phi$$

But in equivalent small angle FM

$$\Delta\phi = \frac{\Delta f}{f_m}$$

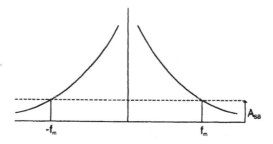

The frequency deviation of the carrier due to the phase noise side-band at f_m is therefore Δf. The mean square deviation due to all phase sideband noise is found from the integral of $\overline{\Delta f^2}$ over all f_m.

If

$$\sqrt{\overline{\Delta f^2}} = 1\,\text{kHz}$$

$$\frac{\Delta f}{f_m} = \frac{10^3}{f_m}$$

and for a $-60\,\text{dBc}$ noise level at f_m

$$\frac{\Delta f}{f_m} = \frac{10^3}{f_m} = 10^{-3}$$

or

$$f_m = 1\,\text{MHz}$$

Similarly at $-40\,\text{dBc}$ noise level $f_m = 100\,\text{kHz}$.

f_m is an equivalent modulation frequency for the 1 kHz FM deviation. But, in fact the 1 kHz deviation is usually quoted as an effect related to the whole sideband noise over the phase sideband frequency range.

The specification shows that, very roughly, $2f_m^2 \mathscr{L}(f)$ is constant with f due to the peaking of phase noise towards the carrier. For the phase-lock option the bandwidth is 100 Hz. At 20 kHz sideband frequency $\mathscr{L}(f) \times \text{BW} \approx 10^{-9}$ ($-90\,\text{dBc}$). Therefore

$$\overline{\Delta f^2} = 2f^2 \mathscr{L}(f) = \frac{2 \times (2)^2 \times 10^8 \times 10^{-9}}{10^2}$$

$$= 4 \times 10^3$$

where f_m has been written as f. The residual FM is

$$\sqrt{\overline{\Delta f^2}} = \sqrt{40} \times 10$$

$$\approx 60\,\text{Hz}$$

which agrees reasonably well with the 50 Hz given in the specification.

Index

Printed in the USA
CPSIA information can be obtained
at www.ICGtesting.com
LVHW021240281023
762436LV00010B/662